Signals and Communication Technology

More information about this series at http://www.springer.com/series/4748

Wynand Lambrechts · Saurabh Sinha

SiGe-based Re-engineering of Electronic Warfare Subsystems

Springer

Wynand Lambrechts
University of Johannesburg
Johannesburg
South Africa

Saurabh Sinha
University of Johannesburg
Johannesburg
South Africa

ISSN 1860-4862 ISSN 1860-4870 (electronic)
Signals and Communication Technology
ISBN 978-3-319-47402-1 ISBN 978-3-319-47403-8 (eBook)
DOI 10.1007/978-3-319-47403-8

Library of Congress Control Number: 2016953320

Printed on acid-free paper

This Springer imprint is published by Springer Nature
The registered company is Springer International Publishing AG
The registered company address is: Gewerbestrasse 11, 6330 Cham, Switzerland

Preface

In electronic warfare (EW), military forces rely on focused electromagnetic energy such as radio waves, infrared radiation, radar and high-powered laser light to intercept, manipulate or disable an adversary's electronic equipment. Successful EW techniques require mature, efficient, high-powered, high-frequency and reliable electronic devices where the choice of enabling technology is application dependent. Soldiers rely on equipment to operate faultlessly in various and often harsh environmental conditions. It is not uncommon for defective equipment to have fatal consequences in certain circumstances. Furthermore, military activity is increasingly becoming more reliant on technology for surveillance, reconnaissance and gathering intelligence. Advances in technology miniaturization and efficiency are ensuring rapid progress in electronic attack (EA), electronic protection (EP) and electronic support (ES), the three major areas of EW. Information warfare is also reliant on electronic equipment to process, store and display large amounts of data instantly, using high-frequency and dependable technologies. High-powered equipment can be used as weapons to destroy electronic equipment and even physically damage vehicles or buildings, or cause harm to humans. The mobility of these weapons is typically problematic, since the electronic subsystems are large and bulky, which is another motivation to miniaturize the electronic supporting circuitry to improve mobility.

Semiconductor processing has also evolved to facilitate complex processing and storage circuits occupying only a few millimetres of area. Integrated circuits (ICs) are the heart of most electronic devices capable of gathering, interpreting and storing digital information or generating, distributing and receiving analog signals. Semiconductor process technologies vary with respect to the application requirement. Inherent trade-offs between process technologies require thorough understanding and research of the requirements, limitations and advantages. Process technologies, such as silicon (Si) variants, which include complementary metal–oxide–semiconductors (CMOS) and Si-germanium (SiGe), indium phosphide (InP) variants, gallium arsenide (GaAs) variants and gallium nitride (GaN), can be used to realize complex, high-frequency and/or high-power ICs. In EW and in space-based applications, reliability in harsh conditions and physical size limitations are frequently required for electronic subsystems. A difficulty that restricts researchers from experimenting and iterating through new ideas and hypotheses regarding the advancement and improvement of EW subsystems is the cost of prototyping and testing submicron electronic circuits. Advantages of Si are a relatively low price to manufacture, high yields and global availability; its caveat, however, is the limitations of its performance. Materials with superior performance in terms of noise, operating speed or output power exist and are available for prototyping, but the higher costs or complex modelling repeatedly deters researchers (and experts) from implementing these technologies.

This book, *SiGe-based Re-engineering of Electronic Warfare Subsystems*, provides a comprehensive and relevant arrangement on the applicability of new generation SiGe electronic subsystems in EW equipment. The definition of EW is expanding, with new innovations and replacement technologies becoming more feasible. Although considered somewhat futuristic, charged particle beam acceleration and laser weapons are becoming increasingly more conceivable, using modern high-powered electronic devices accompanied by rapidly switching supporting circuitry. The basis of these technologies is not new, but their application as offensive or defensive weapons is still at the experimental stage. The fundamental principles of charged particle beam acceleration and lasers are reviewed and presented in this book, highlighting the roles that microelectronic components play in facilitating these weapons and applications, such as laser rangefinders. The generation of laser light also allows optoelectronic communication and microwave photonics, transmitting data by using light as the modulating medium. Optoelectronic communication presents superior performance in bandwidth and low-noise operation, compared to traditional copper and microwave transmissions. High-speed signal processing and complex ICs are requirements of the supporting circuitry to enable optoelectronic communication to operate at its potential maximum bandwidth, ideally using low-cost and mature technology. In addition, the possibility of realizing optic and electronic circuitry on a single die makes SiGe (in view of its Si substrate) a dominant contender in this research discipline. The characteristics of SiGe permit its use in EW, and this book exploits its advantages and provides an in-depth review of the subsystems where SiGe presents the ideal replacement technology for previous and current generation systems.

SiGe offers additional advantageous characteristics compared to traditional CMOS technologies for high-radiation environments, such as in low-earth orbit where space EW offers tactical advantages in EA, EP and ES. An in-depth review of the semiconductor component construction in SiGe is presented in this book to provide better understanding of the inherent radiation-hardened characteristics of SiGe, accompanied by a review of the mechanisms that lead to single event upsets or permanent material damage due to radiation.

SiGe-based Re-engineering of Electronic Warfare Subsystems concludes in the final two chapters by reviewing the potential future of EW and the potential contributions that SiGe can offer. Numerous technologies such as traditional Si, wide bandgap materials such as GaN and silicon carbide and high-speed technologies such as InP and GaAs exist and are being used for various applications in the military, space and commercial sectors. The authors have taken it upon themselves to write this book specifically to address these research topics and empower the scholarly community through this research—integrating EW.

Johannesburg, South Africa

Wynand Lambrechts
Saurabh Sinha

Acknowledgments

We would like to acknowledge Dr. Riëtte de Lange, director of Research Productivity & Improvement of Staff Qualifications, Strategic Initiatives & Administration, University of Johannesburg, South Africa for her coordination.

We would furthermore like to acknowledge the support of technical reviewers as well as language and graphic editors who have all contributed to this process. We value the system of scholarly peer review and the approach that the same adds towards producing research text that adds to the body of scientific knowledge.

Contents

About the Authors

Dr. Wynand Lambrechts, MIEEE obtained his B.Eng., M.Eng., and Ph.D. degrees in electronic engineering from the University of Pretoria (UP), South Africa. He received his M.Eng. with distinction. He has authored 2 publications in peer-reviewed journals, published a book and has presented at various local and international conferences. He currently holds a position as an electronic engineer at Denel Dynamics, a state-owned company in South Africa and is serving as a part-time research associate at the University of Johannesburg (UJ).

Prof. Saurabh Sinha obtained his Ph.D. degree in electronic engineering from the University of Pretoria, South Africa. He has authored or co-authored over 90 publications in peer-reviewed journals and at international conferences. In addition, he is the managing editor of the South African Institute of Electrical Engineers (SAIEE) Africa Research Journal. Prof. Sinha served University of Pretoria for over a decade, his last service being as director of the Carl and Emily Fuchs Institute for Microelectronics. Since 2013, he has been executive dean of the Faculty of Engineering and Built Environment at the University of Johannesburg. He was the 2014–2015 vice-president of IEEE Educational Activities and serves on the IEEE Board of Directors.

Abstract

SiGe-based re-engineering of Electronic Warfare Systems provides a comprehensive and relevant arrangement on the applicability of new generation silicon-germanium (SiGe) electronic subsystems in electronic warfare (EW) equipment. Taking forward research to date, including the state of the art, this work is throughout supported by mathematical derivations, theoretical background and informative illustrations. The aim is to provide thorough context of electronically oriented methods used in EW attack, protection and support with focus on the speed and reliability of current generation SiGe circuits. EW systems vary in application, capabilities and strategies depending on its intent, and this book addresses elements of implementation, research, design, development and innovations from a technical and non-technical standpoint.

The text simplifies key technological principles in the EW environment. Resulting from the seemingly constant improvements in technology, the decrease in cost and the increase in performance of SiGe as underlying technology, there is a gradual shift towards implementing SiGe in EW countermeasures and counter-countermeasures. Shared information on EW implementations remains limited due to its implicit military activity, and the authors envisage this book as an engineering guideline in EW, since convenient availability of information is a defining factor of progress and innovation.

EW uses focused electromagnetic (EM) energy, also known as directed energy, such as radio waves, infrared, radar or laser light to communicate with allied forces or to obscure or disable an adversary's electronic equipment. Military activity is increasingly becoming more reliant on technology to reduce casualties through intelligence, surveillance and reconnaissance. Most electronic circuits used in EW already exist in literature and in practice, and this book aims to present these circuits and its comprising technologies as building blocks of EW. Relevance of these technologies is found in equipment such as radar jammers, low-noise communications and high levels of radiation interference. This book then explores SiGe as an enabling technology for maritime, land and airborne (including space) EW. Many emerging economies do not have free and readily available access to information to research these technologies, and a condensed source for EW technology is required.

Wynand Lambrechts
Saurabh Sinha

Chapter 1
SiGe Based Re-engineering of Electronic Warfare Subsystems

1.1 Introduction to Electronic Warfare

Electronic warfare (EW) uses focused electromagnetic (EM) energy, also known as directed energy (DE), such as radio waves, infrared (IR), radio detection and ranging (radar) or light amplification by stimulated emission of radiation (laser) light, to communicate with allied forces or to obscure or disable an adversary's electronic equipment. EW has changed over the decades and communications, device miniaturization and microelectronics, automation, electric motors, radar power output, three-dimensional printing, chemistry, global positioning, metallurgy and sensors and actuators are only a few disciplines in which technology advances have contributed to commercial and military system enhancements. The key term in the definition of EW, especially concerning this book, is *directed energy*. This book delves beyond the resulting application of DE and investigates the underlying technologies used to achieve modern-day integrated systems. DE is defined by Ellis (2015) as "*an umbrella term covering technologies that produce concentrated electromagnetic energy and atomic or subatomic particles. A DE weapon (DEW) is a system using DE primarily as a means to incapacitate, damage, disable or destroy enemy equipment, facilities and/or personnel. DE warfare is military action involving the use of DEW, devices and countermeasures to incapacitate, cause direct damage to or destroy adversary equipment, facilities and/or personnel or to determine, exploit, reduce or prevent hostile use of the electromagnetic spectrum through damage, destruction and disruption. It also includes actions taken to protect friendly equipment, facilities and personnel and retain friendly use of the electromagnetic spectrum*". From the definition adapted from Ellis (2015) it is clear that DE and DEW are predominantly weapons that make use of 'invisible' ammunition, contrary to classical bombs, bullets and missiles. EW, however, incorporates DEW into its means of attack, but encompasses a large variety of disciplines, skills, technologies and strategies that will be discussed in this book, with reference to electronic and microelectronic technology throughout. EW can be

W. Lambrechts and S. Sinha, *SiGe-based Re-engineering of Electronic Warfare Subsystems*, Signals and Communication Technology,
DOI 10.1007/978-3-319-47403-8_1

considered a rather taboo topic in many circumstances and conversations and in this book, the topic is discussed while always maintaining a strong cross-reference to the technology that enables it. The goal is therefore to educate the reader on old and new techniques, enhancements and possibilities of technology. As will become evident in this chapter and throughout the book, strong focus is also placed on microelectronic components, especially components made of silicon-germanium (SiGe) compounds and the benefits that these components hold in unforgiving environments, generally associated with military activity. However, before these discussions take place, we will first reconsider the definition of EW and expand on it.

EW is a term that encapsulates various areas of electronic measures for offensive and defensive application. In general, EW can be divided into three major areas, namely

- electronic attack (EA)—an offensive strategy,
- electronic protection (EP)—which can be offensive or defensive, and
- electronic support (ES)—a defensive strategy.

EA aims to disrupt communication signals of the enemy, whereas EP is used to shield allied communication from electronic attacks such as jamming. ES refers to action taken to identify and classify sources of intentional and unintentional EM interference as preventative measures to protect the integrity of communication systems of allies and in some instances, civilians. These main areas of EW and examples of each are depicted in Fig. 1.1.

In Fig. 1.1, EW is graphically divided into its three major areas, being EA, EP, and ES, as already defined. EA is further subdivided into categories of attack, which include electronic countermeasures (ECM), deception, intrusion, jamming, pulsing and probing. In this book, each of these categories is discussed and examples of electronic circuits that enable such tactics are given, accompanied by detailed descriptions of the circuits. Following the same strategy, the other two categories will also be covered. As seen in Fig. 1.1, EP consists mainly of electronic counter-countermeasures (ECCM), flare rejection, spread spectrum technologies, a joint restricted frequency list (JRFL), emissions control (EMCON) and stealth. ES is broadly divided into signal intercept and intelligence (SIGINT). SIGINT is a relatively broad term and can be further divided into three sub-categories, namely electronic intelligence (ELINT), communications intelligence (COMINT) and foreign instrumentation signals intelligence (FISINT). For convenience and for later use in this book, the categories of SIGINT are depicted in Fig. 1.2.

From Fig. 1.2, ELINT and electronic support measures or electronic surveillance measures (ESM) are grouped together, since both techniques use various electronic surveillance systems to provide intelligence and information required to perform ES (its parent category in Fig. 1.1). ELINT and ESM broadly comprise electronic intelligence-gathering techniques, radar and EM signal matching, electronic signal

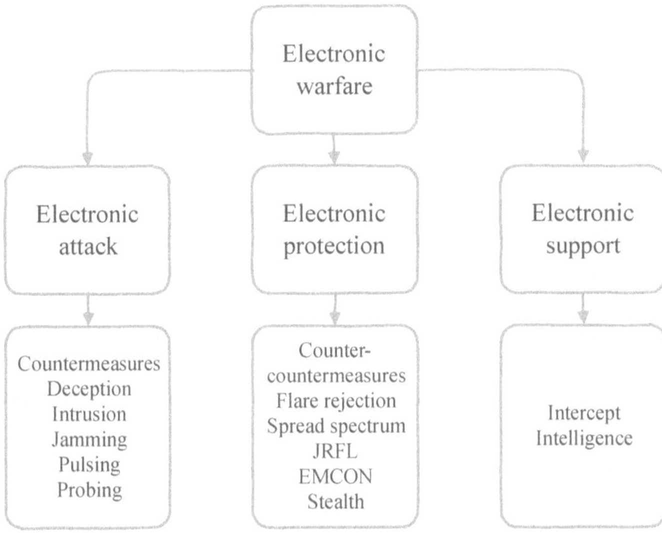

Fig. 1.1 Major areas and subdivisions of electronic warfare: electronic attack, electronic protection and electronic support

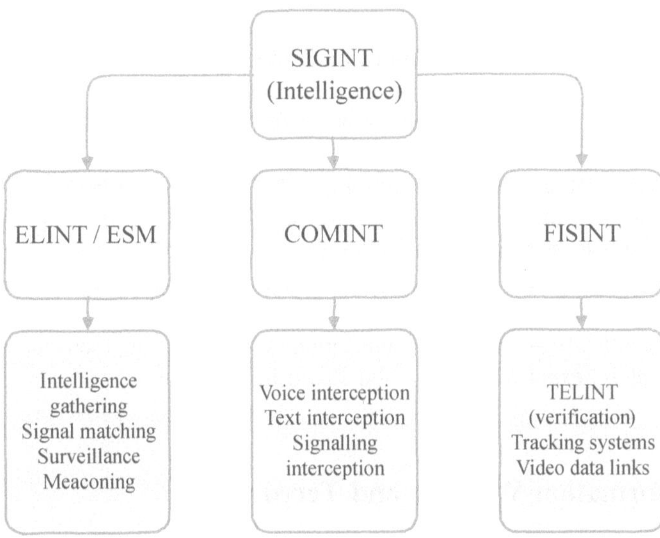

Fig. 1.2 Signal intelligence measures sub-categorized within the ES tactics

surveillance and meaconing,[1] a procedure to learn characteristics of enemy navigation systems such as radio beacons. Also part of SIGINT is COMINT, which focuses mainly on signal interception of foreign communication. The third category of SIGINT is FISINT, associated with interception of foreign EM signals. The difference between COMINT and FISINT is in the content of the communications. COMINT deals with interception of messages between human operators, i.e. messages comprising interpersonal/relational communications, whereas FISINT deals with machine interactions and digital instructions. These instructions include relaying of telemetry data and verifications (TELINT), tracking system data and video data links. These definitions and categories of EW are used to distinguish between the intent of the technology and tactics and give designers and researchers a narrower goal when investigating or experimenting with new technologies to improve EW. Practically the terms are also used throughout military activity and gives military personnel, soldiers and engineers clear and defined intent of operations.

EW applications are found in almost all military environments: maritime (naval EW), on land (ground EW) and in the air (airborne EW and space EW). Vessels may be manned or unmanned and targets include communication systems and radar where the intent is either offensive or defensive. EW systems vary in application, capabilities and strategies, depending on the environment in which they are used. Factors such as vehicle speed and maneuverability, offensive and defensive capabilities and potential risks to humans are considered when implementing EW systems.

EA and EP rely on effective countermeasures and counter-countermeasures to deceive, infiltrate and confuse electronic communication and detection signals such as radar, sonar, laser and IR. Countermeasures of electronic signals are heavily dependent on first identifying an enemy's electronic capabilities and then the implementation of comparable technology that is able to operate at the same standards (for instance when considering high-frequency (HF) and high-power enemy equipment) as that of its adversaries.

The arena of EW has changed over the years and since we have entered the information age, EW has been becoming more important than ever. Historically, EW was specifically aimed at communication systems, but the focus has shifted, or more accurately, expanded to data and stored information.

1.2 Information Warfare and Terrorism

As the world entered the information age (computer/digital age) by shifting from the industrial age to an economy that is essentially based on information and digitization, EW shifted. Historically, cryptography was considered the only

[1]Meaconing is the electronic interception and re-broadcast of navigation signals and is derived from the word beacon.

tangible association between EW and computer security, but these two *disciplines* are fusing at an increasing rate. EW now encompasses many facets, including computer intrusion, computer-based attacks, physical destruction of digital equipment, EW in its classical sense, deception, security system intrusion and psychological operations (Cilluffo and Gergely 1997). Information warfare (IW) is differentiated from other types of warfare through its intention and capabilities to focus on information systems and the information these contain. The power of information (in any context) cannot be underestimated, and in wars, information can help to anticipate enemy movements, execute cyber attacks and falsify digital information. The line between pure IW and EW is becoming increasingly blurred. There are typically five major elements associated with IW, either defensive or offensive, and these are depicted in Fig. 1.3.

Similar to EW and becoming more so, IW places strong focus on COMINT—intercepting digital information and using it as an advantage, ideally unnoticed by the enemy. Destruction and exploitation of an adversary's information systems also fall under the umbrella of IW, along with protection of own/ally information through encryption. Anderson (2008) draws some parallels between EW and computer security, two terms that were not thought to be related decades ago. Anderson (2008) highlights the fact that most people regard computer security as primarily concerning confidentiality, integrity and availability of information and EW intends the exact opposite, since it prioritizes

- denial of service (DoS) (jamming, mimicry and physical attack),
- deception (targeted at automated systems or individuals), and
- exploitation (eavesdropping and obtaining sensitive information from the enemy).

The information that is digitally stored nowadays holds data of almost every facet of a country's infrastructure, including banking and finance, health services,

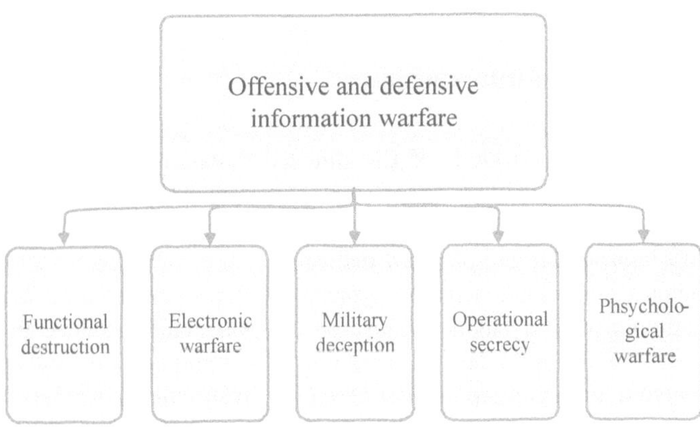

Fig. 1.3 Categories of offensive and defensive IW (functional destruction, EW, military deception, operational secrecy and psychological warfare)

electric power and distribution, transportation and defense infrastructures containing political, economic and military secrets (Cilluffo and Gergely 1997). The availability of low-cost and encrypted communication devices such as cellular telephones enables terrorist activity on a broad scale and eliminates the historic requirement for unique radio equipment and secured radio channels.

An example of terrorist activity making use of low-cost commercial communication devices dates back to 26–29 November 2008 when terrorists used smartphones designed for secure and encrypted commercial communication to access emails, global positioning systems (GPS), voice-over-internet-protocol (VoIP) and telephone calls to successfully coordinate an attack in Mumbai, India—killing 164 civilians and injuring more than 600 people.

In 2014, violent conflict between Russia and Ukraine broke out and became another benchmark for case studies aimed at hybrid conflict where cyber attacks and IW shadowed traditional tactics. Reports of tampering with enemy fiber-optic cables and parliamentarian candidates' cellular phones and more common malicious activity, such as distributed denial of service (DDoS) and web defacements, distinguished this conflict from many others.

The future of EW will definitely have a large IW component and research into countermeasures and new attacks enjoy high priority worldwide. Topics and ideas are shared not only privately, but also publicly at international conferences such as the Australian Information Warfare and Security Conference, the International Conference on Cyber Warfare and Security, the International Conference on Information Warfare and Security, the European Conference on Cyber Warfare and Security and the Information in Warfare Conference, to name a few. Noticeable from these conferences is that many were established in the early 2000s, indicating the relatively 'new' discipline that is IW.

A tactic that has been around for quite some time though, is the use of electronic equipment that generates EM radiation as countermeasures against attacks by enemies.

1.3 Electronic Countermeasures

ECM are focused on two kinds of EM-directed radiation. The first targets communication systems, including voice, text and data links, whereas the second focuses on sensors and actuators of enemy systems, also therefore technically divided into *information* and *physical* damage.

ECM against sensors and actuators typically receives priority compared to ECM against communication systems because of the potential immediate threat of an enemy's intent. Disruption and interception of communication systems lead to delayed reaction or misinterpretation of enemy instructions, whereas destroyed sensors and actuators can have an immediate and fatal effect.

To apply ECM, two classes of ECM are identified, with various implementations of each. These classes are radar jamming and deception. Figure 1.4 highlights the types of radar jamming and deception techniques.

According to Fig. 1.4, radar jamming is divided into electronic jamming and mechanical jamming. Electronic jamming can either be by brute power (obvious) or subtle. Brute methods include high-powered concentrated stepped tones that interfere in the EM spectrum of interest, random noise interference through high-powered transmitters, radio pulsing and signal distortion. Electronic jamming methods that are intended to interfere with the EM spectrum imperceptibly (subtly), primarily through manipulation of the received energy, are squelch capture, modulation intercepting, handshaking intercepting and continuous channel occupation, which is especially effective for technologies that rely on open channels to transmit and receive, such as Wifi and Bluetooth. This kind of electronic jamming is also referred to as digital radio-frequency memory (DRFM) or repeater jamming and is discussed as a separate topic, since it comprises various sub-divisions. A DRFM implementation has a simple premise though; its (simplified) integration into a communication system is show in Fig. 1.5.

As shown in Fig. 1.5, the DRFM circuitry is placed between the incoming radio-frequency (RF) signal preamplifier module and the outgoing power amplifier (PA). DRFM systems are used for the imitation of complex and coherent signals generally related with pulse-compression radars. The expanded functional components and supporting circuitry of a DRFM radar jammer are shown in Fig. 1.6.

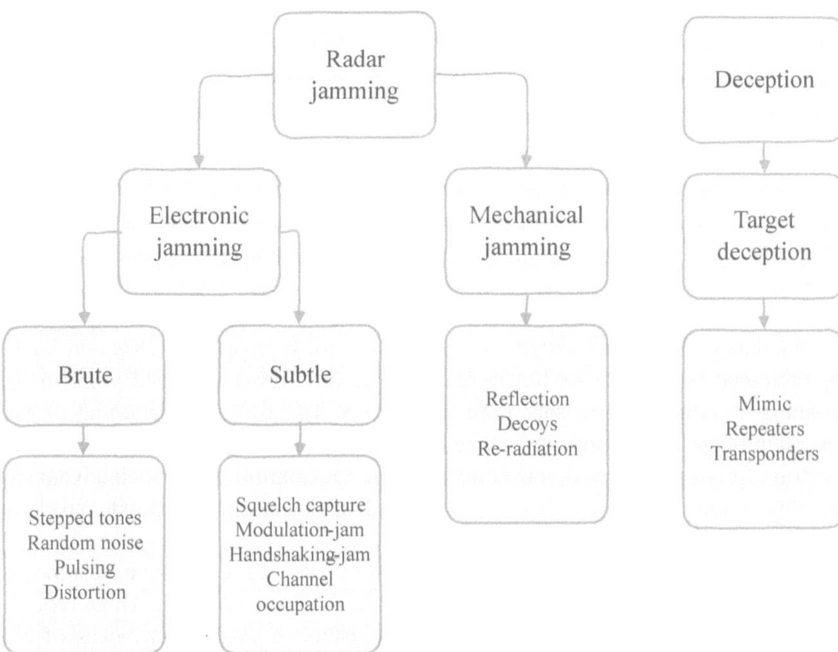

Fig. 1.4 Radar jamming and deception techniques classed under ECM

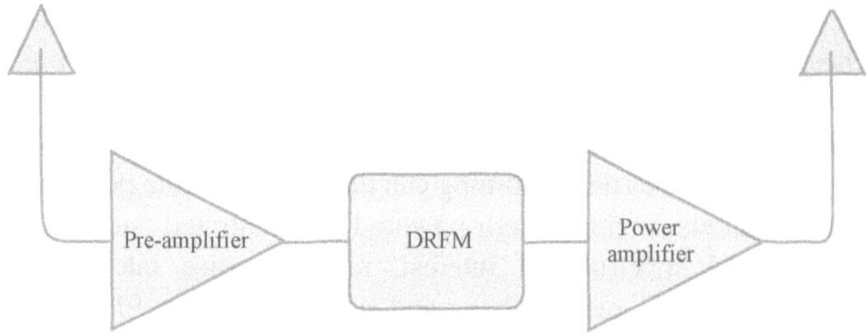

Fig. 1.5 Simplified functional placement of DRFM radar jamming device

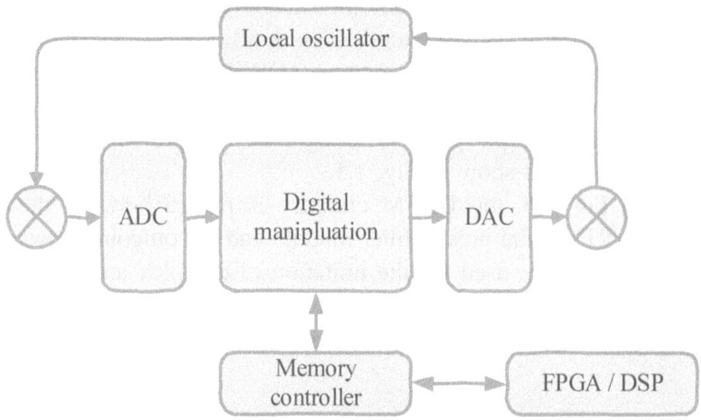

Fig. 1.6 Functional components of a DRFM baseband processor radar jammer

To achieve high-speed DRFM, its functional components are generally designed on a single application-specific integrated circuit (ASIC) to avoid interconnect delays and improve on its modularity and cost-effectiveness. Required components within the ASIC design include preamplifier stages, analog-to-digital converters (ADCs), correction networks, encoding, de-multiplexing, pipeline delay implementation, multiplexing and digital-to-analog converters (DACs). In recent years, the popularity of SiGe as underlying technology for high-speed ADCs and DACs has increased because of its functionality, capability, robustness and performance advantages compared to earlier implementations. To implement jamming protection, significant processing gain, in direct proportion to an increase in available spectrum for signal processing is required. SiGe for example offers such advantages and this technology is detailed and discussed in this chapter and extensively in future chapters of this book.

Conversely, mechanical jamming is caused by physical equipment designed to reflect or re-radiate electronic signals or act as false targets (decoys). Three types of mechanical jamming devices include chaff (clusters of small, thin pieces of aluminum, metallized fiberglass or plastic) appearing as multiple primary targets on enemy radar, corner reflectors and sacrificial decoys. Decoys can in addition implement corner reflectors and chaff to confuse the enemy's electronic radar acquisition and targeting equipment further.

Referring back to Fig. 1.4, deception techniques form a separate class of ECM, mimic radar echo and deceive enemy radar into interpreting received signals as echoes from its targets (or even allied vehicles). Deception techniques will not only mimic radar echoes, but also supply incorrect information on the range and bearing of the proposed target, ideally accomplishing this without the knowledge of the enemy operator. There are two ways of achieving deception: through repeaters or transponders. Repeaters receive the incoming radar signal, manipulate the information and re-transmit it back to the sources, whereas a transponder plays back a pre-recorder duplication of the radar signal when triggered.

In EW, offensive tactics are also required and weapons designed for this specific purpose can become complex and expensive. Depending on the power output of such a weapon, it is generally considered safe for human targets; however, extremely high-powered laser devices can cause injury to humans. These types of weapons are discussed in the following section.

1.4 Directed Energy Weapons

Various forms of DEW have been proposed and used for the last five decades since their invention and constant improvement in technology has ensured continual research and application of DEW. During the last two decades, there has been a significant shift towards interest in moving from medium-energy laser (MEL) to

- high-energy laser (HEL),
- high-power microwave (HPM),
- electromagnetic pulse (EMP), and
- millimeter-wave (mm-wave) technology research (Ellis 2015).

The definition/classification of laser weapons in terms of their energy rating differs between sources, countries and military institutions, but a rough approximation is given by Weise et al. (2004) and adapted in Table 1.1:

There are four fundamental approaches to achieve either high- or medium-power laser energy weapons (GlobalSecurity 2016). These approaches/categories are

- chemical lasers,
- diode-pumped solid-state (DPSS) lasers,
- fiber lasers and
- free electron lasers (FELs).

The laser energy weapons (chemical, DPSS, fiber and FEL) are described further in Table 1.2, with a brief discussion of the basic operation of each. The electronic requirements and technical detail of these laser weapons are discussed in

Table 1.1 Approximate categorization of laser energy rating (Weise et al. 2004)

Energy rating	Directed energy (kW)	Energy density	Applications
Low-energy	<1	<1 mJ/cm^2	Jamming or destruction of sensors
Medium-energy	10–100	>1 J/cm^2	Surface destruction of optronics
High-energy	>100	>1 kJ/cm^2	Destruction of material structures

Table 1.2 Four categories of high- and medium-power laser energy technologies

Laser energy technology	Comments
Chemical lasers	Examples include chemical oxygen iodine laser (COIL), hydrogen fluoride laser and deuterium fluoride (DF) laser
DPSS	Operates by pumping a solid gain medium such as a ruby or neodymium-doped yttrium aluminum garnet (YAG) crystal with a laser diode
Fiber lasers	A laser with an active gain medium in an optical fiber doped with elements such as erbium, ytterbium, neodymium, dysprosium, praseodymium or thulium. High output-power and reliability and low maintenance cost. Used in semiconductor saturable absorber mirrors (SESAMs)
FEL	Uses a relativistic accelerated electron beam (e-beam) injected into a periodic, transverse magnetic field as the lasing medium as opposed to bound molecular or atomic states

subsequent chapters and Table 1.2 aims to provide a brief introduction to the types of lasers. Laser energy weapons are considered an intricate part of future DEW and are therefore especially highlighted in this chapter and expanded in this book.

Other examples of DEW include long-range acoustical devices (LRAD), laser-induced plasma channel (LIPC) and pain beams (Saeed et al. 2008). Any DEW operates at a specified frequency, effective against enemy communications and electronics based primarily on its operating frequency and power level.

Microwave frequencies include electromagnetic radiation with wavelengths ranging from 100 cm (300 MHz) to 0.1 cm or 1 mm (300 GHz), whereas mm-wave frequencies, a subset of microwave frequencies and also referred to as extremely high frequencies (EHF) are defined between 1 cm or 10 mm (30 GHz) to 300 GHz. DEWs are generally defined by their effective range and power output (adapted from Ellis 2015). Table 1.3 summarizes the range and output of HEL, HPM and mm-wave DEWs.

DEW can be fired from various launchers and vehicles. Vehicles that are unmanned are becoming more commonplace owing to advantages as described in the following section. Launching DEWs from these types of vehicles offer similar advantages, mainly the lack of a human operator and the risk of fatalities during missions where DEWs are used.

Table 1.3 Effective range versus effective power output of DEW (adapted from Ellis (2015))

Technology	Effective range	Effective power output
Millimeter-wave technologies	1 km	W to kW
Solid-state and fiber lasers	10 km	High-W to mid-kW
Chemical lasers and diode-pumped alkali and free electron lasers	100 km	kW to MW
1990s "e-bomb" explosively driven HPM	100 m	Mid-kW to GW
Modern electrically or explosively driven HPM	above 1 km	MW to mid-GW
High-altitude nuclear EMP	1000 km	MW to TW

1.5 Unmanned Aerial Vehicles in EW

Unmanned aerial vehicles (UAVs) (also referred to as unmanned aerial systems (UAS) or the mostly military connotation, drones[2]) are, as the name suggests, aircraft with no human pilot on board to steer and maneuver the vehicle. These aircraft can be controlled autonomously through software algorithms to achieve, for instance, marker recognition (such as gray scaling, contour detection, corner calculations, normalization or template matching (Kim et al. 2015)) with various sensors and actuators to allow autonomous flight. Another method of control is exercised remotely by a human operator (possibly to Nikola Tesla's dismay after he demonstrated the first radio signal remote-controlled boat in 1898, realized the military applications of such a remotely manipulated device and chose not to be associated with such activity.) UAVs have many applications. With a decrease in their price, more people (militaries and commercially) are gaining access to such vehicles and applications are expanding rapidly. Not all applications are necessarily tied to military use and commercial use of such vehicles is also growing fast. UAV classification typically falls into one (or more) of six main categories (DHL Trend Research 2014), namely:

- Civil and commercial applications in industries such as mining, civil construction, energy transmission and distribution, water, agriculture, oil and gas, energy services, emergency response, film and photography and telecommunication.
- Logistic support in the service delivery and transport sector.
- Research and development for weather patterns, environmental effects due to global warming, sustainable development and movement of ocean water.
- Armed UAVs used in combat situations.
- Exploration of enemy areas to gain reconnaissance information.

[2]More recently, since the introduction of service drones such as the Amazon® Prime Air delivery drone, the word 'drone' has shed some of its seemingly negative connotation with military activity.

- Copying the flight and physical profile of aircraft to confuse the enemy's air defense systems, thus being used as a decoy.

Apart from the application categories of UAVs listed above, because of the diverse range of systems it is becoming difficult to classify these systems into a standardized set. The US military, for example, uses a tier system with prerequisites pertaining to range, endurance and application. UAVs can, in addition, be classified by their physical dimensions, maximum take-off weight, service ceiling, price and build types (DHL Trend Research 2014). UAV build types include

- fixed-wing,
- tilt-wing,
- unmanned helicopter, and
- multi-copter.

Each of these build types has its own set of advantages and disadvantages, discussed in subsequent chapters of this book. Engine types of UAVs are either electric (these engines are quiet but are heavy owing to battery packs) or internal-combustion engines (which have a long range and are relatively inexpensive to manufacture, but become loud and obtrusive in populated areas). Hybrid systems using electric engines for quiet/covert operations and internal combustion engines to extend their range are also becoming more commonplace, but are still an expensive alternative to pure internal-combustion options. With advances in technology, research into electric multi-copter systems is receiving most attention because of their quiet operation and high maneuverability. Historically, there are various examples of the use of UAVs, or at least some form of aerial vehicles used to replace piloted aircraft.

Arguably, the first real reference to UAVs dates back to August 1849 when Austria launched (number unconfirmed) pilotless balloons, armed with bombs and timed fuses, against the city of Venice (Italy) after the Italian revolt against Austrian rule in March 1848. This was the first attack using UAVs to deliver devastating blows to ground-based targets.

The Royal Flying Corps (RFC) merged with the Royal Naval Air Service on 1 April 1918 after existing as the air arm of the British Army since April 1912. Importantly, the RFC assisted the British Army during the earlier part of World War I through artillery cooperation and photographic reconnaissance. By September 1914 the RFC was able to take aerial photographs from 15,000 to 16,000 ft, which could cover a 2-by-3-mile frontline, giving valuable information about the enemy's movement and being used for planning offensive strikes.

UAVs (or drones) have been prevalent in military activity ever since, with the General Atomics MQ-1 Predator, a medium-altitude, long-endurance unmanned aircraft system, being one of the most well-known military drones. The MQ-1 Predator was introduced in July 1995 and conceived through joint operations between the US Central Intelligence Agency (CIA) and the United States Air Force (USAF) awarding the development contract to General Atomics Aeronautical Systems (GA) in January 1994. It is still in use (2016) by the USAF and has completed missions (with 90 % success rate) over Afghanistan, Pakistan, Bosnia, Serbia, Iraq, Yemen, Libya, Syria and Somalia.

In an article by Keller (2014), mention of an era dawning where small UAVs are used in EW to enable warfighters operating on the forward edge of the battlefield to jam enemy radar, communications and other RF systems is an accurate representation of drone use in emerging military activities. In October 2014, Northrop Grumman Corporation's Aerospace Systems segment demonstrated an internal miniature EA payload on its UAV during testing at China Lake Naval Air Weapons Station in Ridgecrest, California (Keller 2014). The company's UAV (the Bat) is a catapult-launched low-flying UAV with a 10 or 12 foot (3 or 3.65 m) wingspan (depending on the exact model) and houses a variety of payloads including an EW jammer, still-image and video cameras, IR and synthetic aperture radar (SAR) sensors, laser range-finders, laser designators, communication equipment, chemical and biological detection systems and flare dispensers (Keller 2014). It is a beyond-line-of-sight (BLOS) system configured for communications relay, SIGINT, EW, search and rescue and port or maritime surveillance. The maximum payload specification for the Bat is 125 lbs (56.6 kg). The amount of technology available in such a relatively small payload shows that the future of UAVs for military use is aimed at taking as much advantage of miniaturization as possible.

Uninterrupted and noise-free communication between devices and humans in military operations is extremely important and radio channels designated for military use are a complex but well-defined facet of EW. Contrary to commercial bands, generally operating in unlicensed bands, the military spectrum is strictly defined.

1.6 Military Spectral Bands

Spectral assignment for military applications should ideally ensure adherence to strict requirements for permanent, uninterrupted and noise-free access for control, surveillance, reconnaissance and reporting. Military activity requires accuracy and quick response between forces and is a vital requirement that is prioritized. Frequency assignment for military activity differs from commercial applications and frequency bands are carefully selected to provide the abovementioned requirements. The need for specialized personnel to manage frequency allocations has become more prevalent in recent times with the increase in technology use in military activities and the complexity of the equipment. These personnel are responsible for identifying possible incompatibilities and interferences between their own communication systems and those of other forces, either for combined and joint operations between allied forces or intercepting enemy communications. Frequency management in military forces is dynamic (changes frequently) to ensure flexibility and unpredicted activity, whereas commercial frequency allocation is usually constant. Globally, the use of RF bands is regulated by governments in most countries. Various regulating bodies, such as the International Telecommunication Union (ITU), are dedicated to standardizing frequency allocation.

Another regulating body is the Electronic Communications Committee (ECC) within the European Conference of Postal and Telecommunications Administrations (CEPT), which provides the European table of frequency allocations in the frequency range 8.3 kHz to 3 000 GHz (3 THz) (ECA table). The key objectives of the ECC, as given in the Emergency Response Communications 2015, are among others, "*to develop European common positions and proposals for use in the framework of international and regional bodies, and to forward plan and harmonize within Europe the efficient use of the radio spectrum and satellite orbits so as to satisfy the requirements of users and industry*".

Similarly, in the United States of America, regulatory responsibility for radio spectrum allocation is shared by the Federal Communications Commission (FCC) and the National Telecommunications and Information Administration (NTIA). The FCC is an independent regulatory agency and administers spectrum for non-federal use such as state, local, government, commercial, private internal business and personal use. The NTIA is an operating unit of the Department of Commerce and administers spectrum allocation for federal use such as for the army, the Federal Aviation Administration and the Federal Bureau of Investigation. The NTIA provides the Federal Government Spectrum Use Reports for frequencies between 225 MHz and 5 GHz. The FCC provides non-federal spectrum allocations in its Table of Frequency Allocations from 9 kHz to 275 GHz and divides its report into an International Table of Frequency Allocations and the United States Table of Frequency Allocations.

During military frequency management, considerations of national and international regulations, technical requirements of the systems and tactical requirements must be carefully considered. Because of the nature of military frequency use, spectra must be capable of handling high capacities of voice, data and command operations. In this book, these considerations are identified and guidelines to understand frequency management and interoperability are provided. The frequency band designations for three frequently used sets of nomenclature are presented in Table 1.4.

In Table 1.4 the military radar band nomenclature includes HF, very high-frequency (VHF), ultra-high frequency (UHF), L, S, C, X, Ku, K, Ka and mm-wave frequencies. The military radar band nomenclature originated during World War II as a secret communication code between scientists and engineers. Ka and mm-wave frequencies were added to the list after the codes had been declassified. The ITU radar bands in Table 1.4 are designated for radar systems and the ECM industry has its own band designations consisting of alphabetical nomenclature from the letter A to M. In Table 1.5, typical applications of frequency bands for military and commercial applications are given.

The typical applications of the frequency bands in Table 1.5 give a high-level summary of the applications in each of these bands. The high data-rate capability of UHF, μw and mm-wave frequencies is inversely proportional to the atmospheric losses and generally require LOS access. In addition, at mm-wave frequencies,

Table 1.4 Frequency band designations of military radar bands, ITU radar bands and ECM radar bands

Military radar (IEEE[a] radar bands)		ITU radar		EU/NATO/US ECM	
Band	Frequency	Band	Frequency	Band	Frequency
HF	3–30 MHz	VHF	138–144 MHz	A	30–250 MHz
			216–225 MHz		
VHF	30–300 MHz	UHF	420–450 MHz	B	250–500 MHz
			890–942 MHz		
UHF	300–1000 MHz	L	1.215–1.400 GHz	C	500–1000 MHz
L	1–2 GHz	S	2.3–2.5 GHz	D	1–2 GHz
			2.7–3.7 GHz		
S	2–4 GHz	C	5.250–5.925 GHz	E	2–3 GHz
C	4–8 GHz	X	8.500–10.680 GHz	F	3–4 GHz
X	8–12 GHz	Ku	13.4–14.0 GHz	G	4–6 GHz
			15.7–17.7 GHz		
Ku	12–18 GHz	K	24.05–24.25 GHz	H	6–8 GHz
K	18–27 GHz	Ka	33.4–36.0 GHz	I	8–10 GHz
Ka	27–40 GHz			J	10–20 GHz
V	40–75 GHz			K	20–40 GHz
W	75–110 GHz			L	40–60 GHz
mm	110–300 GHz			M	60–100 GHz

[a]The Institute of Electrical and Electronic Engineers

Table 1.5 Typical applications in military and commercial frequency designations adapted from Adamy (2006)

Frequency range	Typical applications
VLF, LF, MF	Very long-range communications (maritime distress and emergency signals), commercial amplitude modulation (AM) radio, non-directional navigation radio beacons, coast-guard and ship-to-shore
HF	Signals reflect from ionosphere for long-range over-the-horizon communications, amateur radio, military and governmental communication, ship-to-shore, maritime distress and emergency signals, automatic link establishment (ALE) links
VHF	Mobile phone communications, television and commercial frequency modulation (FM) radio, air traffic control, amateur radio, air navigation beacons, maritime mobile. High losses beyond line of sight (LOS)
UHF	High-speed mobile telecommunication and digital television, UHF stealth tracking radar. High losses beyond LOS
μw	High-definition television and telephone links, satellite communication, radar, spectroscopy, satellite navigation, radio astronomy. LOS is required
mmw	Radar and high-speed data links, radio astronomy, fire-control radar, security screening. LOS is required. High oxygen absorption

centered around 60 GHz, oxygen absorption losses are high and attenuate radio signals at high rates. Oxygen absorption researchers were especially active during the 1960s and 1970s when satellite communications became increasingly popular, and many of the early findings are used in propagation modelling today. Several methods of estimating atmospheric absorption exist, and these methods are briefly discussed by Hill (1987). The Van Vleck absorption coefficient estimates oxygen absorption attenuation at around 10–15 dB/km (0.01–0.015 dB/m) for a 60 GHz transmission, depending on the altitude. Free-space losses and attenuation from the oxygen absorption band account for losses that occur regardless of the path followed by the signal (assuming LOS).

Frequency management and allocation can only really be effective if the equipment that needs to perform functions at these frequencies is also effective. Transistors are known to be HF and potentially high-powered components and are the key active component in practically all modern electronics. The fundamental operation of transistors as switching devices is well defined, but their capabilities in terms of speed, power consumption, noise and size vary greatly with fabrication techniques (process parameters) and material composition. Bipolar transistors have emerged as one of the most sought-after technologies when designing high-speed circuits requiring high levels of gain, without compromising in terms of noise. Initial applications were largely based on technologies such as gallium-nitride (GaN), aluminum-gallium-nitride (AlGaN), indium-phosphide (InP) and silicon-carbide (SiC). These technologies, although still in use for lower frequency and cost-effective applications, have been suffering since the introduction of gallium-arsenide (GaAs) and SiGe. This book focuses on the current use and future possibilities of SiGe as the enabling technology in EW.

1.7 SiGe Overview

SiGe heterojunction bipolar transistors (HBTs) have one major advantage that has led to their success. This is their low noise capability in HF circuits (Niu 2005) due to their inherent good gain and current densities at a fraction of the cost compared to other specialized materials (discussed in Chap. 10). In EW, for example, a noisy circuit can lead to poor and compromised communication and can have devastating consequences. In commercial applications, low noise capability is also sought, but the effects of a noisy channel are generally more of an inconvenience as opposed to a life-threatening compromise.

Digital wireless communications are based on the principle of modulating baseband data onto an HF carrier in the RF domain (modulation) and down-converting the signal at the receiver towards its original baseband signal (demodulation). The modulation-demodulation process is the origin of the term modem, a device that people have become accustomed to as something that 'provides the internet'. During this modulation-demodulation process, however, the transmitter, the wireless channel and the receiver add noise to the signal. When the noise becomes comparable to the amplitude of the signal itself, the probability of an error in the received signal becomes high.

A parameter to define the possibility of an error at the receiver is called the bit error rate (BER) and is generally plotted against the signal-to-noise ratio (SNR) of the signal. Several techniques of advanced modulation and demodulation, such as quadrature phase-shift keying (QPSK) and orthogonal frequency division multiplexing (OFDM), can decrease the BER for a constant SNR. The wireless channel (in most cases the atmosphere) cannot be manipulated and is an inevitable source of noise and signal degradation. Mobile phones, for instance, have to demodulate signals as weak as −100 dBm (10^{-13} W) (Niu 2005) and it is therefore crucial that the receiver front-end is able to process the incoming RF signal while adding minimum noise to the already weak signal. RF carrier frequencies have increased in recent years because of their advantageous bandwidth characteristics, meaning that the active electronic components such as transistors must operate at these high frequencies with minimum compromise in noise. HF operation is proportional to noise and SiGe technologies provide a much needed benefit to these devices. To illustrate the need for high-speed active components, consider the simplified operation of a radio receiver.

A digital communication receiver mixes the received and modulated RF signal with a local oscillator (LO) signal, which is typically generated at the output of a phase-locked loop (PLL) frequency synthesizer. A typical representation of a frequency synthesizer using a PLL is given in Fig. 1.7.

The selectivity of the receiver and the integrity of the signal are affected by this frequency synthesizer, resulting in its phase noise, also referred to as spectral impurity. Up-conversion of low-frequency signals in the voltage-controlled oscillator (VCO), $1/f$ flicker noise, thermal noise and transistor terminal current noise all add up to reduced signal integrity.

In RF engineering, the minimum noise figure (NF_{min}) can be reduced by using transistors with small base resistance (r_b), high current gain (β) and high cutoff frequency (f_T) (Johnson 1965). SiGe HBTs have a superior NF_{min} because of their inherent characteristics of low r_b, high β and high f_T. Cutoff frequencies in the hundreds of GHz range at 300 K with half-THz maximum oscillation frequencies

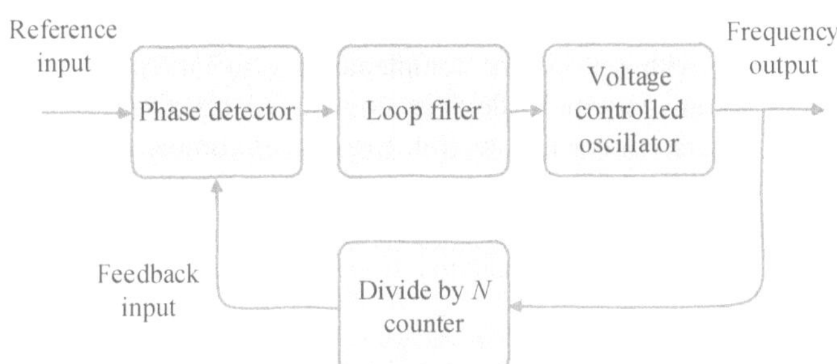

Fig. 1.7 Functional components of an RF synthesizer using a PLL

(f_{max}) have been reported by Jungemann and Hong (2013) and f_{max} of 480 GHz by Ulusoy et al. (2014), whereas f_{max}/f_T of 798/479 GHz SiGe HBTs operating at 4.3 K has been demonstrated by Chakraborty et al. (2014). A low-noise amplifier (LNA) is typically used to facilitate the quantization of device noise parameters in low-noise circuit design (Niu 2005). Electronic equipment in EW devices favors the low levels of $1/f$ noise, RF noise and phase noise of SiGe HBTs.

1.8 SiGe Integration into EW

In Schuh et al. (2010), the advantages of and common use of GaAs and GaN as enabling technologies for transmit/receive (T/R) modules in X-band (8–12 GHz used in radar and countermeasures) defense electronics are described. Output power, broad bandwidth and increased power-added efficiency (PAE) are among the advantages listed for GaAs and GaN monolithic microwave integrated circuits (MMICs) by Schuh et al. (2010). GaN, for example, has a higher breakdown voltage compared to GaAs; this is considered advantageous since the power supply voltage could be increased, leading to lower required gain and system current and hence less thermal loss and noise in the system. It is used in various high-power and high-temperature microwave applications, with a Johnson's figure of merit[3] (JFM) of 27.5—much higher than the JFM of 2.7 of GaAs. This characteristic becomes especially vital in military equipment, as device cooling is challenging in harsh environments. GaN in addition provides high power density and high gain capabilities, coupled with good efficiency, robust operation and long pulse capabilities. An active electronically scanned array (AESA), also referred to as active phased array radar (APAR) is a phased array radar where its transmitter and receiver functions are composed of numerous smaller solid-state T/R modules (TRMs). GaAs was the enabling microelectronic technology introduced in the 1980s, which led to a drastic decrease in the physical size of ASEA receiver modules, succeeded by the binary III/V direct bandgap GaN in the 1990s.

Not only in commercial applications, but also in EW, there is a considerable trend towards higher levels of integration of electronic components and circuits on single integrated circuits (ICs). Integration of multiple functions on a single chip becomes especially popular with technologies such as SiGe; the ability to combine high-speed bipolar with low-power complementary metal-oxide semiconductors (CMOS) leads to an extremely flexible mixed-signal BiCMOS circuit. These integrated systems are compact, powerful, low-power compared to mixed circuits on ceramic substrates and have a lower manufacturing and assembly cost. Future generation ASEAs and current research are pushing towards ASEA antennas with various operating modes in the antenna front-end. These modes include radar,

[3]The Johnson's figure of merit is the product of a semiconductor's charge carrier saturation velocity and the electrical breakdown field.

communication (data links) and EW through jamming (Rieger et al. 2014). SiGe BiCMOS and GaN semiconductor materials are extremely capable of complex, low-noise, high-power and high-speed MMICs and could potentially completely replace GaAs in the RF front-end of TRMs. The SiGe BiCMOS process on its own has also advanced to a point where it is capable of delivering the required output power levels at X-band while also preserving the cost and yield advantages of CMOS (Dinc et al. 2014). In 1999, the ability to fabricate HBTs to execute multi-GHz digital processing algorithms and wideband analog functions on the same IC with low-power, high gate-count CMOS devices for intensive digital signal processing (therefore leading to high-temperature requirements) was successfully demonstrated (Fritz et al. 1999). This work was specifically supported in part by the Microsystems Technology Office of the Defense Advanced Research Projects Agency (DARPA/MTO)—an indication of the military's interest in the technology.

EW is not confined to ground and air anymore and space EW is quickly becoming a trend for militaries. SiGe, conveniently, also offers advantages when used in the harsh space environment and is superior to GaN and GaAs in this respect.

1.9 SiGe and Radiation (Space EW)

The distinctive features of SiGe bandgap-engineered devices offer various key merits to operate these devices in extreme environments such as space. These key elements are evident in SiGe HBTs with very little to no process modifications and provide a broad spectrum of applications in commercial and defense technologies (Cressler 2013). Applications where devices operate in extreme environments, as outlined in Cressler (2013), include satellite systems in earth orbit supporting the global communications infrastructure, on-engine automotive electronics, various weapon systems, space exploration platforms, various sensor systems, space-based remote imaging systems for climate and resource mapping, energy exploration, particle physics detectors and health imaging systems.

Space is considered an extremely unforgiving environment and electronic components exposed to ionizing radiation and large temperature variations could possibly fail when ionized. Digital electronic equipment can experience phenomena such as bit-flips, a single-event upset (SEU) causing the hardware to enter an unknown state and leading to system failure. For example, if a satellite is launched into orbit and the antenna deployment hardware needed to initiate instructions is corrupt, the antennas could fail to deploy, rendering the entire satellite 'lost in space'. These errors, generally attributed to single-event effects (SEE) and displacement damage (DD) events, are especially likely in standard CMOS devices, since these are lateral transport devices. Without special radiation hardening techniques, susceptibility to ionizing radiation is high.

Hardening techniques include hardening-by-process where changes are made in the underlying fabrication techniques of the devices, or hardening-by-design where devices, circuits and/or systems are optimized, generally through additional layers of redundancy. Research by various authors and a discussion by Cressler (2013) have proven that SiGe HBTs, because of being vertical transport devices as opposed to CMOS, which is a lateral transport device, have a built-in tolerance to total ionizing dose (TID) radiation and a unique ability to operate across large temperature ranges. (Being a vertical transport device rather than a lateral one is not the only reason, but an important consideration in the processing of components and its dependence on lithography resolution.) Using SiGe HBTs for analog and RF high-speed digital circuits, combined with CMOS for high-density and slower sub-circuits (such as memory), enables cost-effective applications in extreme environments. Reconsidering the fact that especially in EW (and in this case space EW) technology failures can be fatal to the operator, such advantages cannot be overlooked. Fleetwood et al. (2015) explore a modest (non-intrusive and no major fabrication modifications) technique to optimize the vertical profile of SiGe HBTs to reduce radiation effects. The profile is optimized for inverse-mode (IM) operation and shows improvement of IM performance while having a minor impact on normal active operation. It involves electrically swapping the emitter and the collector terminals and has been shown to improve the radiation tolerance of SiGe HBTs to single event transients (Fleetwood et al. 2015).

1.10 Radar and Countermeasures

Radar uses radio waves and echoing to determine the range, angle or velocity of objects. Radar in EW systems must detect and characterize sources of electronic noise such as RF jamming or co-location antenna interference and adapt performance accordingly to compensate for that interference. Radar has been an intricate part of military activity since shortly before World War II and continues along this path today. Its applications in EW are crucial and it is used in air defense systems, surveillance and guided missile target locating systems (tracking). In principle, radar operates based on the fact that light travels approximately 300 mm each nanosecond (1×10^{-9} s), measuring the time elapsed between the transmission of the pulse and reception of the same pulse, to determine the distance of the object. The received signal will undergo changes in amplitude, noise and possibly phase and the receiver electronics are responsible for decoding and interpreting the received signal with high-speed ADCs, for example. Generally, the transmitter and receiver are co-located and co-axial on the same optical path. The design of a radar in terms of its power output and frequency of operation requires parameters concerning the SNR of the received signal, which gives an indication of the detection range of the radar, assuming several background radiance and detector sensitivity parameters.

To determine the power returned to the receiving antenna in a radar configuration, the radar range equation can be used. This equation can be written in many forms, depending on the parameter of interest, and (1.1) shows the range equation used to determine the received signal (*SE*) energy in watt-second,

$$SE = \frac{P_{ave} G^2 \sigma \lambda^2 T_{OT}}{(4\pi)^3 R^4} \tag{1.1}$$

where P_{ave} is the average transmitter power, G is the gain of the antenna in watt, λ is the wavelength of the transmitted signal in meters, σ is the radar cross-section of the target in square meters, T_{OT} is the time that the pulse illuminates the target and R is the range to the target in meters. In EW, however, it is more practical to consider the received power in the radar receiver and this equation is generally used when determining the jamming-to-signal ratio. The equation is given by

$$P_R = \frac{P_T G^2 \sigma \lambda^2}{(4\pi)^3 R^4} \tag{1.2}$$

where P_R is the received power and P_T is the transmitted power. Important to note from (1.1) to (1.2) is the raised fourth power in the range, R^4, in the denominator. It shows that the amount of received power at the receiver attenuates at a very high rate owing to the distance between the transmitter and the receiver. This is because the signal must travel from the transmitter to the object and back again towards the receiver. The received power can be increased by increasing the transmitted power, the transmitting antenna gain, the radar cross-section or the wavelength (thus lowering the frequency). Frequency shift also occurs due to the motion (moving object) relative to the radar and the target, which changes the number of wavelengths between the reflective object and the radar transmitter/receiver (transceiver). This change in the number of wavelengths due to motion is called the Doppler effect and can be digitally corrected using electronic circuitry. This circuitry should ideally operate at high speed (higher than the transmit frequency and exceeding the Nyquist[4] frequency) to offer real-time corrections and therefore accurate analysis of the received signal.

Using radar techniques, there are four basic approaches to threat guidance: active, semi-active, command and passive guidance.

For active guidance, a radar must be located on the missile as well. In active threat guidance the missile is launched in the general direction of the target and uses its on-board radar seeker to locate and guide itself to the target. There are several advantages to using this technique. Firstly, the launch vehicle can leave the danger area after the missile has been launched, since it is not used to guide the missile towards impact. Secondly, with the large dependence of the received energy on the

[4]The Nyquist frequency in discrete-time systems is half of the sampling rate (or twice the operating frequency) in signal processing.

distance to the target, as seen in (1.1) and (1.1), the radar becomes more accurate and difficult to jam as it approaches is target (owing to the R^4 dependency). These guidance systems, however, require large radar payload transmitters and receivers and limit the explosive payload of the missile.

If only a receiver is located on the missile, it is a semi-active threat guidance system. The launch vehicle transmits radar signals to the target, which are then picked up by the receiver on the missile as it guides itself to the target. The launch vehicle must therefore be in the vicinity of the target and its dependency on the distance to the target is a function of the movement of the launch vehicle, which would ideally aim to leave the danger area. These systems are generally used in air-to-air attacks where a missile is launched from an aircraft.

Command threat guidance only requires that a radar sensor tracks and predicts the path of the target and relays the information to the weapon. The weapon therefore has no spatial information about the target and only follows the directions received from the tracking radar, located on for example the launch vehicle. Command threat guidance is generally found on surface-to-air missiles.

Finally, passive threat guidance relies only on the radiated signals from the target and has no radiating transmitters of its own or on the launch vehicle. These radar sensors track either IR radiation from heat signatures or the radiated energy from the target's radar. The launch vehicle is also able to leave the area once it has launched the missile and the missile will follow the radiation emitted from the target. Passive systems rely heavily on signal processing of the received radiative signature to identify and follow their target. This is an important aspect of EW: the identification and image-processing characteristics of the radiated target signature. The radar scan characteristics of threat radar are designed with purpose and intent in mind and include features such as radar scan, antenna beam width and antenna beam pointing. Radar scan to an EW receiver is the time history of the signal strength of the received signal (Adamy 2006) caused by the shape of the radar antenna beam and its angular movement relative to the EW receiver. The radar uses a narrow beam width to determine the azimuth and elevation of the target, where the accuracy of the target acquisition is proportional to the beam width (narrower beam width for more accuracy). The antenna beam pointing is related to the angular pointing width towards the target. The angular area is high when locating the target and once located, the beam-pointing area decreases to track the target.

Also covered in detail in subsequent chapters of this book are the threat radar modulation techniques used in EW radar. Modulation is typically divided into three main categories: pulse radar, pulsed Doppler radar and continuous wave (CW) radar. The modulation techniques and carrier frequency are dependent on the technology used in the digital circuitry. SiGe again offers unprecedented speed advantages to increase the carrier frequency into the high-GHz domain without allowing noise to dominate the signals. Figures 1.8, 1.9 and 1.10 depict the simplified general operation of pulsed radar, pulsed Doppler radar and CW radar as adapted from Adamy (2006).

In Fig. 1.8 short pulses are generated by a central processing unit and modulated onto a high-power RF output signal. Since the system only generates short pulses,

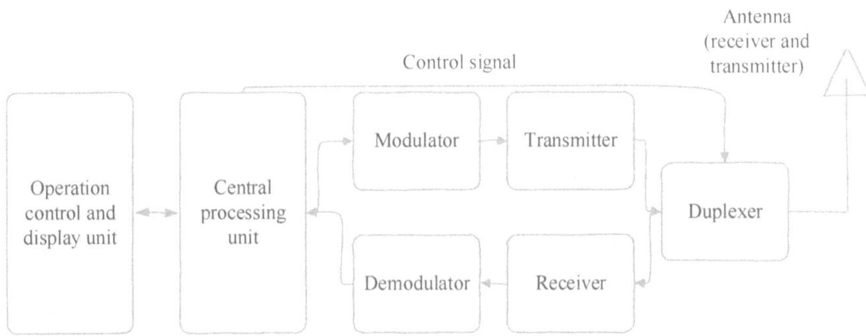

Fig. 1.8 Functional components of a pulsed radar modulation system

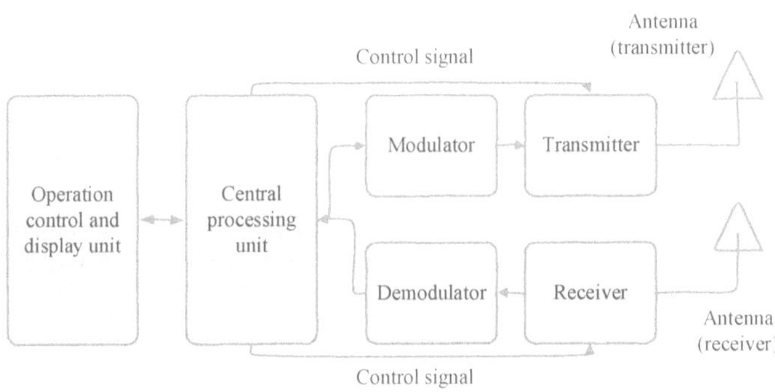

Fig. 1.9 Functional components of a pulsed Doppler radar modulation system

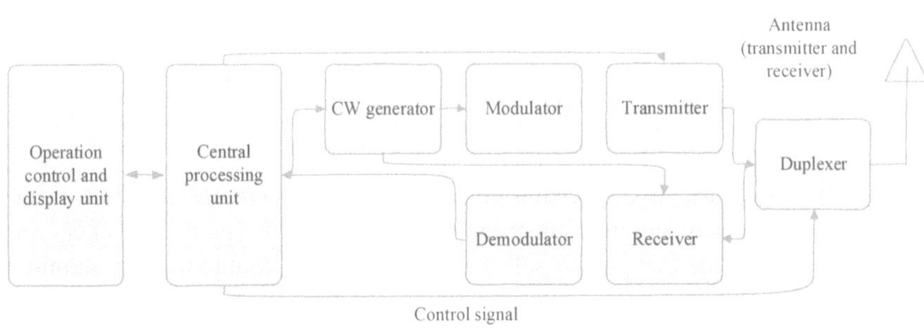

Fig. 1.10 Functional components of a CW radar modulation system

one antenna is used for both transmitted and received signals, which are shared in the time-domain. The high-powered RF signal and received signal are separated and sent to either the antenna or the receiver by the duplexer. Received signals are passed back to the processor to update tracking information by analyzing the incoming signal amplitude and keeping the weapon tracked to the target. A separate control line in the system handles operating modes and target selection. Information about the target is displayed on a control and display unit. Doppler pulsed radar operates similarly to pulsed radar and its simplified operation in shown in Fig. 1.9.

The difference in operation of Doppler pulsed radar as shown in Fig. 1.9 as opposed to traditional pulsed radar is its ability to maintain phase consistency by using coherent transmitted pulses. The transmitted pulses are therefore a continuation of the same signal to keep consistent phase information. This type of modulation technique requires an additional antenna, one for transmitting and another for the received signal. CW radar simplified operation is depicted in Fig. 1.10.

CW radar as shown in Fig. 1.10 also requires two antennas, one for the transmitted CW signal and one for the received signal, since the transmitted signal is a continuous wave and is therefore constantly present. The receiver compares the frequency of the received signal to the frequency of the transmitted signal by also having access to the CW generated signal and determines the Doppler shift as a result of the relative movement of the target. The transmitted signal is modulated onto a known HF carrier for information about the number of wavelengths received back and therefore determining the range of the target. A processor is also present in this system for control functions and used as an interface between the controls and the displays of the system.

Further detail and mathematical derivations are presented in subsequent chapters with the focus on SiGe as a potential underlying technology of radar and countermeasures primary and supporting electronic circuitry.

This book often refers to weapons used in EW, especially the missile, and it has consequently been decided to present a brief introduction to missiles. This introduction does not reveal abundant detail regarding the mechanical and technical functioning of such a device, but aims to present a general understanding of the subsystems located on a missile and how these interoperate.

1.11 The Missile and EW

A missile is a self-propelled precision-guided munition system using various tactics and a large portion of engineering principles to achieve its goal. A rocket, conversely, is an unguided or guided self-propelled craft not containing any munition and has different applications, such as space exploration. There are four main system components in a missile: targeting and guidance, flight mechanics, engine and warhead (payload). The purpose of the missile depends on the intent of the application and is determined by its operation: surface-to-surface, air-to-surface, surface-to-air, air-to air and anti-satellite weapons. To present the simplified workings of a missile graphically, consider Fig. 1.11.

The missile in Fig. 1.11 consists of various subsystems including its launch motors, link antennas, flight mechanics, the warhead (primary payload), a

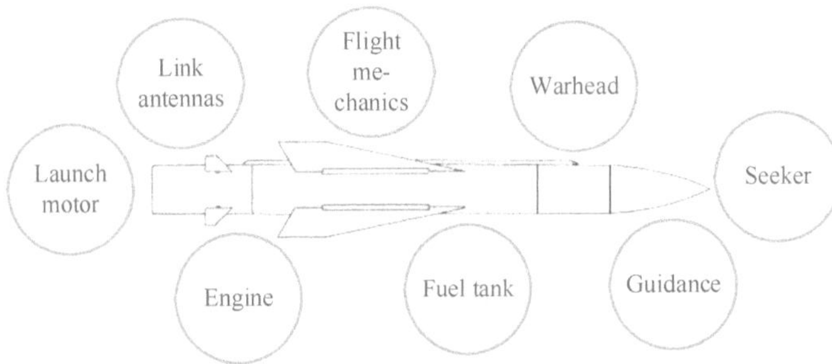

Fig. 1.11 Functional components of a CW radar modulation system

seeker-head, guidance and targeting electronics and microelectronics, the fuel tank and of course its engine. Effective missile defense involves the detection, tracking, interception and ultimately the destruction of an attacking missile. EW missile defense systems use jamming, decoys, anti-radiation missiles and DE to defend against enemy missiles and other weapons such as mortars. These techniques, if not destructive such as anti-radiation missiles, interfere with the seeker, guidance and link antennas, as seen in Fig. 1.11, to confuse, disrupt or redirect the enemy missile, using its anti-electronics payload. Scientists at the Georgia Tech Research Institute (Georgia Tech) are actively researching ways to incorporate SiGe in phased-array radar systems such as theater high-altitude area defense radar to reach the same level of functionality as current GaAs implementations. According to Georgia Tech, current GaAs systems use vast amounts of energy to power many modules and ICs to direct radar beams electronically. Their philosophy of *smaller is better* aims to improve on currently available consumer electronics using SiGe and to miniaturize these EW systems physically, increase functionality per die area, lower power consumption and also the cost compared to GaAs and InP technologies, which have long been used in radar systems. SiGe is also compatible with conventional silicon IC manufacturing, which in addition decreases the cost penalty of migrating to a new technology. The major drawback of using SiGe in radar systems, according to Georgia Tech, is the total amount of power it can generate. SiGe can generate approximately one-tenth the power of traditional GaAs technology. These limitations can, however, be overcome when considering the increase in the number of individual elements that can be manufactured on an equal-sized die. In missiles, for example, power capabilities are limited by the size of the radar payload and the brute computing performance gained by using SiGe, at low power consumption, can lead to smarter and more intelligent radar systems, as opposed to following a strategy of constantly increasing radar output power.

A technology that has shown much promise in the field of EW is using an optical carrier in radar and EW applications, a research field in microwave photonics.

1.12 Microwave Photonics

"*Microwave photonics is an area that studies the interaction between microwave and optical waves for applications such as radar, communications, sensor networks, warfare systems [EW] and instrumentation*" (Yao 2012). Yao (2012) identifies four topics in microwave photonics,

- photonic generation of microwave signals,
- photonic processing of microwave signals,
- photonic distribution of microwave signals, and
- photonic ADC conversion.

Fiber-optic functionalities in laser radar and EW can be found in various subsystems and on system level in EW. Driven by the communications sector and because of the constant evolution of technology, microwave photonics promises (and has occasionally delivered) some cost versus performance benefits compared to traditional analog RF carriers. These advances have been used in multifunctional AESA (M-AESA) antennas for the timing and control functions and the digital beam-forming networks (D-BFN) wideband radar systems. Very short-haul optical digital transport technology offers high compliance against electromagnetic interference (EMI) and high data rate signal generation in excess of 10 Gbps (Pierno et al. 2015). The potential of photonics in EW can be realized using these systems and present high power and linearity benefits and low noise operation. Advances in photonics relate to decrease in cost, which is a large driver to realize microwave photonics-based systems in EW.

1.13 Conclusion

This chapter introduces aspects of EW related specifically to the underlying technology, which enables fast and powerful ECM and offensive electronic attacks. The chapter provides an introduction to EW and lists major aspects concerning its implementation. The central theme of this book is re-imagining and re-inventing the seemingly *old space* of engineering in EW by introducing leading-edge technologies that are currently researched and developed. The worldwide shift towards the information age has changed EW in many aspects compared to its traditional existence; the movement of information to a digital age has created additional facets of EW, one such being IW. The war of information is not only focused on maintaining one's information safely, but also on techniques to intercept, acquire or corrupt the information of the enemy.

More traditional EW techniques are also presented in this chapter; ECM, DEW and UAVs (albeit a more recent addition) are introduced and the basic principles of each of these are discussed. Categories of ECM such as radar jamming (electronic as well as mechanical) and target deception are rapidly changing with advances in technology. High-speed, high-power and low noise electronic circuits are the driver towards more powerful ECM. The electronic weapons space is also evolving and EHF applications are introducing new and more powerful ways to generate vast amounts of energy towards enemy electronics and even radiation levels harmful to humans. Launch strategies of these weapons are shifting towards UAVs and similar applications. The ability to launch high-powered electronic weapons from a safe distance from the target has many benefits that decrease the risk of human casualties. These weapons, as well as communication systems, command stations and remotely controlled vehicles, all require specific and licensed frequency usage. This chapter briefly introduces the availability and designation of military spectral bands to give an idea of the types of equipment used in EW and its frequency requirements.

A central theme in this book is using newer and more advanced technologies to design HF devices. Traditional radar and electronics in EW have been using technologies such as GaN, superseded by GaAs, but SiGe offers advantages that cannot be achieved by the older generation technologies. High speed and device miniaturization are among these advantages, and this book aims to identify techniques that can harness the power of SiGe and overcome its shortages, such as its lower power output capabilities. This chapter introduces SiGe and highlights parameters applicable to improving EW through effective radar and countermeasure tactics.

The missile is an important offensive and defensive weapon in any military and is also a large driver towards developing improved EW. The missile in its simplified representation is also given in this chapter and subsequent chapters explore this component in further detail to expose its strengths and weaknesses to EW. Finally, microwave photonics, deemed an upcoming technology that is also set to change EW in many ways, is discussed in this chapter, as it is set to become an integral technology in building subsystems of EW equipment, especially where even high-speed analog electronics such as SiGe cannot compete. Interconnection between fiber optics and analog circuitry is another active research field that is discussed in this book.

References

Adamy, D. (2006). EW 102: A second course in electronic warfare. *Artech House*. ISBN 1580536875, 2004.

Anderson, R. J. (2008). *Security engineering: A guide to building dependable distributed systems* (2nd ed.). Wiley. ISBN 0470068523.

Chakraborty, P. S., Cardoso, A. S., Wier, B. R., Omprakash, A. P., Cressler, J. D., Kaynak, M., et al. (2014). A 0.8 THz fmax SiGe HBT operating at 4.3 K. *IEEE Electron Device Letters, 35* (2), 151–153.

Cilluffo, F. J., & Gergely, C. H. (1997). Information warfare and strategic terrorism. *Terrorism and Political Violence, 9*(1), 84–94.

Cressler, J. D. (2013). Radiation effects in SiGe technology. *IEEE Transactions on Nuclear Science, 60*(3), 1992–2014.

DHL Trend Research. (2014). Unmanned aerial vehicle in logistics: A DHL perspective on implications and use cases for the logistics industry. *DHL Customer Solutions and Innovations*.

Dinc, T., Kalyoncu, I., & Gurbuz, Y. (2014). An X-band slow-wave T/R switch in 0.25-μm SiGe BiCMOS. *IEEE Transactions on Circuits and Systems-II: Express Briefs, 61*(2), 65–69.

Ellis, J. D. (2015). *Directed-energy weapons: Promise and prospects*. Center for a New American Security.

Fleetwood, Z. E., Wier, B. R., Raghunathan, U. S., Lourenco, N. E., Oakley, M. A., Joseph, A. J., et al. (2015). Optimizing the vertical profile of SiGe HBTs to mitigate radiation-induced upsets. In *2015 IEEE Bipolar/BiCMOS Circuits and Technology Meeting (BCTM)* (pp. 72–75).

Fritz, K. E., Randall, B. A., Fokken, G. J., Walters, W. L., Lorsung, M. J., Nielsen, A. D., et al. (1999). High-speed, low-power digital and analog circuits implemented in IBM SiGe BiCMOS Technology. In *21st Annual GaAs IC Symposium* (pp. 41–44).

GlobalSecurity. Directed energy weapons. Retrieved January 8, 2016 from http://www.globalsecurity.org

Hill, R. J. (1987). Absorption by the tails of the oxygen microwave resonances at atmospheric pressures. *IEEE Trans. on Antennas and Propagation*, AP-*35*(2), 198–204.

Johnson, E. O. (1965). Physical Limitations on Frequency and Power Parameters of Transistors. *1958 IRE International Convention Record*. 13, 27–34, 1965.

Jungemann, C., & Hong, S.-M. (2013). Investigation of electronic noise in THz SiGe HBTs by microscopic simulation. In *2013 IEEE Bipolar/BiCMOS Circuits and Technology Meeting (BCTM)* (pp. 1–8).

Keller, J. (2014). Small UAVs begin taking on role of electronic warfare aircraft. Retrieved January 11, 2016 from http://www.militaryaerospace.com

Kim, J., Lee, Y. S., Han, S. S., Kim, S. H., Lee, G. H., Ji, H. J., et al. (2015). Autonomous flight system using marker recognition on drone. In *21st Korea-Japan Joint Workshop on Frontiers of Computer Vision (FCV)* (pp. 1–4).

Niu, G. (2005). Noise in SiGe HBT RF technology: Physics, modeling, and implications. *Proceedings of the IEEE*. Invited Paper, *93*(9), 1583–1597.

Pierno, L. Fiorello, A. M., Secchi, A., & Dispenza, M. (2015). Fibre optics in radar systems: Advantages and achievements. In *2015 IEEE Radar Conference (RadarCon)* (pp. 1627–1633).

Rieger, R., Schuh, P., & Oppermann, M. (2014). SMTR® module—Evolution towards airborne applications. In *2014 International Radar Conference* (pp. 1–6).

Saeed, F., Zulfiqar, R., Aziz, A., & Zaman, N. (2008). Fighting terrorism using DEW. *IEEE International Multitopic Conference, 485–488*, 2008.

Schuh, P., Rieger, R., Fleckenstein, A., & Oppermann, M. (2010). T/R-module technologies today and future trends. In *2010 European Microwave Conference (EuMC)* (pp. 1540–1543).

Ulusoy, A. C., Schmid, R. L., Zeinolabedinzadeh, S., Khan, W. T., Kaynak, M., Tillack, B., et al. (2014). An investigation of fT and fmax degradation due to device interconnects in 0.5 THz SiGe HBT Technology. In *2014 IEEE Bipolar/BiCMOS Circuits and Technology Meeting (BCTM)* (pp. 211–214).

Weise, T. H. G. G., Jung, M., Langhans, D., & Gowin, M. (2004). Overview of directed energy weapon developments. In *12th Symposium on Electromagnetic Launch Technology* (pp. 483–489).

Yao, J. (2012). *A tutorial on microwave photonics* (pp. 4–12). Canada: Microwave Photonics Laboratory Research Laboratory, School of Electrical and Computer Science, University of Ottawa.

Chapter 2
Charged Particle-Beam Acceleration and Lasers: Contextualizing Technologies that Shaped Electronic Warfare

2.1 Introduction

In 1931, Nikola Tesla, famous inventor and the father of alternating current (AC), publicized at a media conference that he was on the brink of discovering and presenting a totally new source of energy. Regarding this invention, he stated: "*The idea first came upon me as a tremendous shock ... I can only say at this time that it will come from an entirely new and unsuspected source*". Nikola Tesla was known to be strongly anti-war and his inventions were inspired by the desire to put an end to warfare. By 1937 war in Europe was inevitable, as tensions were rising, and World War II started in September 1939. A discouraged Tesla distributed his technical paper titled, "*New Art of Projecting Concentrated Non-Dispersive Energy through Natural Media*" to the United States, Canada, England, France, the Soviet Union and Yugoslavia to generate interest and financing for his idea. This paper described a method (although never implemented) to direct high-velocity air through an open-ended vacuum chamber; today the basis of charged particle beam weapons. The first research conducted to realize a DEW, a particle beam accelerator aimed at incoming missiles, was undertaken in 1958 by the Advanced Research Projects Agency (ARPA, founded in 1958), two years before the successful demonstration of the laser built by Theodore H. Maiman at Hughes Research Laboratories (the research arm of Hughes Aircraft, established in 1960). A charged particle-beam accelerator, as the name suggests, uses EM fields to accelerate ionized particles to nearly the speed of light and to contain these particles in a narrow, defined beam. A laser emits coherent radiation in a narrow and diffraction-limited beam.

This chapter describes the principles of operation of charged particle-beam acceleration and its potential applications in EW, followed by the principles of operation of lasers and its uses in EW and other military applications.

W. Lambrechts and S. Sinha, *SiGe-based Re-engineering of Electronic Warfare Subsystems*, Signals and Communication Technology,
DOI 10.1007/978-3-319-47403-8_2

2.2 Charged Particle-Beam Accelerator

The history of charged particle-beam accelerators (shortened to particle accelerators in this book) arguably began with the evacuated tube and spark coil of W.K. Roentgen, which accelerated electrons and resulted in the discovery of X rays in 1895 (Woodyard 1948). In 1897, J.J. Thomson discovered the electron using a particle accelerator known as a cathode ray tube by measuring the mass of the cathode rays and showing that the rays consisted of negatively charged particles smaller than atoms. Ernest Rutherford discovered in 1911 that an atom has a positive center, which contains most of its mass, and in 1920 proposed that the particles be called protons. A few years later, in 1932, James Chadwick was able to prove that the nucleus of an atom also contained a neutral particle, later called the neutron. Charged particle-beam accelerators are designed to accelerate protons to speeds approaching the speed of light; primarily to collide the particles against one another and study the collision (particle physics). Particle accelerators can be classified into three categories by adapting the work of Woodyard (1948), namely firstly according to the form factor of the accelerator, secondly according to the electric principles used to obtain an electric field in the generator, and thirdly according to the types of particles that are accelerated. These categories are depicted in Fig. 2.1.

As seen in Fig. 2.1, each category can be subdivided into two types. The form factors of the typical generators include linear and circular devices. To obtain a strong electric field, either direct current (DC) static fields or AC electric fields can be used. Distinguishing between the two types of particles that can be accelerated involves using light or heavy particles.

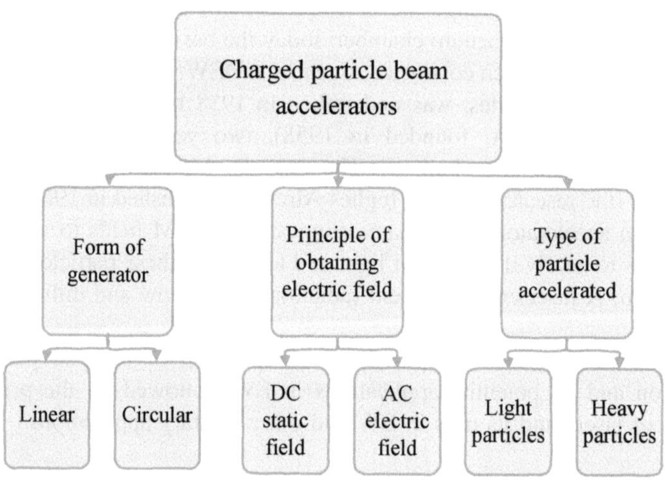

Fig. 2.1 Classifying the categories of charged beam particle accelerators (adapted from Woodyard 1948)

Linear accelerators push charged particles along a straight line through an electric field. The energy transferred to the particle is related to the length of the device and the voltage slope along its path. The Van De Graaff generator is an example of a linear particle accelerator.

Circular accelerators work on a similar principle, except that the particles travel around a circular path, allowing for smaller devices' routing particles to complete many revolutions within the same path. Electromagnets guide the particles in circular paths. Circular accelerators include the cyclotron, betatron and synchrotron. Cyclotrons, first conceived in Germany in the late 1920s, are the earliest operational circular accelerators. The first operating cyclotron was invented by E. O. Lawrence in 1929 at the University of California, Berkeley. Cyclotrons were the most powerful particle accelerators until the 1950s, superseded by the betatron and synchrotron. Particles are kept on a spiral trajectory by a DC static magnetic field and accelerated by a high-frequency RF electric field. The RF field is applied to two D-shaped sheet metal electrodes inside a vacuum chamber. These electrodes are placed face to face with each other with a narrow gap between them. Charged particles are injected into the center of this space. Electromagnets perpendicular to the electrode plane cause the particles to bend in a circle because of the Lorentz force perpendicular to their direction of motion. Synchrotrons synchronize to a particle beam of increasing kinetic energy to bend the charged particles into a closed path. A synchrotron therefore varies the strength of the magnetic field in time according with the particle gain energy, as opposed to varying the field in space. Synchrotrons typically receive pre-accelerated proton particles from cyclotrons or lower-energy synchrotrons. Betatrons, invented in 1940, accelerate the charged particles by induction from an increasing (similar to synchrotrons) magnetic field. The particles behave as if they are the second winding of a transformer and accelerate owing to the change in magnetic flux through its circular orbit.

A guideline of the required kinetic energy (K) in eV to accelerate a proton with rest mass of 0.938 GeV/c^2, where c is the speed of light (299,792,458 m/s), to a percentage of the speed of light is given by the European Organization for Nuclear Research (CERN), in a summary of the accelerators used at CERN (2009). These values are given in Table 2.1.

In Table 2.1, the linear accelerator 2 (Linac 2) at CERN generates 50 meV kinetic energy to accelerate proton particles to 31.4 % of the speed of light. The proton synchrotron booster (PS booster) is able to accelerate particles to 91.6 % of

Table 2.1 Relationship between the required kinetic energy and speed of a proton, achieved by CERN machines only (adapted from CERN 2009)

Kinetic energy of a proton (K)	Speed (percentage of speed of light)	Accelerator
50 meV	31.4	Linac 2
1.4 GeV	91.6	PS Booster
25 GeV	99.93	PS
450 GeV	99.9998	SPS
7 TeV	99.9999991	LHC

the speed of light using 1.4 GeV energy. The proton synchrotron (PS) accelerates proton particles either received from the PS booster or heavy ions from the low energy ion ring (LEIR) to reach 99.93 % of the speed of light by applying 25 GeV of kinetic energy. The super PS provides particle beams for the large hadron collider (LHC) at 450 GeV to achieve 99.9998 % of the speed of light, amplified to a total of 7 TeV in the LHC to reach 99.9999991 % of the speed of light. It is theoretically not possible to accelerate a particle with mass (m) to the speed of light. This can be mathematically presented by using Einstein's theory of relativity, where a particle's energy (E) is described by

$$E = \frac{mc^2}{\sqrt{1 - v^2/c^2}} \tag{2.1}$$

where v is the velocity of the particle along the axis of its movement (along a vector $\boldsymbol{r}$). If the velocity of the particle is zero, (2.1) simplifies to Einstein's equation of the theory of relativity for a particle at rest, given as

$$E = mc^2. \tag{2.2}$$

No particle with a non-zero mass can reach the speed of light, no matter how much energy is added to the particle. Famously, Einstein (1905) proposed that "*If a body gives off the energy L in the form of radiation, its mass diminishes by L/c^2*". Einstein used the letter L for energy and in subsequent work the symbol E was used to rewrite it in the form $E = mc^2$. In (2.2), c is constant, therefore adding energy to the particle increases its mass. By theoretically increasing the energy infinitely the mass of the particle would proportionally also increase infinitely, which is practically impossible and it cannot reach the speed of light by applying this principle. If the particle is moving, it has kinetic energy (K) and the total energy of the particle is the sum of its rest energy and its kinetic energy, given by

$$E = mc^2 + K. \tag{2.3}$$

The kinetic energy of the particle can be calculated by rearranging (2.3) to

$$\begin{aligned} K &= E - mc^2 \\ &= mc^2\left(\frac{1}{\sqrt{1 - v^2/c^2}} - 1\right). \end{aligned} \tag{2.4}$$

If the velocity of the particle is much lower than the speed of light ($v \ll c$), then the kinetic energy in (2.4) can be simplified to

$$K \approx \frac{1}{2}mv^2 \tag{2.5}$$

which will be transferred to any object that it collides with at this speed, called the translational kinetic energy (E_t). Nature produces the highest energies compared to any anthropogenic particle accelerators in the form of cosmic rays. The most energetic cosmic ray ever observed was a proton accelerated in excess of 1×10^{20} eV and the source of such ultrahigh energy cosmic rays is still being investigated. In nuclear and particle physics, particle accelerators have contributed in many inventions, discoveries and applications.

The LHC beneath the border between France and Switzerland built by CERN is a 17 mile (27 km) long circular tunnel with depths ranging from 164 to 574 ft (50–175 m) underground. It uses superconducting magnets and is the largest particle collider in the world. The LHC can potentially generate a total of 14 TeV (two high-energy particle beams generating 7 TeV of energy in ultrahigh vacuum each) of energy to accelerate and collide protons. Initial tests of the LHC were conducted at lower energy levels following a magnet quench incident, which damaged the superconducting magnets. The proton collisions are studied at four research stations of the LHC, named CMS, ATLAS and LHCb to investigate the particles that are detected during the collisions. One of the largest drives for building the LHC was the aspiration to find and identify the Higgs boson, which would prove or disprove theories such as supersymmetry and the existence of the multiverse. A Higgs boson of mass 125.3 GeV was identified on 4 July 2012 through experiments (initially estimated at 115 GeV to support the theory of supersymmetry or at 140 GeV to support the theory of a multiverse) on the LHC; however, scientists are still confirming whether this discovery is the true Higgs boson or if there is a possibility that multiple Higgs bosons may exist.

In Barbalat (1994) several examples of applications of particle accelerators are described. These applications are still in use today, with additional uses especially in scientific research emerging. These applications are listed in Table 2.2.

The applications and benefits of particle accelerators, as shown in Table 2.2, have been expanding with advances in technology in recent years. The operating principle of particle acceleration remains constant. Current research focused on electron and proton/ion particle acceleration through lasers to unprecedented energies in the high GeV range is receiving high amounts of attention. Research activity in this field aims to accomplish the development of high-quality and low-cost, versatile and small form-factor proton sources for research, medical applications and EW. In an article published by Extreme Light Infrastructure, the following abstract is taken from the website: "*... the dramatic rise in attainable laser intensity has generated an even more dramatic emergence and now evolution of the fields of research associated with non-linear laser-matter interaction. Production and acceleration of electrons up to 1 GeV over accelerating distances around 1 mm (100 meters for conventional accelerators) and hadron acceleration to 100 meV, are the clearly visible results of this evolution*". The increases in brightness and decrease in pulse duration of particle beams are methods to significantly change and improve on the investigation of particle matter, which will also lead to improvements and new applications of particle acceleration. Berkeley Labs published a world record in December 2014 for a compact particle accelerator

Table 2.2 Applications of particle accelerators (adapted from Barbalat 1994)

Field	Examples of applications
Scientific research	Nuclear physics (exploring the nucleus) Particle physics (building blocks of matter) Cosmology and astrophysics (cosmic observations) Atomic physics Condensed matter physics Chemistry/biology (chemical characteristics of molecules/cancer therapy)
Industrial applications	Semiconductor ion implanting Micro-photolithography Material surface modification Plasma-etching Scanning electron microscopes
Medicine	X-rays Nuclear magnetic resonance (used in X-rays) Gamma-ray therapy Neutron/heavy charged particle therapy Positron emission tomography Deoxyribonucleic acid (DNA) research
Radiation	Food preservation Sterilization of toxic waste Polymerization of plastics Explosive detection
Power engineering	Nuclear waste disposal Inertial confinement fusion Plasma heating

using a powerful laser (Berkeley Lab Laser Accelerator—BELLA) to accelerate subatomic particles to 4.25 GeV in a 3.54 inch (9 cm) long tube. The traditional means of accelerating particles by use of electric fields have a limit of approximately 100 meV per meter, as is the case for the LHC at CERN. Laser acceleration of subatomic particles have clearly overcome this limitation by an order of magnitude. Laser accelerators (also referred to as laser-plasma accelerators) inject a laser light into a short and thin cavity, which contains a specific plasma. The beam effectively creates a channel through the plasma as well as waves that trap free electrons and accelerate them to high energies. The acceleration gradient, $E_{A,}$ for a linear plasma wave can be determined by

$$E_A = c\sqrt{\frac{m_e \rho}{\varepsilon_0}} \tag{2.6}$$

where c is the speed of light in a vacuum, m_e is the mass of an electron, ρ is the plasma density in particles per cubic meter and ε_0 is the permittivity of free space.

As an introduction to DEW and particle-beam weapons, the concept of lasers is discussed in the following sections. A primary goal of this book is to explore and research the advantages and possibilities of using SiGe in various aspects of EW.

Its significance becomes apparent when considering the principles behind the technologies that enable it to make a contribution in these areas. Radar principles are also crucial to understand, interpret, highlight and identify the possibilities of SiGe as enabling technology and will be discussed in the following chapter. The following section first describes highlights in the history of the laser.

2.3 The History of the Laser

Historically, before the invention of microwave/molecular amplification by stimulated emission of radiation (maser) in 1953, usable and practical light emanated predominantly from spontaneous and scattered emissions from sources such as the incandescent bulb (first demonstrated in 1878) and other types of electromagnetic radiation such as IR, ultraviolet (UV) or gamma rays (Townes 1965). The maser was envisioned as a concentrated light source through research conducted by Charles H. Townes, Nikolay Basov and Alexander Prokhorov, leading to its invention in 1953. The maser is based on the theoretical principles of *stimulated emission* proposed by Albert Einstein in 1917, suggesting that excited atoms can radiate at specific wavelengths, including at visible light. To generate microwave energy for the maser, positive feedback by a resonant circuit is required to ensure that the gain in energy is greater than the circuit losses (Townes 1965). Masers are still used in various applications today, including satellite communication, radio telescopes, radar tracking, as oscillators in low-noise applications and as high-precision frequency references for atomic clocks. Masers, however, are historically associated with unwieldy machines requiring active cooling and producing small amounts of usable power as light. Maser research is still conducted; for example, in August 2012 Oxborrow et al. (2012) reported new interest in maser technology when scientists succeeded in exciting pentacene ($C_{22}H_{14}$) molecules by using a high-power laser and releasing microwave particles.

The maser is the predecessor of the laser. "A laser consists of an amplifying atomic medium occupying all or part of the volume of a suitable resonator. The role of the resonator is to maintain an electromagnetic field configuration whose losses are replenished by the amplifying medium through induced emission" (Yariv and Gordon 1963). In 1960, Theodore H. Maiman succeeded in operating the first pulsed ruby (chromium in corundum; Al_2O_3:Cr) laser. The original paper in August 1960 (Maiman 1960), a single-page article in Nature, reported that "an optical pumping technique was applied to a fluorescent solid resulting in the attainment of negative temperatures and stimulated optical emission at a wavelength of 6943 Å[1]" (694.3 nm). The first CW laser, a gas-based laser using helium (He) and neon, was demonstrated by the Iranian physicist Ali Javan in January 1961. The results are recorded in Javan et al. (1961).

[1]1000 angstroms (1000 Å) = 100 nanometres (100 nm).

During the 1960s researchers focused largely on the mechanisms to improve ruby-pulsed lasers and gas-based CW lasers. The US Army Electronics Research and Development Laboratory developed and demonstrated a ruby laser Q-switching technique involving the use of a revolving mirror mechanism, which generated single pulses of megawatt power, in August 1961. In Porto (1963) a technique to measure ruby laser outputs by reflecting the laser light in a barium sulfate ($BaSO_4$) diffuse reflector was presented and led to improvements to measure laser output wavelength. Bhawalkar et al. (1964) presented work on investigation of the spike pulses from the relaxation oscillator in a pulsed laser and Evtuhov and Neeland (1965) reported on measurement and interpretation of the frequency separation between transverse modes in ruby lasers. Military interest in using laser rangefinders was mentioned as early as 1965 (Benson and Mirarchi Benson and Mirarchi 1964) in a publication in which the limits to the range and accuracy of early lasers because of random and low-energy pulsed energy in these systems are addressed. The principle of the laser was already well-established in the 1960s and as seen from these early research papers, focus on its improvement received much attention from scientists. From the first pulsed ruby laser, many variations and different types of lasers have emerged, each with its own set of advantages, disadvantages and intriguing characteristics. Before discussing the types of lasers used in commercial, research and military applications and research, the basic principle of laser physics is presented.

2.4 The Basic Principles of Laser Physics

In the most general terms, a laser uses a mechanism that amplifies and guides light signals at a specific wavelength through a narrow and concentrated beam or cavity. Light is guided by either stimulated emissions or optical feedback, where oscillations of the light beam are typically provided by reflector mirrors. A simplified schematic representation of a laser system is given in Fig. 2.2.

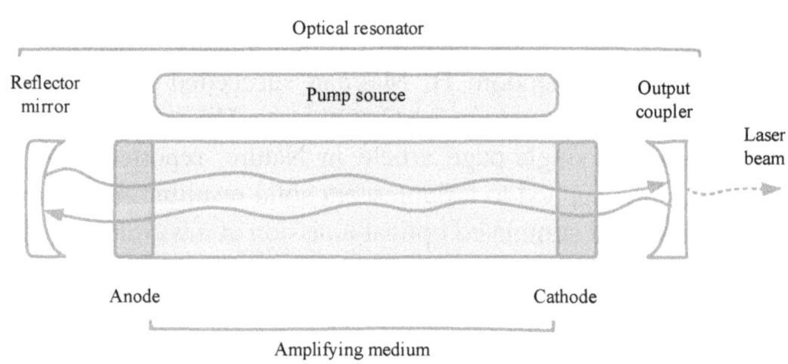

Fig. 2.2 Simplified representation of a laser system and the path of the light beam

As shown in Fig. 2.2, the laser consists of a gain or amplifying medium, also called the lasing material, to energize the light. The lasing material can be a crystal, gas, semiconductor or liquid dye and this is where stimulated emission occurs. A pump source, connected electrically by its anode and cathode, adds energy to the lasing material and the electrons that are added are excited to a higher energy level in a process known as population inversion. The process that induces population inversion in one form or another is called pumping. Pumping can be supplied by an electric current, by a light at a different wavelength than the generated output light or by another laser. Excited electrons then reside in an unstable condition and rapidly decay into its original energy state. The decay of these electrons occurs in two ways, either falling to its ground state while emitting randomly directed photons (light) or from spontaneously decaying electrons striking other excited electrons, causing them to descent to their ground state. The stimulated change releases energy as photons of light that travel at the same wavelength and phase as well as in the same course as the incident energies. Emitted photons move back and forth in an optical resonator and a set of mirrors feeds the light back into the amplifier to increase the developing beam in the optical cavity continuously. At the back of the laser is a fully reflecting mirror to ensure that the beam cannot escape the device at this end and in the front, a portion of the beam is emitted out of the device and a portion is reflected back into the loop through an output coupler. This systematic approach eliminates the possibility of the generated light radiating in all directions from the source and ensures that it has a focused trajectory in a single direction.

Optical amplification in the gain medium can be described by its energy level diagram, showing how the pump source induces transitions of ions from an excited state towards lower states. Quantum mechanical forces in atoms force electrons to take on discrete positions in orbitals around the atom nucleus. A simplified representation of the quantum mechanical effects that induce emission of light is given in Fig. 2.3.

Figure 2.3 shows the three distinguished states of a stimulated electron and the stages of spontaneous emission. If the electron is excited by an external source it takes on an excited state, depending on the energy applied to the system (ΔE). It therefore moves from a ground state E_1 to the excited state at E_2. The change in energy intensity is due to the incident photons directed at the material, such that

$$E_2 - E_1 = \Delta E = hv \tag{2.7}$$

where h is Planck's constant (4.135×10^{-15} eV s = 6.626×10^{-34} J s)[2] and v is the velocity of the incoming light. The energy added to the system can be either through incident photons (light) or incident heat/radiation (phonons) and the energy is transferred to the electrons, while in the excited energy state the electrons must naturally decay back to the ground state if the incident energy is removed or lowered. This fact highlights the reason why in 'pulsed' lasers, electrons are given a

[2] 1 Joule (1 J) = $6.242 \times 10^{+18}$ electron-volts ($6.242 \times 10^{+18}$ eV).

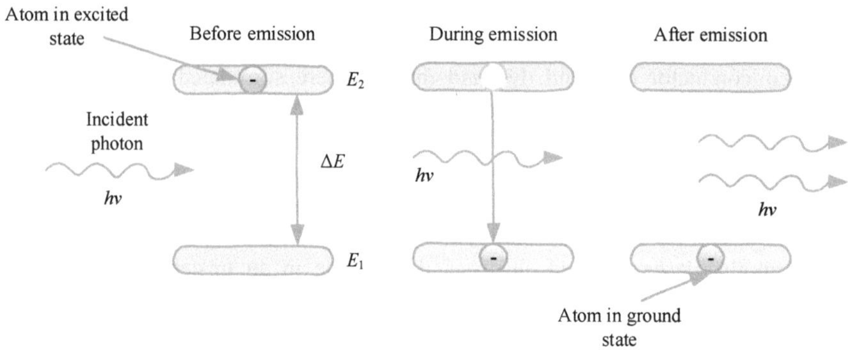

Fig. 2.3 Quantum mechanical effects showing the electron states before, during and after stimulated emission

fair chance/time to decay. During this stage, as the electrons decay, emission of light occurs with a wavelength dependent on the gas, solid or liquid material. The pulse of incident energy must be longer than the inherent time constant of the decaying transition to allow the electrons to reach the lower energy level. If the electron finds an unoccupied state in the lower energy level it emits a photon and this is called spontaneous emission. This photon again has an energy of hv and is spectrally defined at a wavelength. The photon exhibits random phase properties and multiple photons that are emitted without any phase relationship, therefore in random directions, are called fluorescence and thermal radiation—a random occurrence.

In Bäuerle (2013) the thermal and non-thermal processes (thermo-physical, thermochemical, photochemical and photo-physical[3]) of the excitation energy are described using the relaxation time τ_T (molecules moving from E_2 to E_1 in Fig. 2.3) and structural rearrangement time τ_R (molecules moving from E_1 to E_2 in Fig. 2.3) of atoms or molecules within the surface of the material. For a thermal laser process it should hold true that $\tau_T \ll \tau_R$ and a source must be used with $hv > E_g$ to excite electrons to a higher state. Laser-driven heat sources follow the same thermo-physical and thermochemical processes as conventional heat sources; a significant difference in these heat sources is the ability to induce temperature increases localized to small volumes in space and time by focused laser beams. Short and high-powered driver pulses add to thermal emissions through kinetic energy transfer as well.

Photochemical lasers operate under the condition that the thermal excitation energy time constant is relatively slow compared to the natural transition times, therefore when $\tau_T \geq \tau_R$. This process assumes that molecules are already absorbed in the top layer of the material and can interact and react with photolytic chemicals

[3]Photo-physical processes are defined as processes where thermal and non-thermal mechanisms contribute to the overall processing rate (Bäuerle 2013).

to transfer charge between states. Under a laser irradiation source these reactions typically show small variations in temperature over time. Bäuerle (2013) presents a model of the total reaction rate as the sum of the thermal channel (W_T) and the photochemical channel (W_{PC}) such that

$$W = W_T + W_{PC} = k_A \tilde{N}_A \left[1 + \frac{k_{A^*}}{k_A} \left(1 + \frac{hv}{\sigma I \tau_T} \right)^{-1} \right] \tag{2.8}$$

where A and A^* characterize the system in the ground state and excited state respectively, σ is the excitation cross-section, k_i is the rate constant of the system state (τ_i^{-1}), I is the intensity of the incident radiation and $\tilde{N}_i$ is the quasi-stationary density of species at its occupying state. The relevance of (2.8) is specifically found in its ability to characterize the process as thermal or photochemical through the time constant τ_T. For very small constants ($\tau_T \ll \tau_R$) the total reaction rate depends primarily on k_A and the thermal channel dominates. As the time constant rises in relation to the rearrangement time, the interaction and competition between thermal and photochemical processes become significant. The heat distribution of fixed laser parameters within the irradiated zone can be modelled by the heat equation, assuming the absence of thermal radiation and heat transport by convection. The heat equation in a Cartesian coordinate system that is fixed with the laser beam is given in Bäuerle (2013) through the first law of thermodynamics (conservation of energy) such that the spatial energy in the system $Q(x, t)$ is given by

$$\begin{aligned} &\rho(T) c_p(T) \frac{\partial T(x,t)}{\partial t} - \nabla[\kappa(T) \nabla T(x,t)] \\ &+ \rho(T) c_p(T) v_s \nabla T(x,t) = Q(x,t) \end{aligned} \tag{2.9}$$

where $\rho(T)$ is the mass density and $c_p(T)$ is the specific heat at constant pressure, v_s is the velocity of the medium relative to the heat source and κ is the thermal conductivity. In the most general case $T \equiv Y(x, t) = T(x_a, t)$ is a function of both the spatial coordinates x_a and the time t (Bäuerle 2013).

To achieve stimulated emission and increase the probability of an atom entering a transition state called absorption where the incident photon is destroyed, an external electromagnetic field at a frequency associated with the transition is added to the system. Population inversion is essentially the redistribution of atomic energy levels in a system and what makes laser emissions occur. In a simple two-level energy system, it is not possible to obtain a population inversion with optical pumping, as the atoms in such a system are in thermal equilibrium and there are effectively more atoms in the lower energy state than in the higher energy state. If atoms in a two-level energy system are excited from the ground state to the excited state, statistical equilibrium will eventually be reached in response to spontaneous and stimulated emissions. Equal population of the two states can be achieved, which results in optical transparency but without remaining optical gain. To achieve non-equilibrium conditions, an indirect method of populating the excited state must be used. The minimum pumping power ($P_{\min}$) for a pumping process that is assumed to be 100 % efficient, i.e. only the energy lost by radiation needs to be replenished, can be determined by

$$P_{\min} = \frac{N_m hc}{\lambda_L t_c} \tag{2.10}$$

where N_m are the finite number of modes of the laser within its finite bandwidth, the energy in the given mode is characterized by the average lifetime t_c and the operation wavelength is given by λ_L. The rate of stimulated emission is proportional to the difference in the number of atoms in the excited state (N_2) and in the ground state (N_1), which depends on the average lifetime of the atoms in the excited state and the average lifetime of the emission in the laser cavity. Mathematically this is represented by

$$\Delta N_m = N_2 - N_1 = \frac{N_m \tau}{t_c} \tag{2.11}$$

where the atoms with an excited state have a lifetime of τ if placed inside a cavity. For an electron bound to a nucleus and residing at an atomic orbital, its energy level is related to its state by

$$E_m = -hcR_\infty \frac{Z^2}{N_m^2} \tag{2.12}$$

where R_∞ is the Rydberg constant ($R_\infty = 1.097 \times 10^7\ \text{m}^{-1}$ and $hcR_\infty =$ 13.605 eV) and Z is the atomic number. If there are more atoms in the upper level than in the lower level, the system is not in equilibrium and the distribution of atoms between the levels is given by Boltzmann's law as

$$N_2 = N_1 x e^{\left(\frac{E_2 - E_1}{kT}\right)} \tag{2.13}$$

and in this case, N_2 is always less than N_1. To create a situation where atoms are not in equilibrium, an external factor is required to vary the energy levels between states. In lasers, this is called the pumping stage (electrical, optical or chemical), raising atoms to an excited, upper energy level. Light can only be amplified if $N_2 - N_1 > 0$. In optical pumping there must be at least three distinct energy levels to create a population inversion. The energy-level diagram of a three-level system is given in Fig. 2.4.

Figure 2.4 shows the pumping transition between E_1 and E_3 and laser transition as the atoms decay from E_2 and E_1. The transition between E_3 and E_2 is a rapid decay from the highest energy state to the metastable energy level but does not contribute to laser radiation. This transition between E_3 and E_2 limits the possible systems that allow population inversion in three-level systems. Although this representation of energy transitions seems simple to achieve, the ground state of the

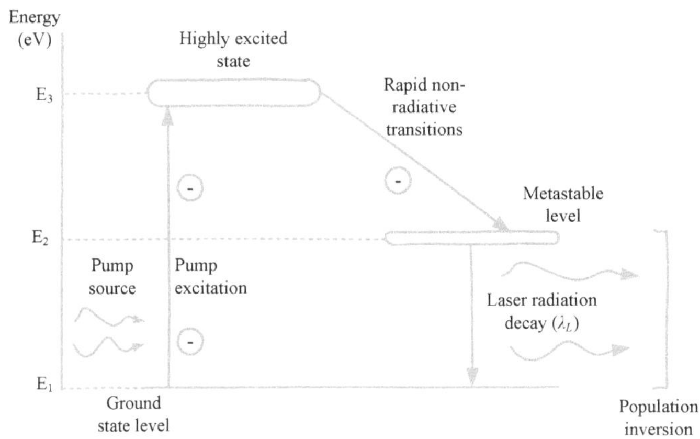

Fig. 2.4 Energy level diagram showing the principle of a three-level system

energy system has a large population of atoms at thermodynamic equilibrium. The practical implementation of achieving such a system is more complex and requires high-powered pumping energies. The metastable energy level (E_2) must be able to store atoms decaying from E_3 and ensure that spontaneous non-radiative emissions become less likely. It also requires great pumping energy to excite atoms from E_1 to E_3. The first ruby laser demonstrated in 1960 used Al_2O_3:Cr, an aluminum crystal matrix doped with Cr^{3+}, whose energy levels match such a three-level system and create the laser effect.

Another example of a spectroscopic system is the four-level laser. The four-level system is depicted in Fig. 2.5.

In the four-level system in Fig. 2.5 the optical pumping transition and the laser transition occur over a pair of distinct levels (E_1 to E_4 for the pumping and E_2 to E_3 for the laser transition). E_2 is chosen to be sufficiently far away from the ground state E_1 to ensure that the thermal equilibrium at thermodynamic equilibrium is negligible. Atoms do not stay in E_4 or E_2 and atoms excited to E_4 rapidly decay to E_3 with non-radiative emissions and from E_2 to E_1 due to natural depopulation, also a non-radiative transition. As soon as an atom moves to E_3, a population inversion occurs and the medium becomes amplifying. To maintain the population inversion, atoms should not accumulate in E_2, but must rapidly decay to E_1. A known compound commonly used that operates as a four-level energy-level system is neodymium YAG, given by Nd:$Y_3Al_5O_{12}$, and emanates at a wavelength of 1064 nm.

An example of a spectroscopic system using an electrical pumping mechanism and energy transitions is a He-neon (HeNe) gas system. Such a system is depicted in Fig. 2.6.

In the HeNe system shown in Fig. 2.6 the neon transitions are used for the laser transitions and He is the intermediary gas capable of transferring energy from the electrons to the neon particles through collisions. It should be noted from Fig. 2.6

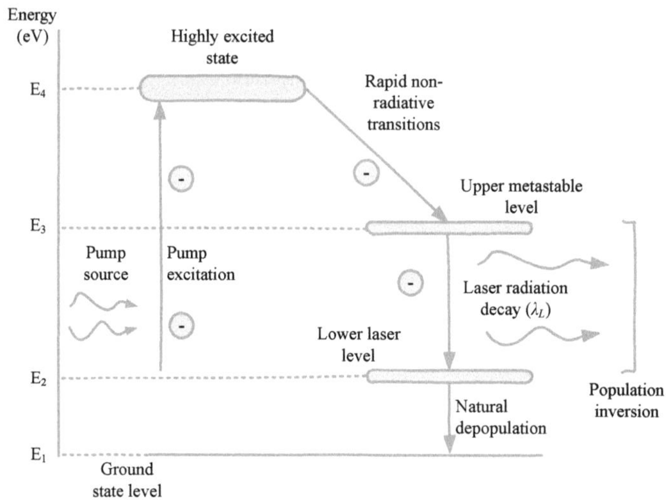

Fig. 2.5 Energy level diagram showing the principle of a four-level system

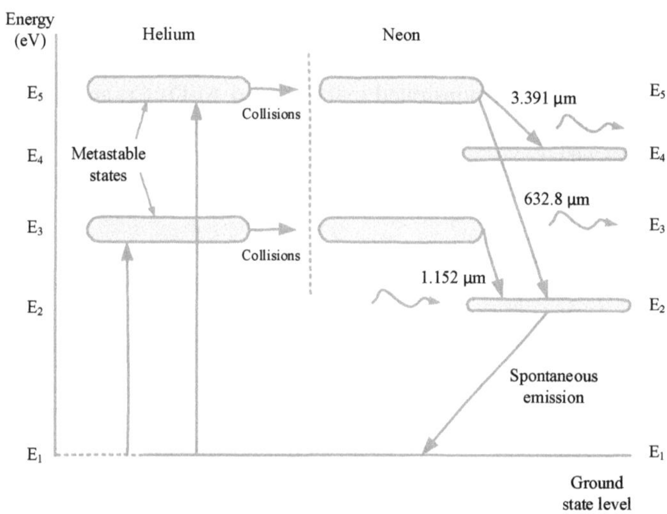

Fig. 2.6 Energy level diagram of a HeNe gas system using electrical pumping to transition between states

that the lower levels of the laser transitions occur relatively far from the ground state and this favors population inversion as opposed to thermal population. He has an additional characteristic in that it has two metastable states where atoms can stay for longer periods of time before falling to the ground state. He atoms are transported to the excited state through collision with electrons. Energy is transferred to the neon particles when the atoms collide because these metastable levels of He coincide with the excited states of the neon. The process of population inversion can simplistically be represented by

$$\mathrm{He}^* + \mathrm{Ne} = \mathrm{He} + \mathrm{Ne}^* \quad (2.14)$$

where He^* and Ne^* represent the atoms at excited levels. The transfer of energy occurs while atoms decay to lower energy levels. HeNe systems can emit radiative laser light at 543, 632.8, 1152, 1523 and 3391 nm; 632.8 nm radiation is most commonly used in commercial applications such as optical laser-disc readers. A list of commonly used laser materials with various energy-level systems and their corresponding wavelengths and applications are given in Table 2.3.

The optical power output of a laser is dependent on the in-lasing threshold. This threshold is the lowest excitation level at which the output is dominated by stimulated emission as opposed to spontaneous emission. This depends heavily on the material used and its band gap (discussed in more detail in the following section). The optical power output of a laser beam is its continuous power output of CW lasers or the average power for a pulsed laser. The energy of a laser (E_{laser}) is the output of a pulsed laser and is related to the maximum attainable power output (P_{pk}), where

$$E_{laser} = P_{pk} \times t \quad (2.15)$$

where t is the duration of the laser pulse in seconds. The average power (P_{avg}) of a pulsed laser is the pulse energy multiplied by the frequency (f) of the laser repetitions per second, such that

$$P_{avg} = E_{laser} \times f. \quad (2.16)$$

Therefore, at constant average power levels, short pulse lengths (pico- to femtoseconds) result in high peak output powers with relatively low pulse energy, whereas long pulse lengths (milliseconds) result in low peak power but high pulse energy.

In Marshall and Burk (1986) the parameters to determine the optical link performance between a transmitter and a receiver through an optical channel are described. In this article the detector sensitivity and required signal calculations are not considered in the optical link performance and are determined purely from analytic models of the system components concerning the laser. The link parameters are listed in Marshall and Burk (1986) as part of a block diagram of an optical communications transceiver consisting of a laser, communications detector, optical assembly and a receiving telescope (optics) and are adapted and given as

- laser average output power (P_{avg}),
- laser wavelength (λ),
- transmit telescope aperture diameter (D_t),
- transmit telescope obscuration ratio (γ_t),

Table 2.3 Commonly used laser materials and their formulas, emitted wavelengths and example applications

Laser material	Formula	Wavelength (λ_L) in nm	Applications
Alexandrite (chrysoberyl)	$BeAl_2O_4$	755	Laser hair removal
Argon		457–528	Photocoagulation in ophthalmology, limited lifetime applications
Argon fluoride (exciplex)	ArF	193	IC production (photolithography), eye surgery
Carbon dioxide	CO_2	10,600	High-power CW applications, cutting, welding, engraving, military range-finding, LIDAR
Carbon monoxide	CO	5000–6000	Glass, metal and ceramic processing, skin resurfacing, surgical cutting
Copper vapor	CVL	510.6/578.2	Analytical instruments, spectroscopy
Erbium:Glass	(CR14 or EAT14)	1540	Skin resurfacing, optical fiber communication
Frequency doubled Nd: YAG	see Nd:YAG	532	Ophthalmology, oncology, laser peeling, laser rangefinders and designators
Helium cadmium	HeCd	325 (UV)/ 442 (blue)	Spectroscopy, stereo lithography
Helium neon	HeNe	632.8 (most used)	Laser disc players, laser alignment, particle measurement, velocimetry
Hydrogen fluoride	HF	2700–2900	High-power CW applications (military and space—missile defense)
Krypton fluoride	KrF	337.5–799.3	Photolithography
Laser diodes	–	405–950	Fiber-optic communication, DVD/Blu-ray readers, laser printing/scanning, laser pointers
Nd:YAG	$Nd{:}Y_3Al_5O_{12}$	1064	Laser-induced thermotherapy, scientific research
Rhodamine 6G	$C_{28}H_{31}N_2OCl$	450–650	Dye lasers, tunable lasers[a] (wide bandwidth research applications)
Ruby	$Al_2O_3{:}Cr$	694.3	Laser range-finding, holographic portraits, hair removal, tattoos
Ti:Sapphire	$Ti{:}Al_2O_3$	690–1100	Scientific research (due to tenability and ultrashort pulses)
Xenon fluoride	XeF	351	Photolithography, micro-machine plastics and composites
Xenon monochloride	XeCl	308	Dermatology, micro-machine organic tissue, precision surgery

[a]Coldren et al. (2004)

- transmitter optics throughput (η_t),
- transmitter pointing bias error (ε_t),
- transmitter jitter (σ_t),
- receiver telescope aperture diameter (D_r),
- receiver telescope obscuration ratio (γ_r),
- receiver optics throughput (η_r),
- narrowband filter transmission (η_λ),
- receiver point bias error (ε_r),
- receiver jitter (σ_r),
- narrowband filter spectral bandwidth ($\Delta\lambda$), and
- detector diametrical field of view (θ).

From the parameters listed above the transmitter and receiver optics throughput, transmitter and receiver pointing bias error and jitter, the receiver telescope aperture diameter and the receiver telescope obscuration ratio can be calculated from the reflection and transmission coefficients of the components in the optical system. Marshall and Burk (1986) state that to determine the link performance accurately, the system operational parameters must also be defined. These parameters include the

- data rate R,
- pulse-position modulation (PPM) alphabet size M,
- PPM slot time τ_s,
- link range L,
- atmospheric transmission loss factor $L_{a,}$ and
- background radiance B.

A convenient method to determine the number of received signal photons per PPM word for a laser pulse N_S as a normalized value of received power P_r in watts and the PPM word rate is presented in Marshall and Burk (1986) as

$$N_S = P_r \frac{\log_2 M}{R} \frac{\lambda}{hc}. \tag{2.17}$$

It follows that the number of received background photons per PPM slot time is given by

$$N_B = P_B \tau_S \frac{\lambda}{hc} \tag{2.18}$$

where the collected background power from point sources such as stars and extended sources such as scattered light is given by P_B and represented as

$$P_B = BL_a \frac{\pi D_r^2}{4} \left(1 - \gamma_r^2\right) \frac{\pi \theta^2}{4} \Delta\lambda \eta_r \eta_\lambda. \tag{2.19}$$

The equations to determine the number of photons per laser pulse assume that the type of laser and therefore the wavelength of the operation are known. The following section highlights the types of lasers used in commercial, research and military applications and the differences in operating modes between the types. Types of lasers including gas, chemical, liquid dye, metal-vapor, solid-state and semiconductor lasers are presented and discussed with precedence given to semiconductor lasers, since the context of this book is predominantly aimed at semiconductors and microelectronic technologies. SiGe can be used as lasers emitting in the THz spectral range (Kelsall and Soref 2003) and the technology is present in fast-switching electronic support circuitry and laser drivers (Knochenhauer et al. 2009 and Moto et al. 2013) discussed in the following chapter. Electroluminescence signals observed from the intersubband transitions in SiGe heterostructure quantum-cascade (QC) lasers have only been realized in the past decade and applications such as free-space communication within the atmospheric window and detection and trace analysis of organic metals (Tsujino et al. 2006) are among the primary fields that can benefit from this technology. Semiconductor lasers are only one category of several types of lasers and these types are discussed below to distinguish between the various alternatives.

2.5 Types of Lasers

Lasers operate on the principle of guided luminescence as a result of stimulation of a material and specifically exclude the luminescence resulting from the temperature (radiation) of the material (Ivey 1966). Luminescence may occur in gaseous, liquid or solid materials; solids offer the most practical means of implementation, primarily owing to the containment methods of the material. The three main categories of laser types are further divided into sub-categories, as shown in Fig. 2.7.

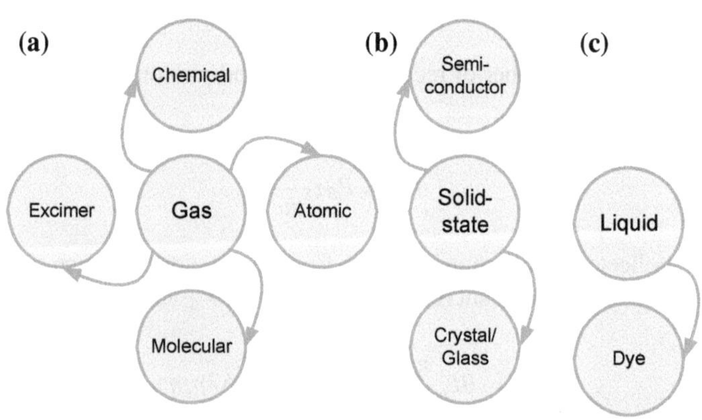

Fig. 2.7 Three main categories of lasers **a** gas, **b** solid-state and **c** liquid lasers and sub-categories of each

In Fig. 2.7 the three main categories of laser types are given, these being gas, solid-state and liquid lasers. Figure 2.8 distinguishes further between some of the most well-known examples of each type of laser.

Types of luminescence are distinguished based on the source of input excitation energy (photoluminescence, cathode luminescence and electroluminescence). Photoluminescence occurs in response to optical radiation, cathode-luminescence is generated by electron beams in cathode rays and electroluminescence is generated by electric fields or currents applied to the material. The source of the excitation process can differ; however, the physics of the transitions as a result of excitation remains the same across excitation methods. Adapted from Ivey (1966), there are three main categories of emitting transitions, which depend on the material and the conditions. These categories are defined by Ivey (1966) as

- transitions involving chemical impurities or physical defects (conduction band to acceptor, donor to valence band or donor to acceptor),
- inter-band transitions (intrinsic emission corresponding in energy to the band gap and higher energy emissions involving hot carriers, also called avalanche emissions), or
- intra-band transitions involving hot carriers (also called deceleration emissions).

It should also be recalled that not all electronic transitions are radiative and are reliant on the radiative qualities of the material. Electroluminescence of semiconductors are generally spontaneous and random, giving incoherent radiation. Emissions of photons initiated by source photons occurs in phase with the stimulating source to generate coherent emissions. For coherent emissions the number of electronic states available for emission must exceed the number of states available for absorption (population inversion of active states). Optical gain is achieved if the stimulated emissions can overcome the optical losses in the material. Radiative semiconductors take advantage of its inherently high index of refraction to reflect light generated inside the material to follow a cavity or optical path naturally. If oscillations in the material take place in response to optical gain, the radiated light

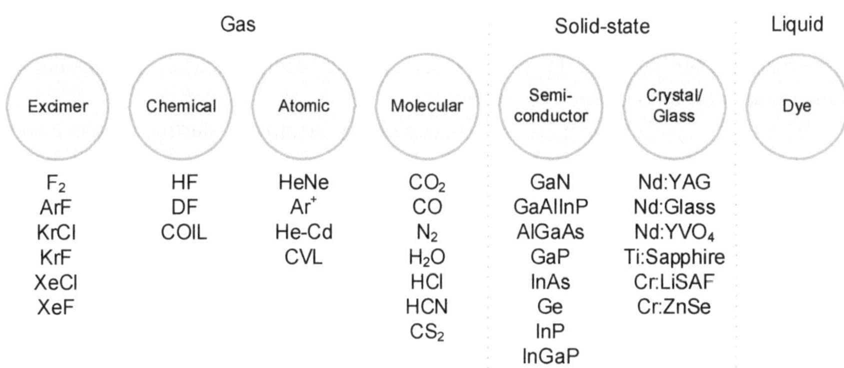

Fig. 2.8 Sub-categories of the types of lasers and common examples of each

will demonstrate spectral narrowing, phase coherence and beam directivity. The earliest recorded laser experiments were achieved in the early 1960s by injecting carriers in a forward-biased *p-n* junction using the direct band gap material GaAs (Nathan et al. 1962; Quist et al. 1962). GaAs was considered the most efficient and most common material used to conduct initial semiconductor laser experiments. Following the experiments conducted on GaAs, Bell Laboratories started researching phase chemistry and liquid epitaxy of $Al_xGa_{1-x}As$ injection lasers. A structure consisting of a layer of GaAs between *n*-type and *p*-type layers of $Al_xGa_{1-x}As$ showed simultaneous characteristics of acting as an optical waveguide and as a confinement region for the carriers injected in the GaAs laser in forward-biased conditions (Patel 1984). This led to the first room-temperature CW semiconductor laser at 880 nm in 1970 (Silfvast 2004). The following sections describe the operation principle of semiconductor lasers, solid-state, gas, chemical and liquid dye lasers in varied degrees of detail. Semiconductor lasers are used as the reference benchmark to elaborate further on the principles of laser operations since solid-state, gas, chemical and liquid dye lasers share many of their operating principles where primarily the pumping method and optical gain differ.

2.5.1 Semiconductor Lasers

Semiconductor lasers operate on the principle of semiconductor gain media, a condition induced by stimulated emission at an interband transition when the conduction band is at a higher carrier density. The gain in a semiconductor is a phenomenon where photon energy above the semiconductor band gap energy can excite electrons into a higher state in the conduction band. As the electrons reach this higher energy level they rapidly decay to states near the bottom of the conduction band. During this period, holes generated in the valence band move to the top of the valence band and electrons in the conduction band can recombine with these holes. During this recombination, photons are emitted at an energy near the band gap energy of the semiconductor. The energetic bands mentioned in the description of gain media above are classified as empty, filled, mixed or forbidden in semiconductor physics. Figure 2.9 depicts the band structures for insulators, semiconductors and conductors (metals).

The energy levels in Fig. 2.9 are occupied by electrons, starting at the lowest energy level. Electrons contributing to electronic conduction are situated in the higher energy bands. Electrons occupying the highest energy band in equilibrium are known to exist within the valence band. The insulator energy band structure in Fig. 2.9 indicates that the energy separation between the valence and conduction band is the largest and these materials do not conduct electrons under normal circumstances. Semiconductors have a smaller band gap between the valence and conduction bands and by applying an energy (electric field) higher that the band gap to the anode and cathode of the material initiates conduction of the material. In conductors such as metal, the conduction band and valence band overlap and there

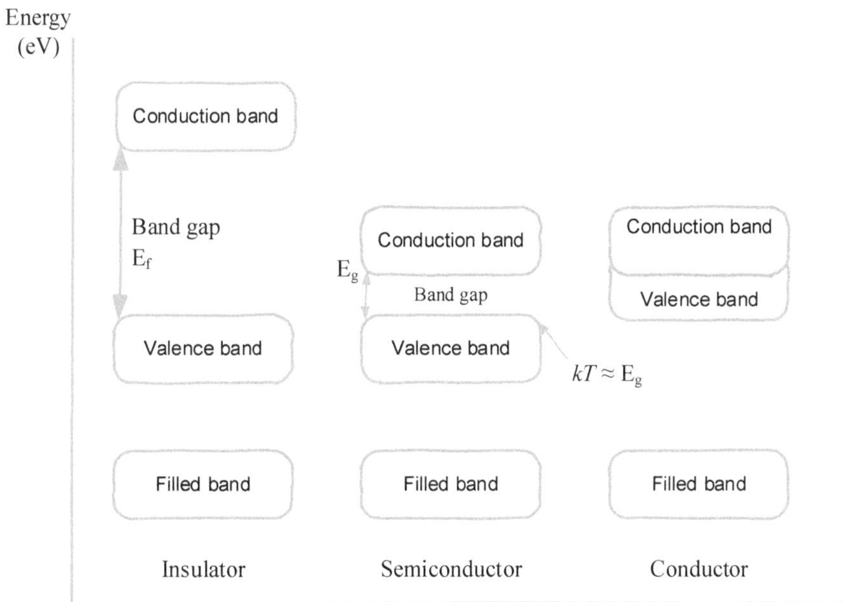

Fig. 2.9 Energy band structure of solid insulators (non-conducting materials), semiconductors and conductors (metal)

is no band gap between these energy levels. These materials conduct electrons under normal circumstances.

In semiconductor lasers the external energy to excite the electrons to higher states are applied by means of an electric field across its *p* and *n* junctions. If the electric field is applied, the electrons in the semiconductor structure acquire directed velocities in the opposite direction of the applied electric field. Each electron has momentum of

$$p = m^{*}v \tag{2.20}$$

where m^{*} is the effective mass of the electron moving in the lattice and having a different mass compared to a free electron due to its additional kinetic energy *E* (and its effective mass may also be negative—Einstein 1905). To relate the momentum, *p,* of the electron to its wavelength, the Broglie relation of

$$p = \frac{h}{\lambda} \tag{2.21}$$

can be used where *h* is Planck's constant and λ is the wavelength. The momentum, kinetic energy and the wave vector **k** can be related by

$$E = \frac{1}{2} m^* v^2 = \frac{p^2}{2m^*}. \tag{2.22}$$

It is shown in Gilmore (2004) that in momentum conservation the wave-function

$$\Phi_1(x) = e^{+ikx} \tag{2.23}$$

represents a particle to the right of a region with constant potential (electric field) $V < E$ with a specific momentum and if

$$k = \sqrt{2m(E - V)/\hbar^2} \tag{2.24}$$

and the reduced Planck constant $\hbar$ is given by

$$\hbar = \frac{h}{2\pi} \tag{2.25}$$

and

$$p = \hbar k \tag{2.26}$$

then (2.22) can be rewritten as

$$E = \frac{\hbar^2 k^2}{2m^*} \tag{2.27}$$

used to draw the energy bands as parabolas on an *E*-*k* diagram. The *E*-*k* diagram (see Fig. 2.10) shows the relationship between the energy and momentum of available quantum mechanical states for electrons in a material. In semiconductor materials it shows the band gap of the material, which is the difference in energy between the top valence band and the bottom conduction band and the effective mass of holes and electrons in the material through the curvature of each band in the diagram. It also indicates how the electron states are equally spaced in the *k*-space—in essence showing the density of states depending on the slope of the curve.

Semiconductors can be classified as direct band gap or indirect band gap materials. Defining the band gap for semiconductors is important in determining physical properties such as photoconductivity (photodetectors) and electroluminescence (lasers).

Semiconductors that can emit photons (crucial for laser operation) fall in the direct band gap category, since the crystal momentum of the electrons and holes is the same in the conduction and valence band. This means that an electron in the lowest-energy state in the conduction band can shift towards the highest-energy state in the valence band with zero change in crystal momentum.

In an indirect band gap semiconductor, the electron must first pass through an intermediate state and transfer momentum to the crystal lattice; photons are not

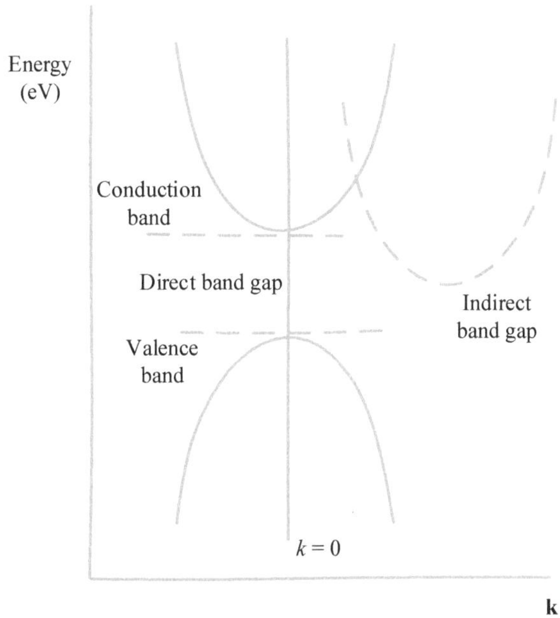

Fig. 2.10 Direct and indirect band gaps and the **k**-vector; showing the change in momentum for indirect band gap solid materials that cannot emit photons freely (adapted from Seo and Hoffman 1999)

likely to be emitted from the lattice. The crystal momentum responsible for the photon emission is described by the **k**-vector in the Brillouin zone defined as the plane boundaries of reciprocal lattice structures. Interactions between electrons, holes, photons and phonons must fulfil the conservation of the energy theorem and crystal momentum principles—thus the conservation of the total **k**-vector (Seo and Hoffman 1999). A photon with energy near the semiconductor band gap has close to zero momentum. Models used to describe free-carrier dynamics in the conduction band include the Boltzmann equation, the Fokker-Planck equation, Monte-Carlo methods, Stuart's single rate equation for the free-electron density and the multi-rate equation model (Gulley 2011). The direct and indirect band gap structures and the **k**-vector are depicted in Fig. 2.10 (Seo and Hoffman 1999).

There are various methods to determine if a solid semiconductor material has a direct or indirect band gap. These methods include plotting the optical absorption coefficient with varied wavelengths, spectroscopy, the *Tauc* plot (Tauc 1968) to determine direct/indirect and allowed/forbidden transitions and the shapes of the highest occupied molecular orbital and the lowest unoccupied molecular orbitals of constituent monomeric conjugated organic molecules.

In principle, a semiconductor/diode laser can be manufactured from any type of direct band gap material, but the efficiency of electrically injected lasers is dependent on precise doping concentrations of the active layers and require lattice-matched materials to facilitate molecule interaction between layers. Most commercial lasers used in for example pumping fiber amplifiers in telecommunications use III-IV compounds such as GaAs and InP heterojunction semiconductors. These materials present good lattice matching and efficient transport. Longer wavelengths of up to approximately 10 μm require other materials such as antimonide (Sb) as an alloy; for smaller wavelengths in the near-UV band (approximately 400 nm) epitaxial layers such as GaN are used. Defense and security applications often use vertical stacked lasers manufactured in II-IV compounds (ZnSe and ZnS for example) with light intensities of up to 2 kW (Laser Enterprise 2016). Common materials used in semiconductor lasers and for other optoelectronic devices are listed in Table 2.4.

Table 2.4 gives an indication of the wavelength of the laser light when implemented in various semiconductor materials. The equation used that relates the bandgap (E_g) and the cut-off wavelength (λ_{co}) is given by

$$\lambda_{co} = \frac{hc}{E_{g,}(eV)} \tag{2.28}$$

where h is Planck's constant and c is the speed of light in free space in m/s. In ternary and quaternary semiconductor compounds such as $Al_xGa_{1-x}As$ as in Table 2.4, the band gap can be varied by changing the content of the materials. The content of the material is defined by x where $1 \leq x \leq 0$. Another example in Table 2.4 is $In_xGa_{1-x}As$, where if $x = 0.53$ then the compound is defined as $In_{0.53}Ga_{0.47}As$ with $E_{g,300}$ of 0.75 eV.

Variant configurations of semiconductor layers produce different laser types; these include single-spatial-mode, multimode, bars and stacks, single-mode distributed feedback (DFB) and less commonly used lasers such as vertical cavity surface-emitting lasers (VCELs), distributed Bragg reflectors (DBRs) and tunable DFB lasers.

The simplest form and also the most commonly used one in commercial lasers is the single-spatial-mode laser. It is also called the single transverse mode laser

Table 2.4 Energy band gap and cut-off wavelengths of commonly used materials in semiconductor direct band gap lasers defined at 300 K

Material	Symbol	$E_{g,300}$ (eV)	λ_{co} (nm)	Band
Aluminium gallium arsenide	$Al_xGa_{1-x}As$	1.42–2.16	575–875	Visible
Gallium arsenide	GaAs	1.42	875	Visible
Gallium nitride	GaN	3.4	365	Near UV
Gallium phosphide	GaP	2.26	555	Visible
Germanium	Ge	0.66	1800	Short IR
Indium antimonide	InSb	0.17	5700	Medium IR
Indium arsenide	InAs	0.36	3400	Medium IR
Indium gallium arsenide	$In_xGa_{1-x}As$	0.73–0.47	1700–2600	Short IR
Indium gallium arsenide	$In_{0.53}Ga_{0.47}As$	0.75	1 655	Short IR
Indium gallium phosphide	InGaP	1.90	655	Visible
Indium phosphide	InP	1.35	919	Near IR

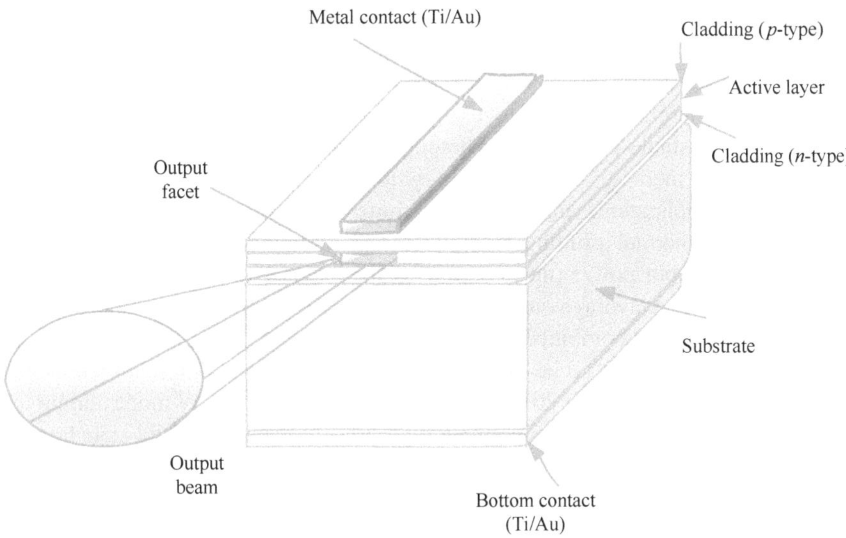

Fig. 2.11 Representation of a double-heterostructure single-transverse-mode semiconductor laser configuration

because of the implication that the beam can be focused to a diffraction-limited spot. Figure 2.11 shows a typical single-spatial-mode laser with a relatively narrow waveguide bounded by cleaved facets (Photonics Handbook 2015).

The single-spatial-mode laser in Fig. 2.11 shows a heterostructure with a narrow waveguide to guide the output laser beam. This waveguide can be achieved by a ridge-structure or a buried index step during processing. A single optical mode is supported in such a structure. The width of the waveguide is typically limited to 2–5 μm, depending on the actual waveguide structure. At the output facet the emitted laser light is typically around 1 μm high and 3–4 μm wide—again depending on the actual structure and application (Photonics Handbook 2015). As the laser light leaves the output facet, it naturally diverges into vertical and horizontal directions. The vertical divergence is perpendicular to the junction plane; it typically diverges about 20–30° and is referred to as the fast axis. The horizontal divergence is approximately 5–10° and called the slow axis. Since the divergence in the two planes is different (anamorphic), the output beam forms an oval shape. The output beam can also be somewhat astigmatic,[4] but both of these effects can be corrected through optics. Single-spatial-mode lasers are ideal when a diffraction-limited focus point is required, such as in optical data storage, thermal printing, fiber communications and laser pointers. Maximum average power output is limited in these lasers by the narrow waveguide generating significant local power density and

[4]Astigmatism in an optical system is when the rays propagating in two perpendicular planes have different focus points.

thermal dissipation. Trading off spatial coherence, to increase the average power output, the width of the waveguide can be increased to several hundreds of microns. These lasers are called broad-area or multimode lasers.

Multimode lasers cannot be focused to a diffraction-limited spot or efficiently coupled into a single-mode fiber (Photonics Handbook 2015). The output power also does not increase proportionally with the increase in the width of the waveguide. Applications that do not require diffraction-limited focus include external-drum thermal printing techniques and pumping solid-state lasers using high average output power (up to 6 watts at the expense of spatial coherence). To achieve even higher average output power in the order of 100 W, bars and stacks of multimode lasers are combined in a single structure and optically corrected at the output. Bar dimensions can be in the order of 1 cm high and 0.5–2 cm wide and a single bar can emit between 20 and 60 W of output power in CW mode and up to a 100 W in pulsed mode. Applications for high-powered pumping solid-state lasers at commercial wavelengths (785, 792, 808, 915 and 940 nm) typically use these configurations. To increase power further, bars are stacked in series and can thus use the same input voltage requirements with matched output impedance to the driver electronics; however, these configurations generally use large and costly high-current sources. These bar-stacked lasers are actively cooled (mostly water-cooled) during operation as a result of their high-current usage, can be used with or without optics depending on the application, found in applications for material processing where high beam quality is not required (welding, hardening, alloying or cladding of metals) and extremely high-power solid-state lasers. Single-mode DFB lasers offer low noise and narrow frequency spectrum characteristics and are becoming more common in optical transmission systems.

DFB lasers do not rely on mirrors to provide optical feedback; they use corrugated structures such as gratings close to the waveguide in the epitaxial layers. Each ripple in the grating offers a small amount of internal feedback and since the device uses a single frequency, each feedback from the gratings adds up in phase, which naturally decreases partition noise found when longitudinal modes in multi-frequency lasers interfere with each other. Noise levels of −150 dB/Hz are commonly seen in these DFB lasers. Applications in fiber-optic transmission systems requiring lower power but also very low-noise characteristics use DFB lasers to match high gain-bandwidth products (GBPs) of fiber amplifiers. DFB lasers are also found in interferometric and spectroscopic applications and have been demonstrated at wavelengths from 633 nm to upwards of 2000 nm. Other less common laser structures include VCEL, DBR and tunable DFB lasers.

In VCSELs, light propagates perpendicularly to the wafer and is emitted from its surface. To implement the required mirrors for optical feedback, mirrors are grown as multilayer dielectric stacks above and below the quantum well active layer.

> VCSEL multilayer stacks must have reflectivity as high as 99.8 % and the layer thicknesses must be precisely controlled to ensure that the resonator frequency falls within the gain bandwidth of the quantum well (Photonics Handbook 2015).

The growing and manufacturing of these devices are simpler compared to more traditional lasers and can be tested at wafer level, therefore making them more cost-effective and easier to control in terms of the quality and yield of the devices. VCSELs have low divergence and nearly circular beam profiles and are ideally used for coupling to optical fibers. DBR lasers provide a higher CW output power compared to DFBs at the cost of introducing mode hops with the drive current (Photonics Handbook 2015). Tunable DFBs can be achieved by having multiple grating structures introduced into the semiconductor, giving it a discrete tunable range, although increasing the manufacturing cost.

2.5.2 *Solid-State Lasers*

Semiconductor lasers also fall into the category of solid-state lasers (Figs. 2.7 and 2.8), but another type of solid-state laser is based on crystals or glass doped with rare earth or transition metal ions for the gain media. Another term for solid-state lasers, which specifically exclude semiconductor lasers, is doped insulator lasers. These can be manufactured in the form of bulk lasers, fiber lasers or other types of waveguide lasers. Commonly used solid-state lasers are mostly optically pumped with flash lamps or arc lamps and can achieve relatively high powers of tens of kilowatts. Drawbacks of solid-state lasers include low efficiency coupled with high thermal losses in the gain medium. As a result of the low efficiency and moderate lifetime due to thermal heating, laser diodes are often used for pumping solid-state lasers and actively cooled. These systems are called DPSS lasers and provide a good beam quality and better lifetime and are relatively compact systems.

Solid-state lasers present long lifetimes of radiative upper-state transitions, leading to efficient energy storage. This characteristic of solid-state lasers makes them suitable for Q-switching lasers that have extremely high peak powers. Solid-state lasers are generally not associated with large tuning bandwidths; rare-earth-doped laser crystals present a low tuning range, whereas rare-earth-doped glass lasers present a larger tuning range.

2.5.3 *Gas Lasers*

A gas laser is identified by the fact that the gain medium in the laser system is a gas that can be pumped by an electric discharge, RF waves, photons or e-beams. The gain medium consists of a narrow tube filled with a combination of gases (for example He and Ne). At its ends, an anode a cathode facilitates a high-current discharge through the tube, while mirrors at the ends of the tube provide optical feedback. Electrons from the discharge collide with the abundant gas atoms and excite these atoms to an upper metastable energy level. Through further resonant impacts the energy is transferred to the minority gas particles and raises these atoms to nearly identical energy levels. These minority particles can then decay back to the ground state to facilitate population inversion. Particles generally follow different routes towards the ground state and as a result of these different paths also generate various wavelengths during radiation emissions. Rare-gas ion lasers achieve stimulated emission when the discharge current is high enough to ionize the gas and through the electron-ion interactions excite the ions from the ground state to the metastable states.

At IR wavelengths it is possible to achieve high-power lasers from molecular gas such as CO_2. These lasers generate stimulated emissions from the low-energy transitions between vibration and rotation states of molecular bonds. High-powered molecular lasers are generally used in electrically discharged pumping techniques but can use RF excitations as well. According to Endo and Walter (2006), some advantages of gas lasers are the ability of the gas to adopt the cavity shape quickly, scalability, long lifetime if in a high-quality vacuum, recyclability of gas, homogeneity, stability, possible use of isotopes to shift the radiation spectrum and the wide spectra covered by varying gas types.

Metal vapor lasers are also gas lasers that vaporize metal (such as copper) at high temperatures and use this vapor as the gain medium in an optical cavity. The high temperatures required to vaporize the metal make it difficult to implement metal vapor lasers practically and require the optical cavity and mirrors to withstand high temperatures. Materials other from pure copper, such as copper nitrate or copper acetylacetonate, can help to achieve lower temperature operation and make these systems realizable at lower cost and with increased modularity. Typically, to achieve population inversion in metal vapor lasers, two successive energy pulses are required to dissociate vapor molecules, followed by a pulse to cause the dissociated ions to lase. There are two main types of metal vapor lasers: ionized metal (such as He-Cd) and neutral metal (such as Cu) vapor lasers; both operate by vaporizing metal in a container.

2.5.4 Chemical Lasers

Chemical lasers use chemical reactions as opposed to light or electrical sources as the pumping source. The COIL laser presents high efficiency and high average power and has proved to be an important laser in antimissile defense systems. A major advantage of chemical lasers is the relatively small power generation required, which is a valuable commodity on mobile air defense systems such as missiles to initiate a chemical reaction. Most of the energy required for the lasing action is stored in the fuel/chemicals themselves. The basic principle of the COIL laser is therefore that a fuel such as Cl_2 gas is mixed with an oxidizing agent such as hydrogen peroxide (H_2O_2) in a combustion chamber and a spark is ignited to activate the reaction. The product of this reaction is the excited molecular oxygen. This reaction is mixed with an iodine vapor and flashed through a nozzle where the components interact to produce excited iodine atoms. The lasing activity is achieved in an optical cavity across the expansion chamber. For atmospheric aerial applications, an iodine laser is ideal, since its 1300 nm wavelength falls in the atmospheric transmission window. For space-based applications, for example, HF lasers are considered to be more efficient.

2.5.5 Liquid Dye Lasers

The gain medium of liquid lasers is optically pumped liquids, typically at room temperature. The most common liquid lasers are dye lasers, which consist of organic dyes in liquid solvents. Dye lasers can generate a large bandwidth of laser light from the excited states of these dissolved dyes. The output types of these lasers can be either pulsed or CW and span across spectra from near-UV to near-IR, depending on the dye used (LaserFocusWorld 1995). The principal operation of dye lasers involves optically pumping the organic molecules to excited states through arc lamps, flash lamps or cascaded laser sources. Typical laser sources used in dye laser pumping include frequency-doubled Nd:YAG lasers, CVL, Ar^+, N_2 and excimer lasers. The dye solution is usually pumped transversely through the laser cavity and is contained by a transparent chamber, which is called a flow cell. The broadband laser emissions originate from interactions between the vibrational and electronic states of the dye molecules, which split the electronic energy levels into broad energy bands similar to those of vibrational lasers. In order to tune the laser to a desired frequency, wavelength-selective cavity optics are used (LaserFocusWorld 1995). Dye lasers have high efficiency, good tuning abilities, broadband operation and high spatial coherence.

2.5.6 Other Types of Lasers

Various other types of lasers exist with variations on pump sources and applications. Such lasers include

- FELs,
- gas-dynamic lasers,
- Raman lasers,
- nuclear-pumped lasers, and
- laser-driven plasma accelerators.

A FEL uses high-speed electrons that can move freely in a magnetic field as its lasing medium. These laser types have the largest tunable frequency range of all laser types and span from microwave, THz-radiation, IR, visible and UV to X-ray. The pump source is a relativistic electron beam with emission wavelength (λ_n) from an undulator[5] with a large number of periods, interference effects in the radiation produced at a large number of essentially co-linear source points in a spectrum with quasi-monochromatic peaks given in Brown et al. (1983) as

$$\lambda_n = \frac{\lambda_u}{2n\gamma^2}\left(1 + \frac{K^2}{2} + \gamma^2\theta^2\right) \quad \text{with } n = 1, 2, 3, \ldots \tag{2.29}$$

where θ is the angle of observation relative to the average electron and direction, n is the harmonic number, λ_u is the spatial period of the magnetic field (the undulator wavelength), γ is the relativistic Lorentz factor and the wiggler strength K is given by

$$K = \frac{eB_0\lambda_u}{\sqrt{8\pi m_e c}} \tag{2.30}$$

where B_0 is the applied magnetic field, m_e is the electron mass, c is the speed of light in a vacuum and e is the electron charge.

Gas-dynamic lasers take a hot and high-pressure mixture of gases, typically CO_2, N_2, H_2O or He, and expand the mixture rapidly through a supersonic outlet. During the expansion the gas is transformed into a laser medium and population inversion is generated. The gas then passes into a laser cavity and the beam is split perpendicular to the flow of the gas through mirrors placed on both ends of the cavity. The gas stream then enters a diffuser where it releases energy and is decelerated to subsonic speeds and exhausted out of the cavity. The non-equilibrium flow through the nozzle and its behavior through the laser cavity determines the distribution of molecules and the conditions where population inversion in the expanding gas occurs. Gas-dynamic lasers are very high-powered and high-efficiency devices with average output power in the hundreds of kWs.

Raman lasers use stimulated Raman scattering to achieve light amplification as opposed to the more traditional electrically induced stimulated transitions. Raman scattering is the inelastic scattering of photons during excitation as opposed to the more general, elastic (Rayleigh) scattering where the photons have the same energy (frequency and wavelength) of the incident photons. Raman lasers are optically pumped, but this optical pumping does not produce population inversion. Pumped photons are absorbed and transitions to lower-frequency laser light. Theoretically, any Raman laser wavelength can be achieved through specifying the pump wavelength, provided both wavelengths are within the transparency region of the material and a sufficiently high non-linearity and/or optical intensity is reached.

If a laser is pumped with energy of fission fragments, it falls under the category of nuclear pumped lasers. The lasing medium is surrounded by a tube lined with uranium-235. To achieve lasing, the medium is subjected to a high neutron flux in a

[5]A periodic structure of dipole magnets forcing electrons to undergo oscillations and resulting in radiated energy.

nuclear reactor core. Nuclear pumped lasers can, for instance, be used to pump X-ray lasers and energize EMP weapons used in EW offensive tactics. Nuclear pumped lasers can scale to very high energy levels, upwards of 100 MW CW beam power, and historically dominates military research. The size and scale of nuclear pumped lasers are extremely large to accommodate the nuclear reactor cores and since these lasers are predominantly CW, the power densities required are very high, but the energy densities emitted are lower compared to electrically charged gas lasers. The most well-known nuclear pumped laser project was Project Excalibur—an X-ray laser used as a DEW for ballistic missile defense. Project Excalibur was officially abandoned in 1992 and research resources were redirected to space-based laser satellite concepts.

Laser-driven plasma accelerators use large electric fields formed within plasma waves to accelerate charged particles to high energies in smaller distances compared to traditional particle accelerators. If a laser pulse propagates through a plasma the non-linear force in an inhomogeneous oscillating electromagnetic field (ponderomotive force) pushes electrons away from the front and back of the pulse and forms a trailing longitudinal density wave. The ponderomotive force in laser-driven plasmas $\mathbf{F}_p$ in newton is expressed by

$$\mathbf{F}_p = \frac{e^2}{4m\omega^2}\nabla\left(E^2\right) \tag{2.31}$$

where ω is the angular frequency of oscillation of the electromagnetic field and E is the amplitude of this field. e and m are the electron charge and mass respectively. These forces allow plasma accelerators to achieve acceleration gradients in the order of 100 GeV/m compared to the approximate 100 meV/m of traditional particle accelerators such as the LHC (Esarey et al. 2009). Calculation of the driven electron plasma waves (wakefields) generated by non-evolving drive laser pulses is presented in Esarey et al. (2009).

Optimization of laser technology involves improving on current technologies and laser types to increase the average power output of devices using the same (or lower) power sources. The following section introduces some optimization techniques based on principles that can be applied to most laser types.

2.6 Laser Optimization

Since the early era of laser research and even today, researchers gave pursued various fundamental goals in improving laser technology. These goals are both technology and application specific and aim to improve the throughput and efficiency of laser pumping through a variety of applications. No single laser can satisfy all the requirements for the vast applications of lasers and research into improving single laser technologies for their associated applications is necessary. A list of general goals of laser advances are improving on

- the type of applications requiring different spectra of luminous light and finding efficient and cost-effective ways to implement it,
- technology advances aimed at increasing the maximum average output power of laser types,
- technology advances aimed at increasing the energy of the peak pulse,
- increasing the output power of the peak pulse,
- improving on the enabling technologies and supporting circuitry to allow higher frequency laser pulse throughput,
- techniques for achieving higher power efficiency, and
- ultimately lowering the cost and decreasing the physical size of lasers.

The classification of lasers differs on the intended end-use of the laser; however, the most general classification that reveals most information is considered to be the state of the lasing medium (gas, solid or liquid). Another consideration when classifying a laser, albeit a somewhat reversed methodology, is classification by its wavelength of operation. This is, however, a common way to determine which laser is right for a certain application when the intended application defines the wavelength required and the type of lasing medium is chosen based on this requirement. Bhatia and Kumar (2002) further classified laser types in terms of their susceptibility to EMI and lists categories that influence interference in operation:

- The physical construction of the laser is decisive; Bhatia and Kumar (2002) stated that more voluminous and longer dimensioned laser tends to be better suited for radiated EMI emissions.
- High-powered lasers contribute to larger conducted noise levels and the operating power as a function of input power should be limited through more efficient lasers.
- Considering the operating mode, CW lasers are generally less noisy compared to their pulsed mode counterparts because of the lower frequency activity in its electrical drivers. Pulsed lasers, especially very high frequency/short pulses, contribute more to noise at the output;
- The optical pumping method influences the generated noise and interference in the laser output and it is generally found that gas lasers, because of their complex electrical requirements and high frequencies, generate higher levels of noise compared to for example flash lamp pumping.
- Any additional accessories used for the laser pumping and supporting circuitry adds to the generated noise levels in the output signal.

Chen et al. (1996) highlight the fact that in optically coupled lasers, the pump source and the laser resonator do not interfere with each other (optical versus electrical characteristics), which lowers the total output noise of the system. Also mentioned is the fact that a fiber-coupled system has significantly lower brightness, since the brightness of the output beam is always lower or at best equal to the brightness of the input beam, measured in W/cm^2.sr (where steradian (sr) is the unit of the solid angle). Chen et al. (1996) found that the optimum focus position of the pumping light in optically coupled systems depends on the pump-beam quality, pump spot size and the laser mode size.

In optically pumped lasers the slope efficiency σ_s (or differential efficiency) is typically used to determine the efficiency of the output power (P_{out}) versus the pump power (P_{in}) of the laser. If third-order effects such as quasi-three-level characteristics of the gain medium or other thermal effects do not dominate laser operation, the slope efficiency plots are generally linear. If the slope efficiency curve displays significant non-linear behavior due to, for example, thermal lensing or overheating of the gain medium, the slope is often determined in pre-defined and normal operating condition sections. The gradient of a non-linear efficiency curve is primarily determined by

- the power of the beam passing through the output mirror compared to the total losses in the cavity, and
- the ratio (between 0.0 and 1.0) of the pump wavelength to the laser wavelength.

Slope efficiency graphs also have a distinct point on the pumping power axis where the output power becomes a non-zero value and this point is referred to as the oscillation threshold (P_{th}). The optimization of the laser output power involves trade-offs between high slope efficiency and low threshold pump power. A typical slope efficiency graph is given in Fig. 2.12.

The efficiency slope as shown in Fig. 2.12 can be determined through

$$\sigma_s = \frac{P_{in} - P_{th}}{P_{out}} \frac{(hv)_{laser}}{(hv)_{pump}} \tag{2.31}$$

where $(hv)_{laser}$ and $(hv)_{pump}$ are the wavelengths of the laser and the pump respectively. The highest theoretical slope efficiency could exist if the pump wavelength and the laser wavelength are exactly equal and the slope in Fig. 2.12

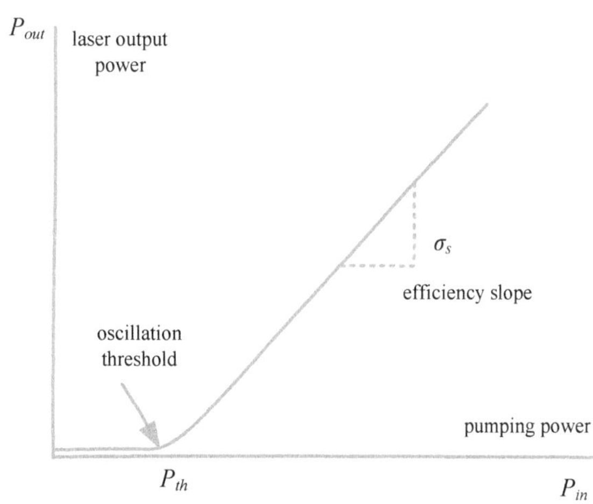

Fig. 2.12 A typical linear slope efficiency curve of pumping power versus output power in optically-pumped lasers

equals 45°. Practically, slope efficiencies are in the range of 5 % or less and can reach up to 30 % for diode pumped lasers. Laser diode performance parameters and experimental procedures to determine and interpret certain parameters are given in Mobarhan (1999). The laser diode parameters include:

Output light versus input current and threshold current: This parameter can be considered one of the most important characteristics of a laser diode, describing the magnitude of the light emitted by the laser as a current injected into the device. The commonly used output light power (P_{out} in watts) versus input current (I_{in} in amperes) (known as the LI curve, which is similar to the efficiency slope) curve is generated by measuring the stimulated radiation to the input current. The LI curve also has a threshold current $I_{th,}$ where the device starts radiating. The slope of the curve is determined by the ratio of $\Delta P_{out}/\Delta I_{in}$ at its nominal operating points, determining its effective slope efficiency in this way. Ideally a diode laser should only require small variations in input current for large variations in output light power, therefore have a large slope efficiency. The LI curve is also typically a linear curve except if third-order effects are dominant during operation at the limits of the device, particularly above the threshold current position and below its maximum input current rating.

Threshold current density: The threshold current of the laser diode, I_{th}, is a factor of the quality of the semiconductor material and the geometric design of the optical waveguide. It furthermore depends on the size and the absolute area of the diode laser—important to consider when comparing lasers, since a higher threshold current could be purely due to a larger device. To avoid confusion in directly comparing threshold current between devices that may occupy a larger or smaller area, the threshold current density J_{th} is used instead. J_{th} is a better indication of the quality of the semiconductor material from which the device is manufactured. The transparency threshold current density (J_O) is used to compare threshold current density between different batches of semiconductor processes. J_O is found by plotting threshold current density versus the inverse of optical cavity length and finding the intersection with the vertical axis of the linear fit line of the data points. J_O can therefore be regarded as the threshold current density of a theoretically infinitely long laser cavity with no losses at its mirror facets used to compare semiconductor process characteristics during diode laser manufacturing.

External differential quantum efficiency: This figure of merit (FOM) parameter of a device indicates the efficiency of the device to convert injected electron-hole pairs to photons emitted from the device (Mobarhan 1999). The external differential quantum efficiency (η_d) can be determined by the slope efficiency with

$$\eta_d = \frac{P_{out}}{I_{in}} \frac{q\lambda}{hc} \tag{2.32}$$

where q is the elementary electron charge (1.6×10^{-19} C). In (2.32) the term P_{out}/I_{in} has units W/A and $q\lambda/hc$ has units A/W—therefore the external differential quantum efficiency is unit-less—a percentage of injected electron-hole pairs converted to actual emitted radiative photons. Mobarhan (1999) also highlights the importance of specifying the type of laser in terms of wavelengths emitted and number of reflective mirrors, since these parameters influence the absolute slope efficiency that can be achieved in a system.

Cavity length dependence on J_{th} and η_d: In order to determine a viable average of the threshold current density and the external differential quantum efficiency for various cavity lengths and to define the material quality and semiconductor process for a batch of laser diodes accurately, a variety of cavity lengths should be measured and compared. Having broad area lasers with clearly defined geometries simplifies these measurements to determine the average values.

Internal quantum efficiency: This parameter, η_i, specifies the efficiency of a diode laser to convert electron-hole pairs into photons within the diode structure itself (not accounting for actual radiated photons) independently of the geometrical properties of the device. The inverse of the internal quantum efficiency is a measure of the injected current converted into other forms of energy, such as thermal energy (heat) and not applied to light photons. To eliminate the influence of cavity length on the efficiency to convert electron-hole pairs to photons, η_i is typically determined by plotting the external differential quantum efficiency versus the cavity length and finding the inverse of the intercept point of the linear fit line from the set of data points with the vertical axis. Graphically, this procedure is presented in Fig. 2.13.

The external differential quantum efficiency is always less than, based on a value of between 0 and 1, the internal quantum efficiency, since not all photons generated in the device ultimately exit the device again as radiated photons and can become trapped or lost inside the device. The ratio of η_d/η_i specifies the number of photons emitted from the laser to the number of photons generated in the laser (Mobarhan 1999).

Characteristic temperature: High-powered lasers inevitably cause the operating temperature of the device to rise. The temperature sensitivity T_0 of the device specifies the amount of change in external differential quantum efficiency and threshold current density with increasing temperature and translates to the thermal stability of the device. T_0 can be determined by

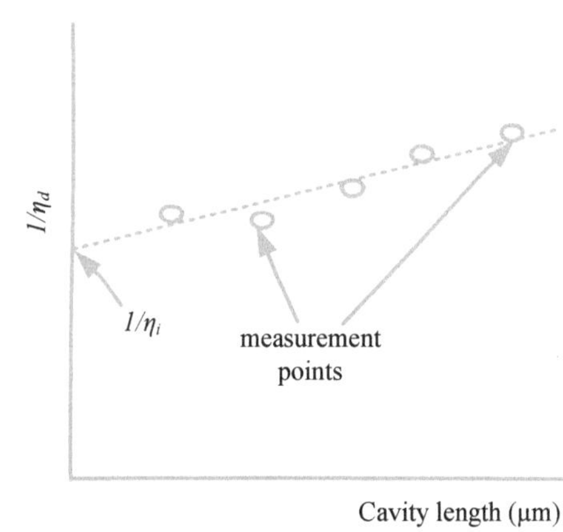

Fig. 2.13 Determining the internal quantum efficiency of a diode laser from the external differential quantum efficiency and the cavity length

$$T_0 = \frac{\Delta T}{\Delta \ln(J_{th})} \tag{2.33}$$

where ΔT is the difference in temperature between two measurements of measured/calculated threshold current density J_{th}.

Spectrum and peak wavelength: The number of spectral outputs that a laser is capable of emitting is a function of the optical cavity structure and the operating current (Mobarhan 1999). As a result, multimode lasers exhibit multiple spectral outputs around their center wavelength. Optical cavity lengths are designed with respect to the output wavelength and are typically chosen at lengths of $\lambda/2$ multiplied by the oscillations, m. The center wavelength of a laser diode is also proportional to the temperature where the output wavelength increases with increasing temperature. This becomes a useful characteristic of laser diodes when tunable lasers are required. The output wavelength can in addition be tuned by changing the input current, generally displaying discrete changes in wavelength. This phenomenon is known as mode hopping and is evident in single-frequency laser diodes.

The following paragraph concludes the discussion on particle accelerators and laser physics, types and optimizations.

2.7 Conclusion

This chapter focuses on particle acceleration technology, its history and its contribution to other technologies such as lasers. Lasers are found in various applications, commercial and military, and this chapter introduces the basic principles of lasers, the types of lasers and laser optimization. In the following chapter, the supporting circuitry to drive lasers is discussed, with the emphasis on the role of SiGe in these devices. The following chapter further investigates the uses of fast-switching, low-noise, compact and efficient SiGe laser drivers in military applications, such as laser rangefinders, laser DEWs, IR countermeasures (IRCM) and laser detonators.

References

Barbalat, O. (1994). Applications of particle accelerators. *CERN-AC-93-04-BLIT-REV*. CERN, 1994.

Bäuerle, D. W. (2013). Laser processing and chemistry. *Springer Science & Business Media*, 29 June 2013.

Benson, R. C., & Mirarchi, M. R. (1964). The spinning reflector technique for ruby laser pulse control. *IEEE Transactions on Military Electronics, 8*(1), 13–21.

Bhatia, M. S., & Kumar, G. (2002). On the EMI potential of various laser types. In *Proceedings of the Electromagnetic Interference and Compatibility* (pp. 3–5).

Bhawalkar, D. D., Gambling, W. A., & Smith, R. C. (1964). Investigation of relaxation oscillations in the output from a ruby laser. *Radio and Electronic Engineer*, *27*(4), 285–291.

Brown, G., Halback, K., Harris, J., & Winick, H. (1983). Wiggler and undulator magnets—A review. *Nuclear Instruments and Methods*, *208*(65–77), 1983.

CERN. (2009). CERN LHC: The guide. Retrieved Jan 21, 2016 from http://cds.cern.ch

Chen, Y. F., Liao, T. S., Kao, C. F., Huang, T. M., Lin, K. H., & Wang, S. C. (1996). Optimization of fiber-coupled laser-diode end-pumped lasers: Influence of pump-beam quality. *IEEE Journal of Quantum Electronics*, *32*(11), 2010–2016.

Coldren, L. A., Fish, G. A., Akulova, Y., Barton, J. S., Johansson, L., & Coldren, C. W. (2004). Tunable semiconductor lasers: A tutorial. *Journal of Lightwave Technology*, *22*(1), 193–202.

Einstein, A. (1905). Does the inertia of a body depend upon its energy-content? *Translated from Annalen der Physik, 18*(639), 1905.

Endo, M., & Walter, R. F. (2006). *Gas lasers*. CRC Press, 26 Dec 2006.

Esarey, E., Schroeder, C. B., & Leemans, W. P. (2009). Physics of laser-driven plasma-based electron accelerators. *Reviews of Modern Physics, 81*(3), 1229–1280.

Evtuhov, V., & Neeland, J. K. (1965). Study of the output spectra of ruby laser. *IEEE Journal of Quantum Electronics*, *1*(1), 7–12.

Gilmore, R. (2004). *Elementary quantum mechanics in one dimension*. USA: JHU Press.

Gulley, J. R. (2011). Modeling free-carrier absorption and avalanching by ultrashort laser pulses. In *Proceedings of SPIE*. 8190 819022-1-112011.

Ivey, H. F. (1966). Electroluminescence and semiconductor lasers. *IEEE Journal of Quantum Electronics*, *2*(11), 713–726.

Javan, A., Bennet, W. R., & Herriott, D. R. (1961). Population inversion and continuous optical maser oscillation in a gas discharge containing a He-Ne mixture. *Physical Review Letters, 6* (106–110), 1961.

Kelsall, R. W., & Soref, R. A. (2003). Silicon-Germanium quantum-cascade lasers. *International Journal of High Speed Electronics and Systems*, *13*(2), 197–223.

Knochenhauer, C., Hauptmann, S., Scheytt, C., & Ellinger, F. (2009). A compact, low-power 40 Gbit/s differential laser driver in SiGe BiCMOS technology. In *2009 European Microwave Integrated Circuits Conference* (pp 324–326).

Laser Enterprise. (2016). Retrieved Feb 4, 2016 from http://www.laserenterprise.com/

LaserFocusWorld. (1995). Retrieved Feb 4, 2016 from http://www.laserfocusworld.com

Maiman, T. H. (1960). Stimulated optical radiation in ruby. *Nature, 187*, 493–494.

Marshall, W. K., & Burk, B. D. (1986). Received optical power calculations for optical communications link performance analysis. *TDA Progress Report*. 42–87, July–Sept 1986.

Mobarhan, K. S. (1999). Application note: Fiber optics and photonics. Test and characterization of laser diodes: Determination of principal parameters> *Newport Application Note*.

Moto, A., Ikagawa, T., Sato, S., Yamasaki, Y., Onishi, Y., & Tanaka, K. (2013). A low power quad 25.78-Gbit/s 2.5 V laser driver using shunt-driving in 0.18 μm SiGe-BiCMOS. In *2013 IEEE Compound Semiconductor Integrated Circuit Symposium (CSICS)* (pp. 1–4).

Nathan, M. I., Dumke, W. P., Burns, G., Dill, F. H., & Lasher, G. (1962). Stimulated emission of radiation from GaAs p-n junctions. *Applied Physics Letters*, *1*, 62–64, Nov 1962.

Oxborrow, M., Breeze, J. D., & Alford, N. M. (2012). Microwave laser fulfills 60 years of promise. *Nature*, *488*, 353–356.

Patel, C. (1984). Lasers—their development and applications at AT&T bell laboratories. *IEEE Journal of Quantum Electronics*, *20*(6), 561–576.

Photonics Handbook. (2015) Retrieved Feb 3, 2016 from http://www.photonics.com/

Porto, S. P. S. (1963). A simple method for calibration of ruby laser output. In *Proceedings of the IEEE*, *51*(4), 606–607.

Quist, T. M., Rediker, R. H., Keyes, R. J., Krag, W. E., Lax, B., McWhorter, A. L., et al. (1962). Semiconductor maser of GaAs. *Applied Physics Letters*, *1*, 91–92.

Seo, D-K., & Hoffmann, R. (1999). Direct and indirect band gap types in one-dimensional conjugated or stacked organic materials. *Theoretical Chemistry Accounts*, *102*, 23–32.

Silfvast, W. T. (2004). Laser fundamentals. In *School of Optics: University of Central Florida* (2nd ed.). Cambridge: Cambridge University Press.

Tauc, J. (1968). Optical properties and electronic structure of amorphous Ge and Si. *Materials Research Bulletin, 3*(1), 37–46.

Townes, C. H. (1965). 1964 Nobel lecture: Production of coherent radiation by atoms and molecules. *IEEE Spectrum, 2*(8), 30–43.

Tsujino, S., Scheinert, M., Sigg, M., Grutzmacher, D., & Faist, J. (2006). Strategies to improve optical gain and waveguide loss in SiGe quantum cascade devices. In 2nd *IEEE International Conference on Group IV Photonics* (pp. 4–6).

Woodyard, J. R. (1948). High-particle accelerators. *Electrical Engineering, 67*(8), 759–767.

Yariv, A., & Gordon, J. P. (1963). The Laser. In *Proceedings of the IEEE, 51*(1), 4–29.

Chapter 3
Electronic Warfare Laser Driver Principles: High-Powered Directed Energy Beam Generation

3.1 Introduction

The ability to control and manipulate the electromagnetic spectrum is a major advantage, both defensively and offensively, for military operations. Modern EW weapons employ radio, radar, IR, optical, UV, electro-optical and laser technologies. Directed energy weapons do not necessarily cause irreversible damage to enemy troops and technologies, as the heat ray targets the skin of enemies and creates unbearable conditions, forcing them to retreat. The heat ray is an active denial system (ADS), a non-lethal DEW primarily aimed at crowd control and developed by the US military. Although its effectiveness is doubtful (because of strong RF dissipation in rain/snow/fog, possibly only effective on exposed skin and its inability to target individuals in large crowds), the principle of using such a DEW in non-lethal operations is still valuable for militaries.

Semiconductor laser diodes offer compact size systems, ruggedness and high efficiency and are expected to have a significant impact on the future of DEWs. Diode-pumped solid-state lasers offer significant potential in optical countermeasure systems in the visible spectrum with higher efficiencies compared to frequency-shifted lasers. Research into identifying improved gain media for these lasers is actively pursued to take advantage of the potentially high-powered laser drivers offered by technologies such as GaAs, InP and SiGe. Energy from HELs can be delivered to a target at the speed of light, much faster than traditional supersonic or subsonic speeds from missiles. Another advantage of HELs is the low shot-to-shot costs involved. Traditional missiles are expensive and only have a single shot before they are destroyed, whereas lasers (that also have high development cost) can be used more than once. A challenge for researchers is creating lasers that use high enough power to destroy a target partially while tracking numerous objects simultaneously (Coffey 2014). In addition, these weapons should be able to withstand environmentally harsh conditions where dust and humidity attenuate the main beam significantly, and to be fired from compact and modular

W. Lambrechts and S. Sinha, *SiGe-based Re-engineering of Electronic Warfare Subsystems*, Signals and Communication Technology,
DOI 10.1007/978-3-319-47403-8_3

platforms. Lasers can target UAV optical sensors used as intelligence surveillance and reconnaissance and can disable these sensors with relatively low power at long range. High power and electrical efficiency are obtainable when sufficient cooling of these devices is available and research projects to develop electric fiber lasers as opposed to conventional solid-state technology are pursued by companies such as Lockheed Martin. The following section briefly introduces ongoing military and defense ventures in high-energy laser research. The applications of these lasers vary; however, a common EW theme is evident throughout these projects.

3.2 Laser Systems Markets in Military and Defense Environment

Laser systems are commonly used for military and defense applications and the global military laser systems market's compound annual growth rate is estimated to be 7.97 % over the period 2014–2019. The leading vendors in laser systems identified by researchandmarkets.com are four companies that have dedicated large amounts of resources to researching and developing new technologies and testing technologies to identify the limitations of current-generation systems. The largest contributors to date are:

- BAE Systems,
- Lockheed Martin,
- Northrop Grumman and the
- Thales Group.

Other prominent vendors in laser systems are also making significant headway in modern laser system research and include a list of companies that are not necessarily military-based, but are also providing valuable insight into a research field that has gained traction in recent years, with semiconductor laser becoming more prevalent. These companies include:

- 3SAE Technologies (USA),
- American Laser Enterprises (USA),
- Boeing (USA),
- Coherent (USA),
- Creaform (USA),
- Electro Optic Systems (Australia),
- Eurolaser (Germany),
- Faro Technologies (USA),
- General Atomics (USA),
- IPG Photonics (USA),
- Jenoptik (Germany),
- Kratos Defense & Security Solutions (USA),
- L-3 Advanced Laser Systems Technology (USA),

- Newport Corporation (USA),
- Trumpf (USA),
- UTC Aerospace Systems (USA),
- Vectronix (USA), and
- Visotek (USA).

Applications of laser systems with the highest levels of interest and within the military operation framework as listed by researchandmarkets.com are

- airborne laser mine detection,
- range finding,
- target designation,
- anti-missile systems, and
- neutralizing enemy weapon systems (EW and DEW).

The growth in international arms trade, varying warfare tactics and preference for reduced sized weapons are among the factors driving the research field and markets of laser energy systems. Laser systems are also becoming prevalent in military communications and laser-based ground-to-air laser communication provides quick mine detection abilities. Coffey (2014) listed numerous ongoing (as of 2014) HEL projects with details on the average output power of each project, either envisioned or practically implemented.

As seen in Table 3.1, these projects include laser solutions using COTS fiber-laser components, diode-pumped solid-state lasers, FELs and liquid lasers. In Coffey (2014) the benchmark and goal for military-based laser applications is given as approximately 100 kW where lower-power but more mobile systems ranging in the 10 s of kW output powers are also being demonstrated. Laser weapons (specifically HEL) have matured in the past decade and have reached various milestones, which are pointing to an increase in power, decrease in size and improvement in efficiency and ruggedness. These systems are not far from being commonly deployed as EW countermeasures, hand-off systems and possibly offensive weapons by military operations.

This chapter will focus on the technologies required and fundamental properties of the semiconductor laser driver used to feed these laser devices with high currents. The enabling technology of such circuits can be crucial in achieving not only high-powered lasers, but also energy-efficiency, compact, rugged and mobile systems. Presenting the laser drivers from an electronic circuit and component perspective enables researchers to identify important characteristics required from the enabling technology, such as SiGe, which provides many of these characteristics as a modern and commercially available, relatively low-cost solution. However, before discussing the laser driver fundamentals, a brief overview of the laser diode and its electronic equivalent model is presented, since it will be referred to during the laser driver discussion.

Table 3.1 Ongoing high-energy laser weapons projects as outlined in an article by Coffey (2014)

Project	Agency	Contractor	Description	Power
Area defense anti-munitions (ADAM)	Commercial venture	Lockheed Martin	Fiber (COTS)	10 kW
Gamma	Commercial venture	Northrop Grumman	Solid-state slabs	13.3 kW per slab
Excalibur	DARPA	Optonicus	Fiber and optical phased array	100 kW
High energy liquid laser area defense system (HELLADS)	DARPA	General Atomics	Liquid (aircraft mountable)	150 kW
High energy laser mobile demonstrator (HEL MD)	US Army	Boeing	Solid-state (grounded)	50/100 kW in development
Joint high power solid-state laser	US Army/Airforce/Navy	Northrop Grumman	Solid-state slab	105 kW
Laser weapon system (LaWS)	US Office of Naval Research	Kratos	Solid-state (IR COTS)	15-50 kW (classified)
Robust electric laser initiative (RELI)	US DoD	Lockheed Martin/Northrop Grumman/Boeing	Fiber	100 kW (goal)
Solid-state laser—technology maturation	US Office of Naval Research	Kratos/Raytheon/Northrop Grumman/BAE systems	Lightweight tactical vehicle-mounted	Design state
High repetition rate advanced petawatt laser system (HAPLS)	Research project	University of Texas	Semiconductor diode and optics	10^{15} W/10 Hz/30 fs pulse

3.3 Optoelectronic Communication

This chapter focuses on the optoelectronic communication components required to achieve high-speed, low-noise and high-powered optical transmission through an optical medium. A typical laser optoelectronic communication link is given in Fig. 3.1.

According to Fig. 3.1, the optoelectronic communication link typically consists of an optical transmitter stage comprising a laser driver and a laser diode, an optical medium, which could be free space, optical fiber or waveguides, and an optical receiver stage comprising a photodetector and a trans-impedance amplifier. Within the optical transmitter, a laser driver is required to convert a modulated source signal into a large current used to modulate the optical source (laser diode). Laser drivers use amplified currents as opposed to voltages to drive the laser diode. The laser diode can then transmit the modulated signal into the preferred optical medium, where guided cavities such as optical fibers or waveguides are generally preferred to converge the signal towards the receiver. The signal will inevitably undergo deterioration, a decrease in amplitude and additive noise while travelling through the optical medium and can be corrected at the receiver. The optical receiver converts the incoming light (photons) at the desired wavelength to electrical pulses using a trans-impedance amplifier (current-to-voltage converter) and feeds the amplified signal for further signal processing. Possible technologies in which high-speed trans-impedance amplifiers with operation above 40Gb/s can be implemented typically include GaAs, InP and SiGe (Weiner et al. 2003). The further processing involves demodulation of the incoming signal with the common carrier frequency and extracting the information as data, sound, video or commands. Depending on the application and wavelength, such a system can be implemented on a single chip, between chips or over longer distances through optical fibers with the help of repeaters. The laser diode and the photodetector can be manufactured from the same material to ensure that both devices operate (peak)

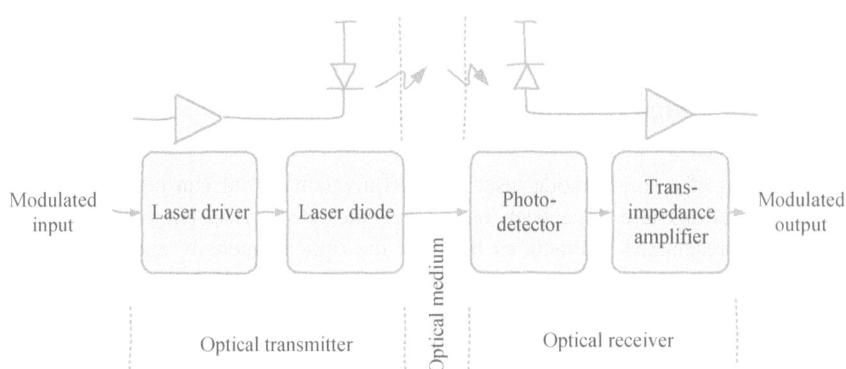

Fig. 3.1 A typical laser optoelectronic communication link showing the optical transmitter stage, optical medium and the optical receiver stage

at similar wavelengths. Material characteristics and peak wavelengths were discussed in Chap. 2. In this chapter the laser driver and the laser diode are discussed, with a brief discussion on the typically expected attenuations in various optical mediums as a function of operating wavelength.

The first discussion presented relates to the laser diode and its equivalent circuit. This discussion effectively introduces the most crucial parameters when considering laser drivers. The optical medium and the photodetector will not be discussed in detail, since they fall outside of the scope of this book; and the trans-impedance amplifier principles are furthermore presented in this chapter.

3.4 Laser Diode Equivalent Models

For most applications of laser diodes, especially high-frequency operation combined with electronic components such as drivers and amplifiers, a precise representation of the laser diode electrical impedance is crucial to match the device effectively to its supporting circuitry. For this reason, modeling the laser diode as an electrical equivalent circuit is the preferred method to determine its impedance characteristics and frequency limitations. Equivalent circuit modeling is used in many electronic designs and gives researchers and designers a baseline to derive the ideal circuit operating parameters and achieve the most efficient transfer of energy. Electronic circuits such as the laser driver can be realized in many CMOS and BiCMOS technologies with varying node sizes providing their own set of advantages and disadvantages. Using small node sizes such as 180 nm and smaller provides fast-switching characteristics and low-power operation, whereas larger nodes such as 350 and 600 nm provide higher power capabilities for the active devices and generally better $^1/_f$ noise performance, since this type of noise is proportional to the gate length of the transistors. Equivalent circuit modeling in addition gives researchers and designers a better understanding of the technology that can be used to achieve the specified results of the system.

3.4.1 The Single Resonance Model

The elements of a single mode laser diode equivalent circuit can be derived from the rate equations and have been used since the late 1970s (Kressel and Butler 1977) to represent the interactions between the optical intensity and the injected carriers to estimate the electrical behavior of these devices. The Fabry-Pérot laser is the most common type of single mode laser and the equivalent model of the single resonance laser model is based on its operation. A Fabry-Pérot resonator is a linear optical resonator (or cavity), which consists of two highly reflective planar mirrors such that the resonator modes extend up to the edges of the mirrors and experience some diffraction losses (Paschotta 2013). The resonance frequency can be tuned by varying the cavity length, therefore the distance between the mirrors, with a piezo actuator, for example. "*When the voltage applied to the piezo is periodically varied, e.g. with a triangular temporal shape, and the transmitted power versus time is monitored with a photodetector and an oscilloscope, the latter can directly display the optical spectrum of the incident light, provided that the spectral width is smaller than the free spectral range and the scan is slow enough to reach a quasi-stationary state of the resonator*" Paschotta (2013). A Fabry-Pérot interferometer is typically used to check if a laser operates in the single resonator mode or as reference cavities. The spectral intensity of the quantum fluctuations of the photon and electron population can be found by adding to the single resonance rate equations Langevin source terms for electrons $f_N(t)$ and for photons $f_S(t)$, assumed to exhibit shot noise features, such that

$$\frac{dn}{dt} = \frac{i}{q} - \frac{n}{\tau_s} - (E_{CV} - E_{VC})s + f_N \tag{3.1}$$

where n and s are the total number of electrons in the gain medium and photons in the single lasing mode respectively (Harder et al. 1982). The pump rate is given by i/q and τ_s is the spontaneous carrier lifetime in the laser. $E_{CV}s$ is the downward stimulated emission rate and $E_{VC}s$ is the upward stimulated transition/absorption rate. The gain in the medium is defined as $E_{CV}s - E_{VC}s$. The rate equation for the spectral intensity of photons is given by

$$\frac{ds}{dt} = (E_{CV} - E_{VC})s + \frac{\beta n}{\tau_s} - \frac{s}{\tau_{ph}} + f_S \tag{3.2}$$

where $\beta n/\tau_s$ is the spontaneous emission coupled to the lasing mode and τ_{ph} is the photon lifetime in the laser (Harder et al. 1982).

Modeling the laser diode can be an effective means to determine some of the most crucial operating parameters of the device, where the current-voltage (*I-V*) characteristics take precedence over many other parameters. Barnes and Paoli (1976) already modeled the double-heterostructure diode model as an ideal diode with an additional series resistance to account for the resistance of the bulk material and the contacts. The *I-V* characteristic of their equivalent circuit is given in an earlier work by Barnes and Paoli (1976) as

$$I = \frac{V}{R_{sh}} + I_s\left\{\exp\left[\left[\left(1 + \frac{R}{R_{sh}}\right)V - IR\right] \Big/ \frac{\eta kT}{q}\right] - 1\right\} \tag{3.3}$$

which can be rewritten in a more suitable form for the measurement of the parameters η and R as (Barnes and Paoli 1976)

$$V = \frac{1}{1 + R/R_{sh}}\left[IR + \frac{\eta kT}{q}\ln\left(\frac{I}{I_s} + 1 - \frac{V}{I_s R_{sh}}\right)\right] \tag{3.4}$$

where V is the applied voltage at the device terminals, I_s is the junction saturation current, η is the exponential parameter characteristic of the *p-n* heterojunction, R_{sh} is the shunt resistance where possible leakage current can occur and R is the series resistance of the bulk and the contacts. To determine the differential resistance at a current I, the differential equation, simplified in Barnes and Paoli (1976), assuming that the effect of the shunt resistance is negligible for currents above 100 pA and assuming $I_s \ll I$, is given by

$$\frac{dV}{dI} = R + \frac{\eta kT}{q}\frac{1}{I} \tag{3.5}$$

which represents a linear plot of dV/dI versus $1/I$ with a slope equal to $\eta kT/q$ and the intercept at $1/I = 0$ is equal to R. The importance of (3.3) becomes apparent when characterizing the current I, which varies exponentially with the applied voltage V by Paoli (1976)

$$I = I_s\left(\exp\frac{q(V - IR)}{\eta kT} - 1\right) \tag{3.6}$$

where its second derivative is given by

$$\frac{d^2V}{dI^2} = -\frac{\eta kT}{q}\frac{1}{(I + I_s)^2} \tag{3.7}$$

assuming that the parameters η, I_s and R are not functions of current flow. Since I_s is much smaller than the current flow I, (3.7) can be simplified to

$$I^2\frac{d^2V}{dI^2} = -\frac{\eta kT}{q} \tag{3.8}$$

and the left-hand side of (3.8) is a direct measure of the effective value of η (since k, T and q are known). By using this approach of characterizing the *I-V* characteristics of a heterostructure diode, Morishita et al. (1979) proposed modeling the laser diode as a parallel resistance-inductance-capacitance (RLC) circuit and this method has been accepted and used since. In Katz et al. (1981) four different parallel RLC circuits representing the equivalent model of a laser diode under different circumstances are given and these are adapted in Fig. 3.2a–d.

According to Fig. 3.2 the four different circumstances of the equivalent model of a laser diode include a basic model where no spontaneous emissions are modeled, (b) the model that includes the effect of spontaneous emission in the laser diode as described in Chap. 2, (c) the model that includes spontaneous emissions and self-pulsations and (d) the laser diode with an externally connected load resistance, also accounting for spontaneous emissions and self-pulsations. The voltage squared across the circuit is representative of the output power/stored energy in the circuit

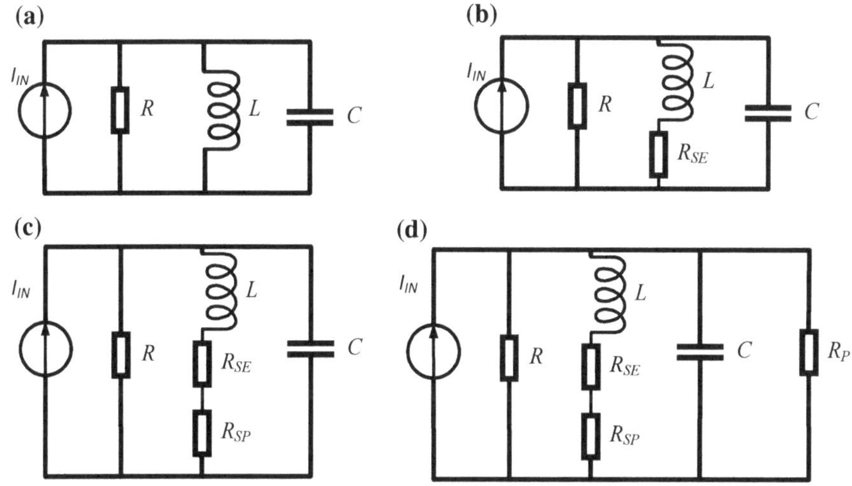

Fig. 3.2 The equivalent circuit models of a laser diode representing **a** the basic model with no spontaneous emission, **b** the model that includes the effect of spontaneous emission, **c** the model that includes spontaneous emissions and self-pulsations and **d** the laser diode with an externally connected load resistance

and the current generator is the equivalent current source, which represents the spontaneous emissions of the laser diode (Irving and Carroll 1990). From the basic model in Fig. 3.2a it can be seen that the diode is represented by three passive components in parallel—a resistance, an inductance and a parallel capacitance. In such a configuration the voltage across the terminals of the three passive components is constant, as opposed to a constant current in a series RLC circuit. The total impedance of such a circuit is calculated using the current of the circuit and the current in each branch where the admittance of each component is calculated as opposed to the impedance with series RLC circuits. The values of each component in Fig. 3.2a can be calculated by the following set of equations:

$$R = R_d\left(\frac{I_{th}}{I^0}\right) \tag{3.9}$$

$$L = R_d\tau_{ph}\left[\left(\frac{I_{th}}{I^0}\right) - 1\right] \tag{3.10}$$

$$C = \frac{\tau_s}{R_d} \tag{3.11}$$

where τ_s is the spontaneous carrier lifetime, τ_{ph} is the photon lifetime, I^0 is the bias current of the laser diode, I_{th} is the threshold current and I_d is a normalized current given by

$$I_d = \frac{N_e^0}{\tau_s} aqd \tag{3.12}$$

where N_e^0 is the DC value of the injected electron density of a *p*-type material, *a* is the diode area, *d* is the thickness of the active region and *q* is again the elementary electron charge. The differential resistance of a diode R_d is given as

$$R_d = \frac{2kT}{q}\frac{1}{I_d} \tag{3.13}$$

and in Katz et al. (1981) the effect of spontaneous emission and saturated absorption or gain saturation and an improved equivalent model to possibly modify the frequency response of the laser diode by electronic means are considered as well. Katz et al. (1981) expand on these equations by taking into account spontaneous emission and self-pulsation. The parallel components as shown in Fig. 3.2b with the additional series resistance R_{se} are given by

$$R = \frac{R_d}{\left(n_{ph}^0 + 1\right)} \tag{3.14}$$

$$L \approx \frac{R_d \tau_{ph}}{n_{ph}^0} \tag{3.15}$$

$$C = \frac{\tau_s}{R_d} \tag{3.16}$$

and

$$R_{se} \approx \beta R_d \frac{n_e^0}{\left(n_{ph}^0\right)^2} \tag{3.17}$$

where n_{ph}^0 and n_e^0 are the normalized steady-state values of the photon density N_{ph}^0 and N_e^0 respectively, given by

$$n_{ph}^0 = A\tau_s N_{ph}^0 \tag{3.18}$$

and

$$n_e^0 = A\tau_s N_e^0, \tag{3.19}$$

where *A* is a proportionality gain constant. For the RLC components in Fig. 3.2c, an additional negative resistance is added in series with R_{se} to accommodate for the self-pulsation of the laser diode and the value of this resistor is determined by

$$R_{sp} \approx -\frac{R_d\delta}{n_s}\frac{1}{\left[1+\left(\frac{n_{ph}^0}{n_s}\right)\right]^2} \tag{3.20}$$

where n_s is the normalized photon saturation density equal to

$$n_s = A\tau_s N_s \tag{3.21}$$

and δ is a parameter denoting the strength of the non-linearity (generally equal to approximately 0.02) and N_s is the saturation density of the photons. Finally, the effect of the parallel resistance R_p is to decrease the resonance peak of the laser diode by a ratio of approximately

$$\frac{RR_p}{(R+R_p)}. \tag{3.22}$$

In order to determine the resonant frequency f_r of the laser diode as depicted in Fig. 3.2a–d, the following equation holds true for a parallel RLC circuit:

$$f_r = \frac{1}{2\pi}\frac{1}{\sqrt{LC}} \tag{3.23}$$

which shows that the resonant frequency of the laser diode can be adjusted or manipulated by changing the values of the equivalent inductance or capacitance. The intrinsic inductance and capacitance are, however, functions of the photon lifetime τ_{ph}, the photon density n_{ph}^0 and the gain coefficient A, which is dependent on the electron density n_e^0 as seen in (3.15)–(3.17). This translates to rewriting (3.23) in a form that identifies the parameters that have a direct effect on the resonant frequency such that

$$f_r = \frac{1}{2\pi}\sqrt{\frac{n_{ph}^0}{\tau_s\tau_{ph}}} \tag{3.24}$$

or in a more convenient form where the gain constant and the steady-state photon density from (3.18) are part of the design equation such that

$$f_r = \frac{1}{2\pi}\sqrt{\frac{AN_{ph}^0}{\tau_{ph}}} \tag{3.25}$$

to represent the equation of the relaxation oscillation frequency and effectively the desired operating frequency of the supporting circuitry, such as the laser driver. Determining the impedance of the laser diode from its equivalent circuit is equally important during the design of the supporting circuitry to ensure effective transfer of power between the electronic circuit and the laser diode. Using the above equations for the effective impedance of the inductor, capacitor and intrinsic resistances of the equivalent model, the complex impedance of the laser diode as a function of operating frequency $Z(\omega)$ where ω is the frequency in radians ($\omega = 2\pi f$) is given in Ozyazici (2004), as adapted from the original article of Tucker and Pope (1983) as

$$\begin{aligned} Z(\omega) &= \frac{V(\omega)}{I(\omega)} \\ &= R_d \frac{j\frac{\omega}{\tau_s} + \frac{1+n_{om}-n_e^o}{\tau_s\tau_{ph}}}{-\omega^2 + j\frac{\omega}{\tau_s}\left[n_{ph}^o + 1 + \frac{\tau_s}{\tau_{ph}}\left(1+n_{om}-n_e^o\right)\right] + \frac{n_{ph}^o + \beta\left[n_e^o\left(1+1/n_{ph}^o\right)-n_{om}\right]}{\tau_s\tau_{ph}}} \end{aligned} \tag{3.26}$$

where $V(\omega)$ is the complex amplitude of the junction voltage and $I(\omega)$ is the complex amplitude of the junction current. The differential resistance R_d in (3.26) is determined by (3.13) and I_d in (3.13) is a normalized current as in (3.12). A similar approach is followed to determine the equivalent circuit for a multiple resonance mode laser diode, which is presented in the following section.

3.4.2 *The Multiple Resonance Model*

The operation of the multiple resonance model for a laser diode is observed by first presenting a circuit structure of a double resonance model, as adapted from Irving and Carroll (1990) and given in Fig. 3.3.

As seen in Fig. 3.3, to determine the effects of a double resonance mode laser, two single resonance mode laser diode equivalent circuits are placed in series and the effects of this resultant circuit are investigated. The output voltage of the circuit in Fig. 3.3 changes in amplitude, phase and rate of phase, since the input current source is varying with these same parameters. The total admittance of the circuit

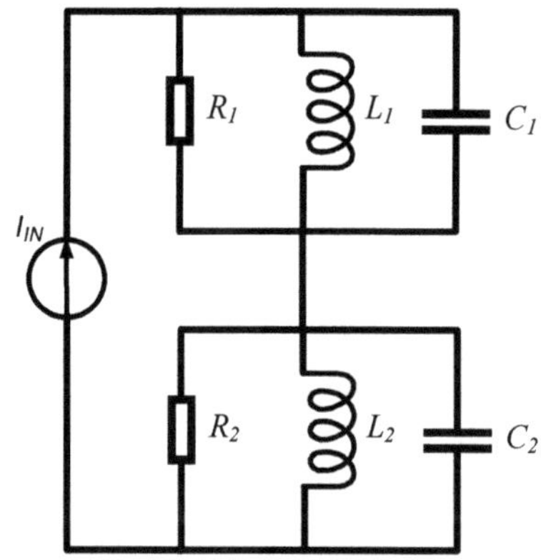

Fig. 3.3 Equivalent circuit model of a laser diode operating in double resonance mode (adapted from Irving and Carroll 1990)

can be determined by solving combined differential equations of the circuit admittance. This theory is then adapted to determine the circuit parameters for multiple resonance models through the electronic rate equations of the equivalent laser diode circuit. The two resonant frequencies of the circuit in Fig. 3.3, f_{r1} and f_{r2} are given by adapting (3.26) such that

$$f_{r1} = \frac{1}{2\pi}\frac{1}{\sqrt{L_1C_1}} \tag{3.27}$$

and

$$f_{r2} = \frac{1}{2\pi}\frac{1}{\sqrt{L_2C_2}} \tag{3.28}$$

and the impedance of the circuit is a series combination of the equation given in (3.26), consequently the addition of each RLC combination. The equivalent circuit of a multiple resonant circuit, including the parasitic components at the input and the additional modes of spontaneous emission and self-pulsation, is given in Fig. 3.4, which is adapted from Kim (2005).

According to Fig. 3.4, L_W represents the bonding wire intrinsic inductance if the laser diode is connected to the supporting circuitry and operating at high frequencies, whereas R_S and C_S are the series resistance between contacts and the contact capacitance respectively. The single resonance parallel RLC sections are depicted in Fig. 3.4 and marked $RLC_1 \ldots RLC_n$, and L_s and C_t represent the small-signal photon storage and active-layer diffusion capacitance respectively. R_{SP},

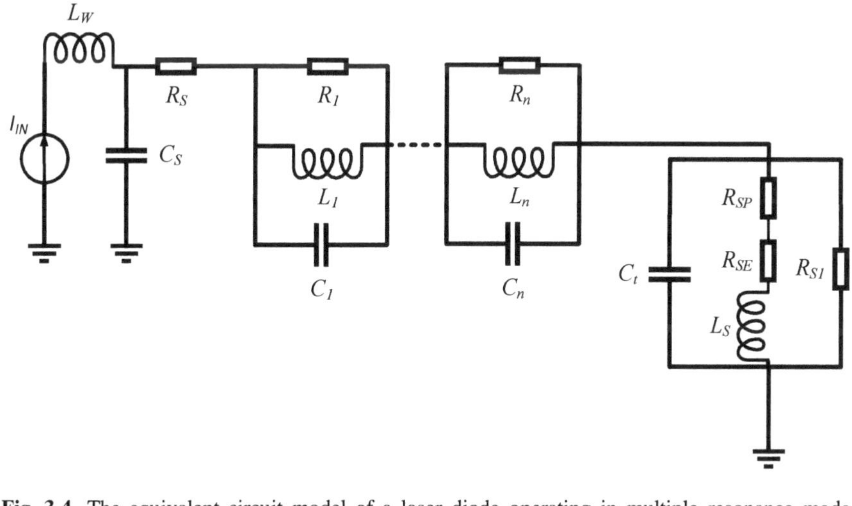

Fig. 3.4 The equivalent circuit model of a laser diode operating in multiple resonance mode, including intrinsic parasitic components and electrical DC characteristics and the optical intensity of the laser diode (adapted from Kim 2005)

R_{SE} and R_I are related to the laser diode electrical DC characteristics and the output voltage across R_{SI} represents the optical intensity of the laser. Each resonant frequency for the n-stages (modes) is then simply determined by

$$f_{rn} = \frac{1}{2\pi} \frac{1}{\sqrt{L_n C_n}} \tag{3.29}$$

where n represents the mode number and the resonance for RLC_n. After considering the equivalent circuit of the laser diode and presenting a brief theory on the design parameters, the laser driver is presented in the following section. The laser driver can only transfer maximum power to the laser diode if it is properly matched and designed in accordance with the operating conditions of the laser diode.

3.5 Laser Drivers

An ideal laser driver can provide a constant, regulated and adjustable current to the laser diode, has linear gain, does not vary with temperature and does not add noise to the signal. Of course this is impossible to achieve practically, but these parameters should be the primary goals in laser driver design. In EW and military operations, the tolerances on these parameters become small and this section highlights these tolerances and considerations for improvements when implementing a laser driver for a specific application. In a typical laser driver, temperature changes from ineffective thermal dissipation (especially in CW lasers) can shift the operating point of the laser diode by shifting the threshold current and cause mode hopping. One key consideration in laser driver circuits is an adjustable current source, primarily because of the large variations in temperature during operation. The output stage is equally important to operate as a low-noise and high GBP linear device. The configuration of the laser diode and the laser driver is the third important consideration when implementing an optical laser transmitter. The laser diode can be situated between the main power supply and the current source (typically p-type laser diode) or between the current source and ground (typically n-type laser diode), and this configuration depends on the application. Laser drivers are essentially trans-conductance amplifiers, amplifiers that convert an incoming voltage signal to a constant current output to drive the laser diode. One such circuit, using a single transistor, is given in the following section.

3.5.1 *Single Transistor Current Source*

Laser diode drivers, which are essentially current sources with optimal frequency response and specific input and output impedance, voltage or current gain and, can be realized with a single transistor, although the overall operation is not optimized for the application. The three most basic active single transistor amplifiers are common-base, common-emitter and common-collector amplifiers and the simplified schematics for each configuration are given in Fig. 3.5.

In Fig. 3.5, the three configurations are shown in their most simplistic form. These configurations are given using BJT transistors, but CMOS implementations have similar topology to realize common-gate, common-source and common-drain amplifiers. For this discussion the BJT implementations will be used. In Fig. 3.5a the common-base configuration requires the input voltage to be applied to the emitter of the transistor and the output voltage is measured on the collector of the transistor. The base of this configuration is grounded. The common-emitter configuration in Fig. 3.5b requires an input signal on the base of the transistor, while the output is again measured on the collector. The emitter is grounded (if no emitter degeneration is implemented). For the common-collector amplifier, the input is again applied to the base of the transistor, but the output is measured over the emitter where the collector is directly tied to the positive supply voltage. Each configuration in Fig. 3.5 is suited for specific applications because of its characteristics of impedance, gain and frequency response. A summary of these characteristics is given in Table 3.2.

The common-base amplifier is typically used as a current buffer since the current gain is approximately unity (Table 3.2) or it is used as a voltage amplifier with an approximate gain of g_mR_C. The common-base amplifier is commonly used for circuits that require a low input impedance; as seen in Table 3.2 the input impedance is approximately equal to the emitter resistance of the configuration, r_e, generally a low value in the 100s of Ohms. This configuration in addition has high isolation between the input and output, which leads to a circuit with relatively high

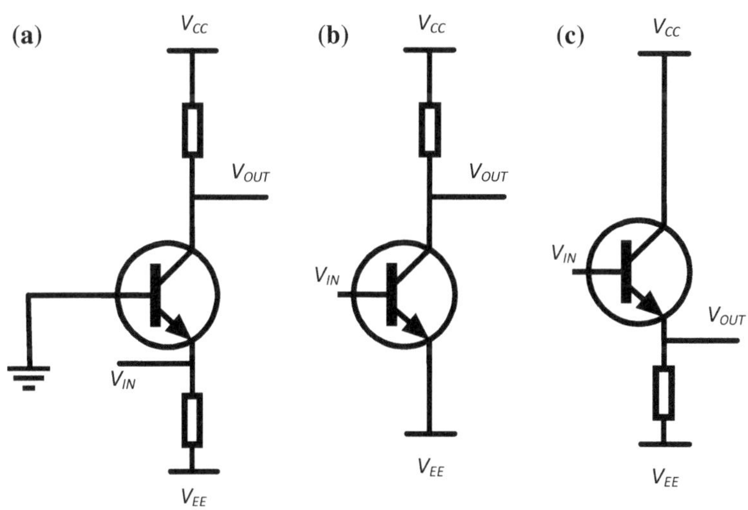

Fig. 3.5 Schematic representation of single transistor **a** common-base, **b** common emitter and **c** common-collector amplifiers

Table 3.2 Voltage and current gain, input and output impedance of the common-base, common-emitter and common-collector amplifiers

	Expression	Approximate	Condition
Common-base			
Voltage gain	$\frac{(g_m r_o + 1)R_C}{R_C + r_o}$	$g_m R_C$	$r_o \gg R_C$
Current gain	$\frac{r_\pi + \beta r_o}{r_\pi + (\beta + 1)r_o}$	1	$\beta \gg 1$
Input impedance	$r_e\left(1 + \frac{r_b}{r_\pi}\right)$	r_e	$r_o \gg R_c \parallel R_L$
Output impedance	$R_C \parallel r_o$	R_C	$r_o \gg R_C$
Common-emitter			
Voltage gain	$-g_m(R_C \parallel r_o)$	$-g_m r_o$	$R_C \gg r_o$
Current gain	$g_m r_\pi$	β_0	constant
Input impedance	β_0 / g_m	r_π	$g_m = qI_C/kT$
Output impedance	$R_C \parallel r_o$	r_o	$R_C \gg r_o$
Common-collector			
Voltage gain	$\frac{g_m R_L}{1 + g_m R_L}$	1	$g_m R_L \gg 1$
Current gain	$\beta_0 + 1$	β_0	$\beta_0 \gg 1$
Input impedance	$r_\pi + (\beta_0 + 1)(R_L \parallel r_o)$	$\beta_0 R_L$	$\beta_0 \gg 1, g_m R_L \gg 1$
Output impedance	$\frac{R_S + r_\pi}{(\beta_0 + 1)}$	$\frac{1}{g_m}$	$r_\pi \gg R_S, \beta_0 \gg 1$

stability from the low levels of feedback from the output (Rogers and Plett 2003). The output impedance is higher than that of other configurations such as the common-emitter, and it is typically used in cascade with the common-emitter amplifier, acting as the first current buffer stage. The common-base amplifier is used ideally in high-frequency systems, since its input capacitance does not undergo the Miller effect, which degrades the available bandwidth of the amplifier, as it does with common-emitter amplifiers, for instance. As a result, the common-base amplifier can be used as a current source whose current is nearly independent of the voltage across its terminals (Gray et al. 2001).

The common-emitter amplifier, as seen in Fig. 3.5b, tends to have a relatively low bandwidth due to the large effect of the Miller capacitance across its collector-base terminals. There are, however, ways to reduce this effect, including lowering the voltage gain by using emitter degeneration, using a common-collector amplifier at the output to reduce the output impedance or a common-base amplifier at its input to lower the input impedance or using a differential amplifier topology. The common-emitter amplifier does, however, present a relatively high input impedance, low output impedance and high voltage and power gains. These characteristics, along with the fact that the current gain from the base to the collector is typically high, make these amplifiers ideal for high-powered devices. For these reasons the common-emitter amplifier is generally used in laser diode drivers and because of its small size and high power output, the circuits are ideal for EW and other military operations requiring modular and lightweight systems. Because of the effect of the Miller capacitance, if used in its basic form, the common-emitter amplifier is generally used in low-frequency and low-noise amplifiers (Mukai and Yamamoto 1982).

The common-emitter amplifier with emitter degeneration is a similar topology as in Fig. 3.5b, but with a resistor placed between the emitter of the transistor and ground. This emitter effectively reduces the trans-conductance, increases the output resistance and also increases the input resistance. These changes are introduced owing to the feedback effect from adding this resistor. As the input signal increases and the transistor conducts more, the collector current increases. As a result, the voltage drop across the emitter resistor increases and this effectively decreases the net value of the base-emitter voltage V_{BE}, which forces the transistor into a less-conducting state and normalizes the collector current flow.

The common-collector amplifier in Fig. 3.5c, also called an emitter-follower, is typically used as a voltage buffer, since its voltage gain is approximately 1, as seen in Table 3.2. The common-collector amplifier has a large input impedance and a small output impedance, making its implementation useful to ensure the amplifier does not load down the circuit it is connected to and it can drive low resistance loads—also required for effective voltage buffer operation. The common-collector has a large current gain between the input and output, as seen in Table 3.2, similar to the common-emitter with a value of β_0. A brief summary of the voltage-gain, current gain, input impedance and output impedance of each of the three amplifier configurations is given in Table 3.3.

From the above discussion it becomes evident that laser drivers, which require a large and constant current to achieve population inversion, can be implemented using either a common-collector or a common-emitter configuration. Since the common-emitter amplifier in addition provides high voltage and power gain, it is generally the preferred choice when implementing laser drivers. Variants of the common-emitter circuit (such as its differential topology) are also used to realize laser drivers and are also discussed in this chapter. Firstly, the basic principles of operation of the common-emitter with emitter degeneration (since this modification is generally added to increase circuit stability) is presented. A schematic representation, which expands on Fig. 3.5b, shows the common-emitter amplifier configuration, seen in Fig. 3.6. Transistor characteristics will be focused on in this discussion in order to determine the viability, advantages and disadvantages of using SiGe as the laser driver enabling technology.

Starting from the left-hand side of Fig. 3.6, the input signal V_{in} is a low-amplitude (typically in mV) AC input signal, which may have a DC offset

Table 3.3 Summary of the voltage gain, current gain, input impedance and output impedance of the common-base, common-emitter and common-collector amplifier topologies

	Common-base	Common-emitter	Common-collector
Voltage gain	High	Medium	Low
Current gain	Low	Medium	High
Input impedance	Low	Medium	High
Output impedance	High	Medium	High

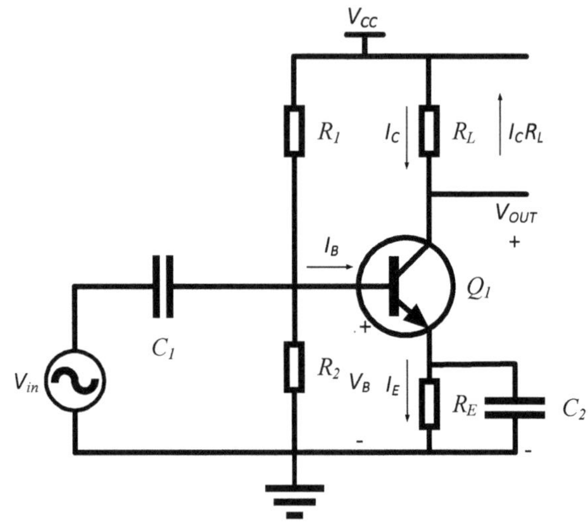

Fig. 3.6 Schematic representation of the single transistor common-emitter amplifier with emitter degeneration used as a simple laser driver

coupled to the signal, depending on the source of the input signal. The frequency of this AC signal determines the speed at which the transistor Q_1 must be able to switch on and off and is called the cut-off frequency f_T.

f_T indirectly gives valuable information about the quality of the vertical doping profile of the transistor. Although f_T is often used as a single figure of merit (FOM) of the capabilities of a transistor, there are various other behaviors describing the practical performance of the transistor, which are not taken into account if one looks only at the cut-off frequency. These include the series resistance of the transistor when a charge is flowing through it, which can have substantial effects on the frequency performance of the transistor in view of its intrinsic *RC*-time constant. Moreover, the transistor presents a specific collector-substrate capacitance based on the specific process and quality of the material and it is important that this value be kept low for high-frequency applications; this parameter is not extracted from only defining the transistor f_T. A better approximation of the transistor performance is its maximum frequency of oscillation, f_{max}. The definition of f_{max} is essentially the frequency of operation where the unilateral power gain of the transistor is unity. By relating the operating frequency of the transistor to its admittance, the approximate solution of f_{max} becomes

$$f_{max} \approx \sqrt{\frac{f_T}{8\pi(r_{bb'} + r_{ee'} + 1/g_m)c_\mu}} \quad (3.30)$$

where $r_{bb'}$ and $r_{ee'}$ are the effective base and emitter resistances respectively and c_μ is the intrinsic collector-base capacitance of the transistor.

The capacitor C_1 acts as a DC-blocker for the DC offset on the source signal to ensure that the transistor is modulated only by the incoming AC signal and not biased by a possibly variable DC value. The biasing of the transistor base is realized by R_1 and R_2 where the voltage V_B across the base is achieved by

$$V_B = V_{CC} \frac{R_2}{R_1 + R_2} \tag{3.31}$$

where V_{CC} is the supply voltage to the circuit. This supply voltage also determines the maximum collector current through R_L, which can be replaced by the laser diode, when the transistor is fully on and the collector-emitter voltage is zero (V_{CE} = 0 V). The transistor has a built-in gain factor β and the collector current (generally required to be high for laser drivers) is the base-current multiplied by this value and given as

$$I_C = \beta I_B \tag{3.32}$$

where I_B represents the base current. β can also be referred to as h_{FE}, which is known as the transistor forward current gain in the common-emitter configuration. The addition of the emitter-resistor, which is used as a feedback to the net base-emitter voltage and the emitter resistor voltage, can be calculated by

$$I_E = I_C - I_B \tag{3.33}$$

by following a simple Ohm's law approach. Another way of determining the emitter current is through

$$I_E = (\beta + 1)I_B. \tag{3.34}$$

The maximum collector current $I_{C(max)}$ that can flow through this configuration when accounting for the emitter-resistor is therefore given by

$$I_{C(\max)} = \frac{V_{CC} - V_{RE}}{R_L} \tag{3.35}$$

where V_{RE} is the voltage across the emitter-resistor, assuming that the transistor operates in saturation mode (V_{CE} = 0). The quiescent current of the common-emitter configuration $I_{C(quiescent)}$, therefore the current flowing with no input applied to the base of the transistor, is determined by

$$I_{C(quiescent)} = \frac{\frac{V_{CC} - V_{RE}}{2}}{R_L} \tag{3.36}$$

which is approximately half of the supply voltage divided by the load resistor. The voltage gain A_V of this configuration at low frequencies is determined by the load resistance, the emitter resistance and the internal resistance of the emitter of the transistor and is given by

$$A_{V(low\,frequency)} = -\frac{R_L}{R_E + R_e} \tag{3.37}$$

where the negative value is due to the fact that the output signal is inverted from the incoming input signal. At high frequencies, the bypass capacitor C_2 in Fig. 3.6 can short out the emitter resistance to ground and the voltage gain is higher, given as

$$A_{V(low\,frequency)} = -\frac{R_L}{R_e} \tag{3.38}$$

since the internal emitter resistance R_e is generally much lower than the added emitter resistor R_E. The output signal is measured at the collector terminal of the transistor.

For laser drivers a constant current at the output is required, which does not vary with small changes in the input voltage V_B, which may occur in practical applications. A method to improve the stability and mitigate the variations on the output current due to input voltage changes is to introduce a constant-voltage reference by means of a diode to the base of the transistor, as shown in the common-emitter amplifier topology in Fig. 3.7.

In Fig. 3.7 a forward-biased diode D_1 is added to the base of the transistor to maintain an approximate 0.7 V bias across the base-emitter junction of Q_1. The exact voltage of the diode is dependent on the current flowing through the diode and the temperature and is evident from the first principles of diode operation where the current through the device is given by

$$I_D = I_S\left(e^{qV_D/nkT} - 1\right) \tag{3.39}$$

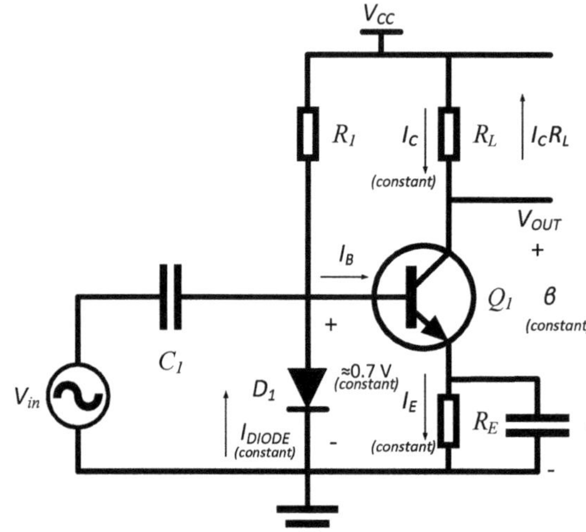

Fig. 3.7 Schematic representation of the single transistor common-emitter amplifier with emitter degeneration and active constant current implementation

where q is the elementary electron charge, V_D is the voltage across the diode, n is the non-ideality or emission coefficient (typically between 1 and 2), k is Boltzmann's constant and T is the temperature of the device in Kelvin. The current through the diode is varied by varying the value of R_I. Any change made to the value of R_{BIAS} will increase or decrease the voltage drop across the transistor base-emitter junction and hence increase or decrease the emitter current by the same proportion. Since the collector current is related to the emitter current by (3.32)–(3.34), this variation will also be seen in the collector current. The collector current therefore essentially mirrors the current through the diode based on the change in output voltage of the diode. The diode can also be replaced by another transistor, which is diode-connected (its collector and base terminals are connected) to increase the matching between the driver transistor and the active current-mirroring diode (transistor).

In order to achieve better temperature stability for the laser driver circuit realized by the common-emitter amplifier topology, an additional transistor (apart from the biasing transistor) should be introduced into the circuit. This changes the topology from a single transistor current source to a dual transistor current source, as discussed below.

3.5.2 Dual Transistor Current Source

The common-emitter amplifier can be modified slightly to present higher temperature stability, generally required for high-powered laser driver systems. The temperature compensation circuit is given in Fig. 3.8.

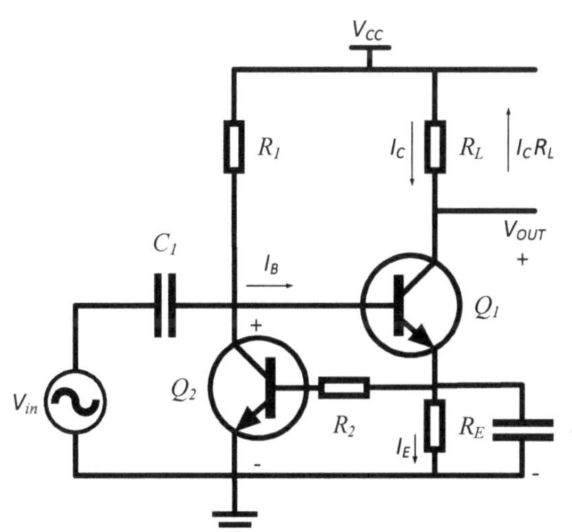

Fig. 3.8 Schematic representation of a temperature-compensated active constant current amplifier circuit

The principle of operation of Fig. 3.8 aims to introduce a means to compensate for any changes in temperature, which should not be noticeable at the output current. For the circuit in Fig. 3.8, when assuming a current is flowing through R_1 and to the base of Q_1, the transistor Q_1 is switched on and allows the output current to flow through the load resistor and the emitter resistor. The current through the emitter resistor induces a voltage drop across this resistor and if this voltage drop reaches a value comparable to the V_{BE} of transistor Q_2, Q_2 will also switch on and start conducting. If transistor Q_2 is switched on, it allows more current to flow through R_1 and the total current available from the supply is diverted between the base of Q_1 and the collector-emitter of Q_2 towards ground. This effectively means that Q_1 will begin to conduct less current through its collector-emitter junction as it starts switching off. This negative feedback reaches equilibrium once the voltage drop across R_E reaches a value of exactly V_{BE} of Q_2—which is also matched closely to Q_1 (through semiconductor layout and process parameter control). The current through Q_2 will be much lower compared to Q_1, since the device is 'barely on' and this means that the power dissipation and eventually the heat generated in Q_2 are very low. This means that the voltage drop across R_E (which is equal to V_{BE} of Q_2) remains constant together with the current through Q_2. The circuit is, however, still sensitive to the ambient temperature, as V_{BE} of Q_2 will vary with the temperature of its environment, but if the ambient temperature is kept constant, Q_2 will not heat up and the current will remain constant since V_{BE} is a function of temperature.

The differential amplifier using two common-emitter amplifier topologies is a useful and commonly used configuration with various advantages over single-ended amplifiers. These advantages include:

- It provides a voltage gain for differential signals on the inputs where the output voltage is $V_d = V_1 - V_2$ and it attenuates interfering common-mode signals with output voltage of $V_c = (V_1 + V_2)/2$.
- It provides the inverting and non-inverting inputs needed for operational amplifiers (op-amps).
- It provides feedback to the input of the circuit to actively correct any variations common to the inputs of the system.

The operational principles of differential amplifiers, since this configuration is typically used in laser drivers, is discussed in the following section.

3.5.3 Dual Transistor Differential Current Source

The simplest form of the emitter-coupled differential amplifier using a common resistor at the emitters of both transistors to induce a current flow through the tail of the configuration is given in Fig. 3.9.

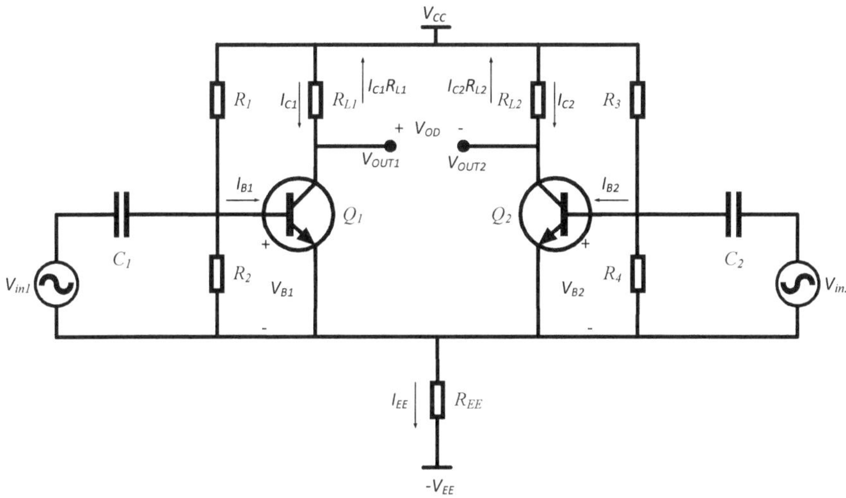

Fig. 3.9 Differential emitter-coupled amplifier with resistive biasing

In Fig. 3.9 there are two *npn*-transistors, each able to receive an AC input signal and each biased to be fully on. The differential amplifier ideally only amplifies the difference between the two input voltages such that

$$v_{od} = A_{dm} v_{id} \tag{3.40}$$

where v_{od} is the differential output voltage such that $v_{od} = v_{out1} - v_{out2}$, A_{dm} is the differential-mode voltage gain and v_{id} is the difference between the two input signals v_{in1} and v_{in2}, therefore $v_{id} = v_{in1} - v_{in2}$. In practice, the differential amplifier will also amplify the average of the two inputs, its common-mode input, given by

$$v_{oc} = A_{cm} v_{ic} \tag{3.41}$$

where v_{oc} is the common-mode output voltage given as $v_{oc} = (v_{out1} + v_{out2})/2$, A_{cm} is the common-mode voltage gain, and v_{ic} is the average of the input signals given as $v_{ic} = (v_{in1} + v_{in2})/2$. The differential amplifier has an important FOM, which is the common-mode rejection ratio (CMRR), and this is given by

$$\text{CMRR} = \frac{A_{dm}}{A_{cm}} \tag{3.42}$$

where a larger CMRR represents a better differential amplifier, which is given in dB. For a purely differential input ($v_{id} \neq 0$ and $v_{ic} = 0$) a virtual AC ground is seen at the transistor pair emitters and the voltage seen at both emitters does not fluctuate. If the input signal is purely common-mode ($v_{id} = 0$ and $v_{ic} \neq 0$) the current flowing through the emitter resistor is double that of the current flowing through each transistor and effectively the impedance seen from each side of the emitter-coupled pair is also double. For purely differential and pure common-mode input signals, the circuit in Fig. 3.9 can be split into two small signal half-circuits representing each mode of operation, shown in Fig. 3.10.

When inspecting Fig. 3.10a, b, it is noticeable that the circuit presented in Fig. 3.10a is a typical common-emitter amplifier without emitter degeneration resistor with input voltage of $v_{id}/2$ and output voltage of $v_{od}/2$. If the output is taken between the two collector terminals, the differential gain or double-ended output of this circuit is therefore given by

$$A_{dm} = -g_m R_c = -\frac{R_C}{re} \tag{3.43}$$

and if the output is taken between either collector terminal of the transistors and ground, a single-ended output is measured and its gain is

$$A_{dm} = -\frac{R_C}{2re} \tag{3.44}$$

which is a half of the gain of the double-ended output. The input resistance in the differential mode is effectively the equivalent resistance seen between the two bases of the transistors. Since a virtual ground exists at the emitters of both transistors, the input resistance is only $r_\pi + r_\pi = 2r_\pi$.

Figure 3.10b represents the common-mode half circuit of the differential amplifier and when investigating the figure, it is noticed that the topology appears

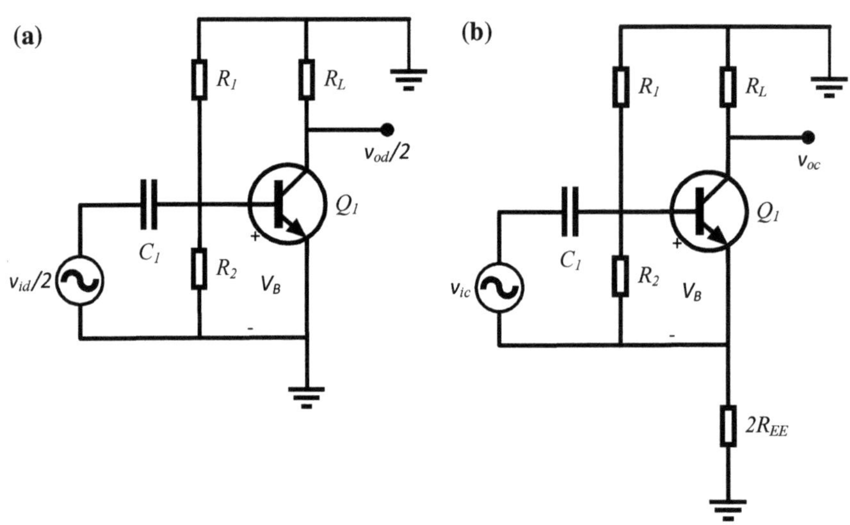

Fig. 3.10 Differential emitter-coupled amplifier with resistive biasing half-circuits representing **a** purely differential mode and **b** purely common-mode

similar to a common-emitter amplifier with emitter degeneration with a value of $2R_{EE}$. Recalling the fact that $2R_{EE} \parallel 2R_{EE} = R_{EE}$, it means that the circuit is not changed and represents one half of the common-mode circuit. The single-ended common mode gain of this circuit is therefore

$$A_{dm} = -\frac{R_C}{2R_{EE}} \tag{3.45}$$

if it is assumed that $2R_{EE} \gg r_e$, which is a fair assumption. The common-mode input resistance can be determined by

$$R_{in(CM)} = 2r_\pi + 2(\beta + 1)R_{EE} \tag{3.46}$$

which can be approximated as

$$R_{in(CM)} = 2\beta R_{EE} \tag{3.47}$$

assuming $\beta >> 1$ and $R_{EE} \gg r_e$. The value of the common-mode input resistance can therefore become very large. By using these results of the differential mode gain and common-mode gain reconsidering the CMRR in (3.42), the expression for the CMRR in terms of circuit components can be derived as

$$\text{CMRR} = \frac{A_{dm}}{A_{cm}} = \frac{-\frac{R_C}{2re}}{-\frac{R_C}{2R_{EE}}} = \frac{R_{EE}}{r_e} = g_m R_{EE} \tag{3.48}$$

which is normally expressed in dB as

$$\text{CMRR} = 20\log(g_m R_{EE}). \tag{3.49}$$

In standard laser transmitters used in light wave communications in electronic surveillance and countermeasures, the laser is driven by a combination of DC bias current, which is adjusted by some feedback control circuit and a modulation current based on the input data signal (Bosch and Swan 1985). In the normal operation the DC bias current keeps the laser just below its lasing threshold, where the laser acts as a light-emitting diode (LED) and the optical power in this current range is low. The modulation current is added to the DC bias so that the laser is turned on when a pulse corresponding to the input data appears. Above the threshold current, lasing will occur and the optical output power will increase linearly with increasing modulation current. The modulation current is generated, in laser drivers, predominantly by using differential amplifiers and it is aimed at operating the amplifier near its emitter-current density (Kaiser and Manz 1994). The differential amplifier and therefore the quality of the driving transistors should be able to switch at high frequencies to allow high-bitrate operation of the modulation current of the lasing diode. It is important to consider transistors that have very small and symmetric rise and fall times, since any discrepancies in the transistor will be reproduced in the output modulation current and in the quality of the optical output signal. The variations that can occur in the total output current generated by the emitter resistor will effectively also change the output level of the modulated current during the on-and-off switching of the dual transistors. For this reason, an improved approach is to use a constant-current source in the tail (or at the top—depending on the application requirements) of the differential amplifier. Such a configuration is shown in Fig. 3.11.

The emitter resistor in Fig. 3.9 is replaced in Fig. 3.11 with a constant-current source using two additional *npn*-transistors at the tail of the differential amplifier. The magnitude of the current is determined by the reference resistor R_{REF} and this current is mirrored from the diode-connected transistor Q_3 towards Q_4, which drives the constant DC bias current of the differential amplifier. Variations of the differential amplifier to improve upon its pulse response and the quality of the modulating current by employing a series of cascaded differential amplifiers have been patented in Kaiser and Manz (1994), for example.

The differential amplifier forms the basis of op-amps, since it provides the ability of having an inverting and non-inverting input to be used as a dual-input, single input, dual output or single output amplifier, depending on the application. Op-amps are therefore a critical building block of laser drivers and based on

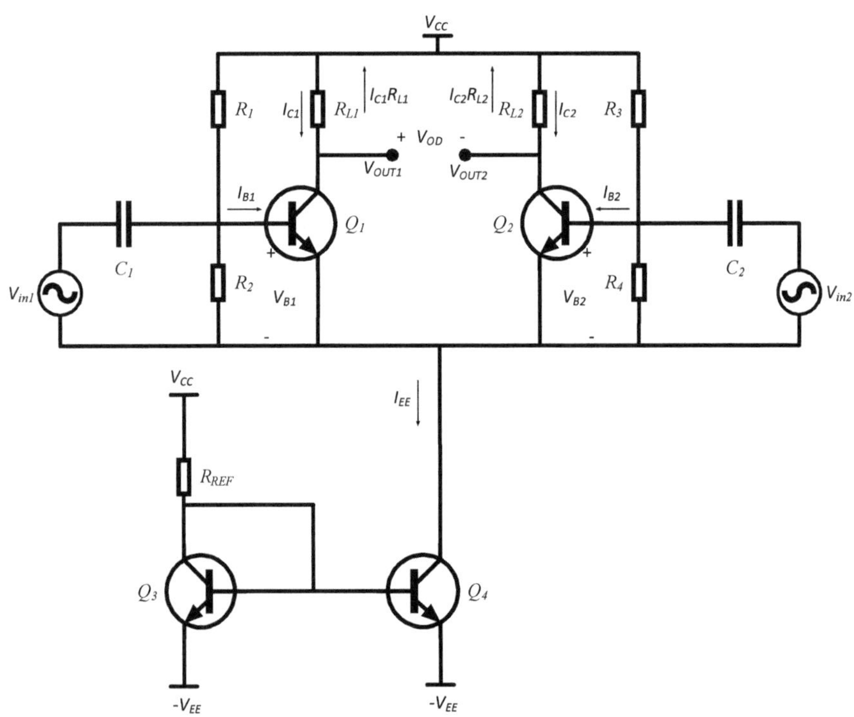

Fig. 3.11 Differential emitter-coupled amplifier with active constant-current biasing

understanding of the basic principles of differential amplifiers, the principles of operation of op-amps are discussed in the following section.

3.5.4 Op-Amp Current Source

Semiconductor diode lasers (laser diodes) are extensively used in a wide variety of laser devices intended for tactical military applications, such as laser rangefinders (Burns et al. 1991), proximity sensors, short-range laser dazzlers and laser aiming devices. To recall, laser diodes primarily comprise a constant-current source that can provide current to a forward-biased laser diode to bias the device just below its threshold current. The modulated current from the intended input signal provides the rest of the current to drive the laser into its lasing mode and provide optical output power synchronized to the input signal. Temperature correction and stability are important factors when designing laser drivers, since thermal variations are inevitable in these high-current inversion-type of applications and should be carefully controlled irrespective of changes in the ambient temperature. Laser diodes generally exhibit reliable operation under ideal operating conditions but are highly susceptible to damage from too high drive currents, thermal transients and electrostatic discharge (ESD), which can all influence the optical output signal. In many military applications, a stable and constant-wavelength optical output is required and beam divergence and failure to induce lasing action can have devastating results.

Laser drivers can be operated in either one of two possible topologies: a constant-current or constant-power output drive. The preferred topology is that of constant-current and precise control of the operating temperature. If no temperature compensation is included in the circuit, the laser output can drop below its lasing threshold and emit spontaneous light only or it can operate past its absolute maximum conditions, damage the device and produce unmodulated optical outputs. Constant-current topologies therefore require a current-sensing element that continuously senses the drive current and with feedback, adjusts the input voltage accordingly, compared to a carefully controlled and precisely maintained reference voltage. The differential amplifier described in the previous section therefore has many attributes ideal for achieving this type of operation. It is able to use a feedback signal at its second input to adjust the output current, has constant output-current capabilities and allows for temperature compensation at its input terminals. The op-amp, which is essentially a differential amplifier, can therefore be implemented to achieve these capabilities as well. A simple op-amp-based constant-current implementation with feedback and reference voltage used to *source* the load with current is shown in Fig. 3.12.

The basic principle of the constant-current mode laser driver in Fig. 3.12 can be described when considering the fact that a feedback voltage is fed to the inverting input of the op-amp. The op-amp is used to amplify the difference between the input signals and this difference can therefore vary as the current through the diode and the driver transistor changes. The voltage-drop across R_1 will change as the load

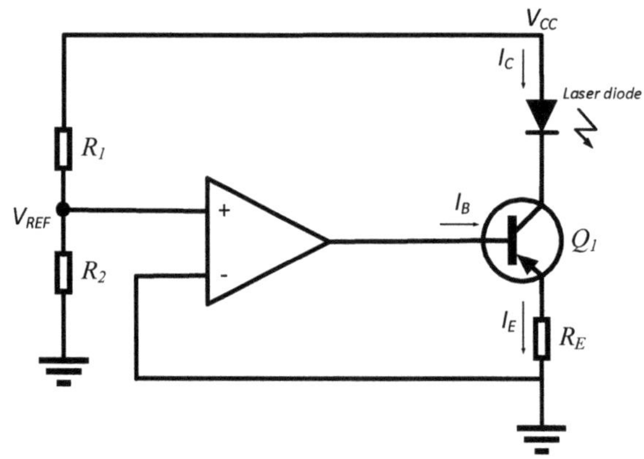

Fig. 3.12 Simple constant-current mode operational amplifier design using a *pnp*-transistor in the output node and a floating load

current changes and this variation in voltage-drop is fed back to the op-amp where the difference between the carefully controlled reference voltage and the feedback voltage determines the output of the op-amp, fed directly to the base of the transistor Q_1. The transistor will therefore switch more on or more off depending on the voltage it receives from the op-amp and its biasing conditions. In Fig. 3.12 the laser diode is connected directly to the positive voltage supply V_{CC}, called a floating load. It can also be connected between R_1 and ground, which is called a grounded load configuration and *sinks* the current through the load. This configuration is shown in Fig. 3.13. Its operational principle is similar to that of Fig. 3.12.

The configuration in Fig. 3.13 operates on a similar principle as the floating load configuration in Fig. 3.12 where the feedback voltage from a current flow through R_E appears at the inverting input of the op-amp. The difference between the feedback voltage and the constant reference voltage V_{REF} on the non-inverting input

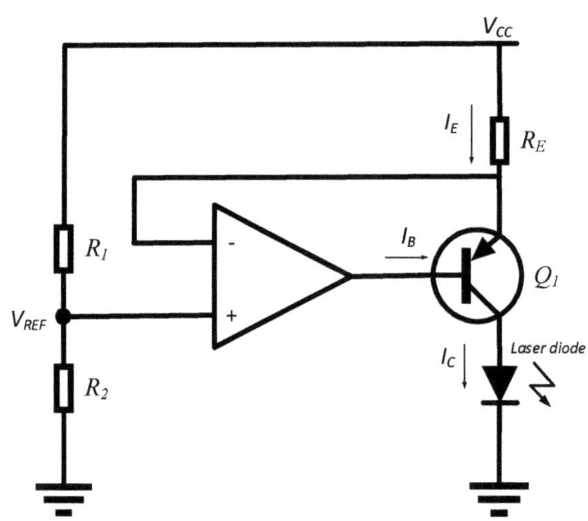

Fig. 3.13 Simple constant-current mode operational amplifier design using a *pnp*-transistor in the output node and a grounded load

biased by R_1 and R_2 is amplified and fed to the base of the *pnp* transistor Q_1. The resultant current flows through the load and the laser diode and is varied based on the voltage supplied to the transistor base.

Several variations on the op-amp configurations for laser drivers can be implemented and can be found readily in literature and as SoC implementations at online vendors. Listing all of these variations would be somewhat impractical to do in this book and there are several resourceful alternatives, which present useful techniques and design modification for improving the performance of op-amp-based laser drivers. Three examples of such sources are Albaugh (2001), Maxim Integrated (2004) and Erickson et al. (2008). A summary of typical performance parameters of laser drivers implemented in different technologies is presented in the following section.

3.6 Laser Driver Performance

Table 3.4 lists various performance characteristics of laser drivers implemented historically in InP, GaAs and more recently, SiGe.

The laser driver implementations in Table 3.4 show the trends of technologies and typical performance attained with each during approximately the past 30 years. GaAs laser drivers were the focus of laser drivers during the late 1980s up to about the early 1990s and their performance characteristics were praised during that time. The typical data rates that were achieved with these devices ranged between about 1.5–5 Gbit/s with power consumption from about 500 mW to around 2 W. It is difficult to report exactly on the power dissipation and other factors such as the area of these devices, since many of the reports have supporting circuitry on-chip, which adds to the total size and power of these devices. Some reports also use multiple lines to achieve the maximum data rates so the actual data rates per channel are not always known. Table 3.4 nevertheless gives a good approximation of the trends and achievements of GaAs, InP and SiGe. GaAs technologies during the 1980s and 1990s used mostly 500 nm–1 μm channel lengths since this was approximately the attainable technology during this time.

InP-based and InP/InGaAs technologies used for laser drivers received attention in research after GaAs-only implementations and as seen in Table 3.4, this happened around the late 1990s to early 2000s. Throughput for laser drivers increased to approximately 10–30 Gbit/s with InP/InGaAs implementations; however, power consumption in these devices was relatively high, ranging between 240 mW and higher than 2 W for active laser drivers. Importantly, transistor f_Ts increased to higher than 50 GHz using this technology, which paved the way for high-frequency devices and circuits in many fields, not only for laser drivers. Power consumption using InP/InGaAs prevented this technology from being widely implemented in mobile and low-power devices, although the advantages of its speed were not overlooked. Research into improving on the technology through improvements in other fields of measurement equipment, process manufacture and advances in

Table 3.4 Technology implementations and performance characteristics of reported works of InP, GaAs and SiGe laser drivers

Reference	Technology	f_T (GHz)	Channel data rate (Gbit/s)	Power dissipation (mW)	Die area (mm^2)
Moto et al. (2013)	180 nm SiGe BiCMOS	200	25.78	160	6.49
Kim et al. (2003)	350 nm SiGe BiCMOS	–	2.5	–	–
Fattaruso and Sheahan (2006)	SiGe BiCMOS	90	4.25	~300	2.66
Morley (2005)	SiGe SOI bipolar	70	10.7	670	2.25
Knochenhauer et al. (2009)	SiGe BiCMOS	180	40	80	0.41
Reyaz et al. (2015)	130 nm SiGe BiCMOS	300	Differential amplifier	122	0.27
Riishoj (1993)	1 μm GaAs MESFET	12	2.5	2000	–
Yamanaka and Takada (1989)	500 nm GaAs FET	–	1.5	927	4
Chen and Bosch (1988)	1 μm GaAs MESFET	(12)	1.7	500	–
Hakamata et al. (1993)	GaAs MESFET	Experiment	5	–	–
Miyashita et al. (1994)	800 nm GaAs MESFET	25	2.5	650	6.16
Meghelli et al. 1997	InP/InGaAs DHBT	56	30	1800	–
Schneibel et al. (1999)	InP/InGaAs DHBT	79	8	>2000	–
Banu et al. (1991)	InP/InGaAs HBT	–	10	1350	–
Pradhan et al. (2002)	InP-based OEIC	45	10	240	–

physics has also produced high-speed InP/InGaAs devices. Kim (2007) reports results of f_T of 112 GHz and f_{max} of 307 GHz achieved in InP/InGaAs heterojunction bipolar transistors.

More recently, from about the mid-2000s until presently, SiGe has become the preferred technology for high-speed and low-powered laser drivers. SiGe achieved throughputs of between 2.5 Gbit/s already in 2003 to upwards of 40 Gbit/s. Power consumption of these circuits is also significantly lower compared to older technologies and ranges between 80 and approximately 700 mW, depending on the structures of the circuits and the supporting circuitry implemented on chip. Die area of these circuits is also inevitably smaller, averaging about 2.5 mm^2 depending on the implementation, where channel-lengths of 130 nm and smaller are not uncommon and have become relatively cost-effective in recent years. To date, the smallest commercially available technology node has a channel length of 14 nm and is produced by Intel® Corporation in the *Skylake* range of digital processors.

3.7 Conclusion

This chapter briefly looks at the market size of laser systems in military applications and the largest contributing companies in this field. A list is also provided of the ongoing research and development of laser weapon projects around the world and the corporations that are involved. The chapter aims to provide better understanding of optoelectronic communication systems and their subsystems to enlighten the reader about the crucial considerations when developing such systems and why technology advancements are playing a large role in the power capabilities and improvements in mobility of high-powered laser weapons. In order to achieve this goal, the chapter gives an overview of the equivalent model of the laser diode to identify the design considerations of the driver circuits. This follows a discussion on laser drivers, typical topologies and first principle design considerations of such systems. Since SiGe presents many advantages in terms of speed, power and size for modern electronics, it can also be considered as an enabling technology for laser drivers, as the subsystems of laser drivers require such components to transfer output power effectively to the optical device. A discussion on basic differential amplifiers is presented and followed by laser drivers achieved by op-amp circuits.

Literally hundreds of op-amp configurations are used in laser drivers and each configuration has its own advantages and disadvantages, depending on the application of the system when choosing the ideal topology. The goal of this book is not to list and describe all of the possible configurations, but rather to describe the basic principles to address when choosing a laser driver. This chapter should therefore be used in conjunction with op-amp design strategies and various other sources when designing a laser driver. The chapter also aims to discuss the building blocks of the laser driver amplifiers and highlight the requirements of the semiconductor technology used to realize the circuits. Important considerations and FOMs to check when operating a laser driver are the input voltage versus output power and output current characteristics and the input impulse response of the circuit, as well as the variation of these parameters with temperature. These parameters will give a good estimation of the performance of the laser driver and its stability under operating conditions.

The chapter furthermore aims to provide background information on the critical requirements of differential amplifiers and op-amps, which will be used in determining the optimum enabling technology for these types of circuits and to determine if SiGe is a well-suited alternative based on its performance parameters, cost-efficiency and operating limitations. SiGe is the fastest growing semiconductor process and continues to provide better and more reliable operation at very high frequencies; military-based electronic components such as radar systems, high-speed analog systems and devices operating in extreme environments can benefit from this technology as much as the commercial sector has already gained from it.

Finally, a list of laser driver performance characteristics is presented for various technology implementations such as GaAs, InP/InGaAs and SiGe to identify the trends and the performance improvements across the implementations.

References

Albaugh, N. (2001). Optoelectronics Circuit Collection. *Texas Instruments Application Report* (SBEA001, 1–18). September 2001.

Banu, M., Jalali, B., Nottenburg, R., Humphrey, D. A., Montgomery, R. K., Hamm, R. A., et al. (1991). 10 Gbit/s bipolar laser driver. *Electronics Letters, 27*(3), 278–280.

Barnes, P., & Paoli, T. (1976). Derivative measurements of the current-voltage characteristics of double-heterostructure injection lasers. *IEEE Journal of Quantum Electronics, 12*(10), 633–639.

Bosch, F. L., & Swan, C. B. (1985). Laser driving means. *Patent US4539686 A*. September 3, 1985.

Burns, H. N., Christodoulou, C. G., & Boreman, G. D. (1991). System design of a pulsed laser rangefinder. *Optical Engineering, 30*(3), 323–329. March 1991.

Chen, F. S., & Bosch, F. (1988). GaAs MESFET laser-driver IC for 1.7-Gbit/s lightwave transmitter. *Journal of Lightwave Technology, 6*(3), 475–479.

Coffey, V. C. (2014). High-energy lasers: New advances in defense applications. *Optics and Photonics*. October 2014.

Erickson, C. J., Van Zijl, M., Doermann, G., & Durfee, D. S. (2008). An ultrahigh stability, low-noise laser current driver with digital control. *Review of Scientific Instruments, 79* (073107). July 2008.

Fattaruso, J. W., & Sheahan, B. (2006). A 3-V 4.25-Gb/s laser driver with 0.4-V output voltage compliance. *IEEE Journal of Solid-State Circuits, 41*(8), 1930–1937. August 2006.

Gray, P. R., Hurst, P. J., Lewis, S. H., & Meyer, R. G. (2001). Analysis and design of analog integrated circuits. *John Wiley and Sons, Inc*. Fourth Edition.

Hakamata, Y., Takemoto, K., Nakanishi, T., & Nakano, J. I. (1993). 5 Gb/s transmission experiment using 1300 nm fabry-perot LD transmitter module with GaAs MESFET LD driver and p-i-n PD receiver. *IEEE Photonics Technology Letters, 5*(2), 251–254.

Harder, C., Katz, J., Margalit, S., Shacham, J., & Yariv, A. (1982). Noise equivalent circuit of a semiconductor laser diode. *IEEE Journal of Quantum Electronics, QE-18*(3), 333–337.

Irving, C. R., & Carroll, J. E. (1990). An equivalent circuit model of a multimode semiconductor laser. *IEE Colloquium on Modelling of Optoelectronic Devices,* 11/1–11/4.

Kaiser, N., & Manz, W. (1994). Driver circuit for a laser. *Patent US5297157 A*. March 22, 1994.

Katz, J., Margalit, S., Harder, C., Wilt, D., & Yariv, A. (1981). The intrinsic equivalent circuit of a laser diode. *IEEE Journal of Quantum Electronics, QE-17*(1), 4–7. January 1981.

Kim, J. H. (2005). Wide-band and scalable equivalent circuit model for multiple quantum well laser diodes. *A Dissertation Presented to The Academic Faculty: Georgia Institute of Technology*. August 2005.

Kim, M. (2007). Self-aligned InP/InGaAs heterojunction bipolar transistors with crystallographically defined emitter contact. *Journal of the Korean Physical Society, 51*(2), 612–615. August 2007.

Kim, Y., Bien, E., Chang, J. J., Robinson, M. A., & Lee, M. (2003). 2.5 Gbit/s/channel laser driver design for four-channel parallel optic transceiver using SiGe BiCMOS technology. *Electronics Letters, 39*(5), 457–458.

Knochenhauer, C., Hauptmann, S., Scheytt, C., & Ellinger, F. (2009). A compact, low-power 40 Gbit/s differential laser driver in SiGe BiCMOS technology. *2009 European Microwave Integrated Circuits Conference (EuMIC)*, 324–326.

Kressel, H., & Butler, J. K. (1977). *Semiconductor Lasers and Heterojunction LEDs*. New York: Academic (Chapter 17).

Maxim Integrated. (2004). Accurate power control of the MAX2740A laser driver. *Application Note 3080*. Available from http://www.maximintegrated.com. August 2004.

Meghelli, M., Bouche, M., Launay, P., & Konczykowska, A. (1997). 30 Gbit/s InP DHBT 2:1 selector-driver IC for external laser modulation. *Electronics Letters, 33*(12), 1057–1058.

Miyashita, M., Andoh, N., Nakano, H., Motoshima, K., Kebukawa, N., Shibao, S., et al. (1994). A GaAs LD driver IC for 2.5 Gb/s optical communications system. *24th European Microwave Conference, 2*, 1649–1654.

Morishita, M., Ohmi, T., & Nishizawa, J. (1979). Impedance characteristics of double-heterostructure laser diodes. *IEEE Journal of Solid State Electronics., 22*(951–962), 1979.

Morley, S. (2005). A 3 V 10.7 Gb/s differential laser diode driver with active back-termination output stage. *2005 Solid-State Circuits Conference (ISSCC)*, 220–221 (Session 12, Optical Communications).

Moto, A., Ikagawa, T., Sato, S., Yamasaki, Y., Onishi, Y., & Tanaka, K. (2013). A low power quad 25.78-Gbit/s 2.5 V laser diode driver using shunt-driving in 0.18 μm SiGe-BiCMOS. *2013 IEEE Compound Semiconductor Integrated Circuit Symposium (CSICS)*, 1–4.

Mukai, T., & Yamamoto, Y. (1982). Noise in AlGaAs semiconductor laser amplifier. *IEEE Journal of Quantum Electronics, 18*(4), 564–575.

Ozyazici, M. S. (2004). The complete electrical equivalent circuit of a double heterojunction laser diode using scattering parameters. *Journal of Optoelectronics and Advanced Materials, 6*(4), 1243–1253. December 2004.

Paoli, T. L. (1976). Observation of second derivative of the electrical characteristics of double-heterostructure junction lasers. *IEEE Transactions on Electron Devices, 23*(12), 1333–1336.

Paschotta, R. (2013). Fabry-pérot interferometers. *RP Photonics*. Available from www.rp-photonics.com. Accessed on February 28, 2016.

Pradhan, S., Bhattacharya, P., & Liu, W. K. (2002). Monolithically integrated 1.55 μm photoreceiver-laser driver optoelectronic integrated circuit. *Electronics Letters, 38*(17), 987–989.

Reyaz, S. B., Malmqvist, R., Gustafsson, A., & Kaynak, M. (2015). SiGe BiCMOS high-gain and wideband differential intermediate frequency amplifier for W-Band passive imaging single-chip receivers. *IET Microwaves, Antennas and Propagation, 9*(6), 569–575.

Risshoj, J. (1993). 2.5 Gb/s laser-driver GaAs IC. *Journal of Lightwave Technology, 11*(7), 1139–1146.

Rogers, J. W. M., & Plett, C. (2003). Radio frequency integrated circuit design. *Artech House*.

Schneibel, H., Graf, C., Bauknecht, R., & Melchior, H. (1999). High current InP double hetero bipolar transistor driver circuit for laser diodes. *1999 International Conference on Indium Phosphide and Related Materials*, 455–457.

Tucker, R., & Pope, D. (1983). Circuit modeling of the effect of diffusion on damping in a narrow-stripe semiconductor laser. *IEEE Journal of Quantum Electronics, 19*(7), 1179–1183.

Weiner, J. S., Leven, A., Houtsma, V., Baeyens, Y., Chen, Y., Paschke, P., et al. (2003). SiGe differential transmimpledance amplifier with 50-GHz bandwidth. *IEEE Journal of Solid-State Circuits, 38*(9), 1512–1517.

Yamanaka, N., & Takada, T. (1989). A 1.5 Gbit/s GaAs four-channel selector LSI with monolithically integrated newly structured GaAs Ohmic contact MSM photodetector and laser driver. *IEEE Photonics Technology Letters, 1*(10), 310–312.

Chapter 4
Electronic Warfare Optoelectronic Receiver Fundamentals: Applications and Research Opportunities

4.1 Introduction

Optical technology in military and aerospace provides unsurpassed advantages in speed, efficiency and radiation hardness. The combination of optical and electronic technologies, generally referred to as optoelectronic systems, combine sensors, communication systems and computing power to form highly reliable and high-data-rate transfer of voice, data, audio, video and commands. The optoelectronic realm and the use of optics in military systems include new and advanced implementations for transmission across optical fibers, signal generation, analog and digital signal processing and optical control of microwave devices. Realizable and practical implementations of systems operating at high-GHz frequencies are possible with optoelectronic combinations. Techniques used in these systems can also be adapted to mm-wave application in military operations and therefore the continuation of research and development in this field is crucial for even higher data rate communication. Primary applications of optoelectronic systems include applications based on enabling technologies such as

- light-emitting diodes for electroluminescence determined by the band gap of the *pn*-junction diode offering high-efficiency and long-lifetime advantages to a multitude of commercial and military applications,
- avalanche photodiodes to convert light into electrical signals through the photoelectric effect, generally coupled with low-noise receivers,
- laser diodes,
- integrated optic circuits or photonic integrated circuits (PIC), which allow integration of optically active and passive functions on a single chip where commonly used materials are InP and InGaAsP alloys based on their wavelength characteristics,

W. Lambrechts and S. Sinha, *SiGe-based Re-engineering of Electronic Warfare Subsystems*, Signals and Communication Technology,
DOI 10.1007/978-3-319-47403-8_4

- integrated optic-electronic circuits, which combine PICs and electronic circuitry to realize full SoC implementations, and
- PIN photodiodes, a *pn*-junction photodetector with an additional intrinsic region that requires a large reverse-bias to cause impact ionization.

These enabling technologies all have specific applications, which are expanding rapidly in various fields. Optoelectronic systems are found in commercial telecommunications, consumer electronics, data communication systems and military and EW advances. Commercial telecommunications include applications such as

- interexchange transmissions by electronic data interchange (EDI) standards on peer-to-peer networking, value-added networks and the internet,
- undersea telecommunications where a submarine communications cable is laid on the sea bed and carries signals between land-based stations either through fiber-optic cabling or copper for electronic transfers,
- long haul fiber-optic networks by using repeater amplifiers and support for high-bandwidth information transfer,
- loop distribution or fiber in the loop, essentially upgrading wired telephone exchange services,
- private network bypassing, and
- upgrading traditional CAT-V coaxial television subscription infrastructures to optic-fiber networks.

In consumer electronic equipment, optoelectronics can be found in

- optical disc readers,
- laser printers, and
- digital cameras.

Optoelectronics in data communication systems are also expanding significantly with the decrease in initial capital to realize systems in the fields of

- commercial user-premise communications,
- intra-systems,
- local area networks,
- machine control,
- automotive applications,
- industrial applications,
- office equipment, and
- equipment interconnects realization.

Future trends in optoelectronic systems, which are showing significant promise for the near future, include optical backplanes, photonic switching and fully optical computers capable of processing data at light-speeds. Finally, and applicable to this book, which investigates the technologies used in military operations and EW, optoelectronic systems are found in applications such as

- shipboard sea systems such as underwater autonomous laser velocimetry, marine particle sensors, underwater acoustic sensors, seismic detection, non-acoustic underwater imaging, non-acoustical submarine tactical communications,
- airborne systems such as fiber-optic weapons sights, passive fiber-optic components to withstand extreme aircraft environments, optoelectronic IR sensors in anti-air operations, airborne imaging spectroscopy,
- ground systems such as ground-based remote sensors, avalanche and transit time devices, missile seeker transmitters,
- EW implementations such as focal place filters for IR detector arrays, low probability of intercept or jammed voice and data links,
- surveillance and detection of wideband radar signals, secondary sensors for anti-radiation missiles,
- navigation applications such as improved airborne infrared search and track (IRST),
- guidance and control such as passive non-cooperative target recognition sensors, and
- various radar applications such as radar cross-section validation or high-effectivity radiated power transmitters.

Photonics essentially involves molding the flow of light by optoelectronic devices through generation, transportation, detection and processing and manipulation of light-waves. To understand and grasp the fundamentals of photonics completely, a clear understanding of the propagation of a photon field, the interaction between photon fields and electron systems in insulators, semiconductors and metals, quantum electronic device concepts and heterojunction devices is required. This book does not aim to provide detailed and first-principle derivations and discussion on each of these disciplines, but rather aims to provide the required fundamentals to apply the physical interactions of photons and electronic devices. This interaction is an important consideration when implementing military and commercial applications and in addition provides benefits when choosing electronic technologies that are capable of manipulating the incoming signals; especially considering long-range communication in military operations where detection of the source is of primary concern. Various sources, online articles, books and journals provide detailed explanations of electromagnetic fields and waves, commonly used equations such as Maxwell's equations, polarization, semiconductor physics, band structures, *pn*-junction theory and physics and the physics of optics and quantum mechanics. Referring to such material during the re-engineering of electronic warfare systems can benefit the reader when identifying and understanding fundamental principles in system design. Light described as EM waves with μm-wavelengths provides valuable insight into small-signal attenuation phenomena, high directionality and efficient antennas by using lenses and mirrors, the availability of sensitive receivers, high-propagation velocities and low dispersion and the availability of low complexity (albeit slower) mechanical modulators such as reticules. Historical pitfalls of light-wave technology are the lack of light sources that could be modulated at high data rates, line of sight (LOS) communication requirements with high attenuation and scattering in dust, fog, rain and vegetation and limitations of compact controlled transmission media such as optical waveguides. Advances in technology in the past decade have improved upon all of these limitations and are steadily providing relatively easy to integrate systems using light-wave technology in conjunction with semiconductor processes. Two key technology inventions of the previous century that have ensured progress and expanding applications for optoelectronic systems are the invention of the laser and low loss and compact optical glass fibers as transmission mediums. Laser diode development has cemented new goals for high-speed communication and research into lower operating currents, longer wavelength operation, >80 GHz bandwidth capabilities for short-distance links and monolithic integration with photodetectors, external modulators and semiconductor optical amplifiers (SOAs). If considering bandwidth in optoelectronic systems, the electronic IC has become the bottleneck for achieving data rates surpassing the most advanced fully electronic systems currently and has become the primary driver for high-speed digital and analog IC technologies. Technology considerations such as SiGe aim to decrease the bottleneck between optical bandwidth and electronic component capabilities. Integrated systems such as low-noise preamplifiers, gain-controlled broadband power amplifiers and driver circuits for power amplifiers for LEDs, lasers and modulators are among the culprits introducing electronic limitations and bottlenecks from intrinsic parasitic characteristics for achieving theoretical speeds. In this chapter the terms photodetector and photodiode are used interchangeably.

4.2 Optoelectronic Communication

This chapter focuses on the optoelectronic communication components required to achieve high-speed, low-noise and high-powered optical transmission through an optical medium. The typical laser optoelectronic communication link as presented in Chap. 3 is shown again in Fig. 4.1 to identify the focus components in this chapter.

The optical medium is the transport vessel for the information signal sent from the transmitting laser diode to the receiving photodetector. The optical medium will introduce signal degradation based on the distance between the transmitter and receiver and the wavelength of operation. A brief introductory discussion on the attenuation in the optical medium is provided in the following paragraph; however, it is not discussed in detail, since this discussion falls outside the scope of this book. The attenuation and additive noise in the optical medium should be corrected at the receiver and the photodetector should be able to distinguish between noise in the system and the information. The considerations when combining a low-noise amplifier with the photodetector receiver are discussed in detail in this chapter. Chapter 3 focused largely on the laser diode and laser driver subsystems in Fig. 4.1, the transmitting side, whereas this chapter focuses on the receiving side. For

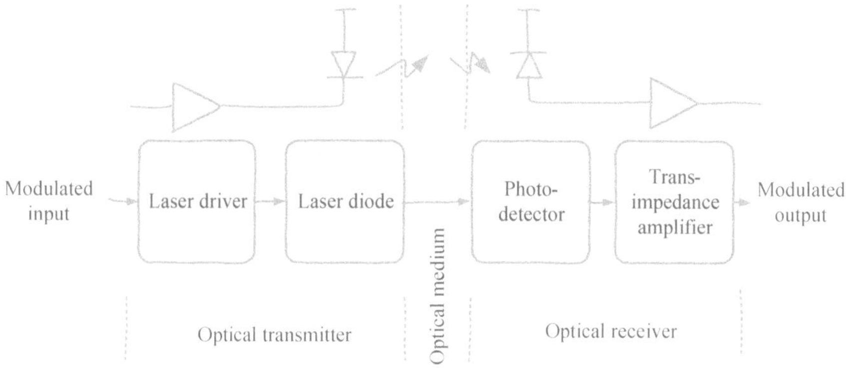

Fig. 4.1 A typical laser optoelectronic communication link showing the optical transmitter stage, optical medium and the optical receiver stage

simplified insight into signal degradation through the optical medium, the following paragraph is presented.

4.3 Optical Medium Signal Degradation

The optical medium in Fig. 4.1 is the material through which the electromagnetic waves in a laser system propagate and is a transmission medium. The electromagnetic waves that propagate through the medium are influenced primarily by the permittivity and the permeability of the transmission medium—if the signal attributes are not considered. The intrinsic impedance η of any transmission medium and the amount of resistance it presents to the flow of electromagnetic waves is given by

$$\eta = \sqrt{\frac{\mu}{\varepsilon}} \tag{4.1}$$

where μ is the permeability and ε is the permittivity of the medium, assuming it has no electrical conductivity and the electric field and magnetic field are negligible. If considering the characteristics of the signal as well, the communication of an optical device works on the same principle as any sound or light transponder system, where the Friis free-space equation can be used to determine the distance and power requirements of the system;

$$P_r(d) = \frac{P_t G_t G_r \lambda^2}{(4\pi)^2 d^2 L} \tag{4.2}$$

where $P_r(d)$ is the power received by the receiving antenna (in watt) at distance d (in m), P_t is the amount of power transmitted (in watt), G_t is the gain of the transmitter and G_r is the gain of the receiver (dimensionless), λ is the wavelength of the transmitted signal (in meters), and L is the system loss factor ($L \geq 1$). A value of $L = 1$ would suggest that there are no losses in the system due to transmission line attenuation, filter losses, or arbitrary antenna losses. It is important to note that from (4.2) the received power is attenuated proportionally to the square of the distance (d) between the transmitter and the receiver. In a passive implementation, G_r would represent a loss (as opposed to a potential gain) within the system. The frequency f in Hz of the optical signal is related to the wavelength in m by

$$f = \frac{c}{\lambda} \tag{4.3}$$

where c is the speed of light in a vacuum in m/s. Electromagnetic radiation interacts with matter in different ways across the frequency spectrum. The scope of this book does not include a detailed discussion on these interactions across the spectrum and focuses instead on the electronic components within the transmitter and the receiver of the optoelectronic communications system. The following section discusses the optical link trans-impedance amplifier as shown in Fig. 4.1, with the focus on considerations based on the technology used.

4.4 Optical Link Trans-Impedance Amplifiers

Using a photodiode in an optical sensor is a common method to transform incoming pulses of data or commands to electrical outputs for light-based sensors and equipment. Applications include absorption and emission spectroscopy, color measurement, turbidity, gas detection and various military and EW applications such as laser rangefinders (Burns et al. 1991) and radar, which are discussed in this book in more detail. A typical goal to achieve in optoelectronic applications and therefore the trans-impedance amplifier capability is the 40 Gbit/s transmission rate as defined by the IEEE 802.3bm-2015 standard. All of these applications rely on precision light measurements from efficient and accurate photodiodes where precision in reproducibility is key. Light striking the active area of the photodiode must be converted to an electrical signal. Normally, trans-impedance amplifiers are used to convert the generated photodiode current to a measureable and low-noise output voltage. The following sections will discuss the operation of the trans-impedance amplifier in terms of its gain, noise and stability to determine the possible benefits of using SiGe BiCMOS technology in high sensitivity photodiode applications. A simplified schematic representation of an inverting trans-impedance amplifier using an op-amp is shown in Fig. 4.2.

In the amplifier in Fig. 4.2 a current is generated in the photodiode D_1, which is operating in photovoltaic mode (zero bias) and the op-amp keeps the voltage across

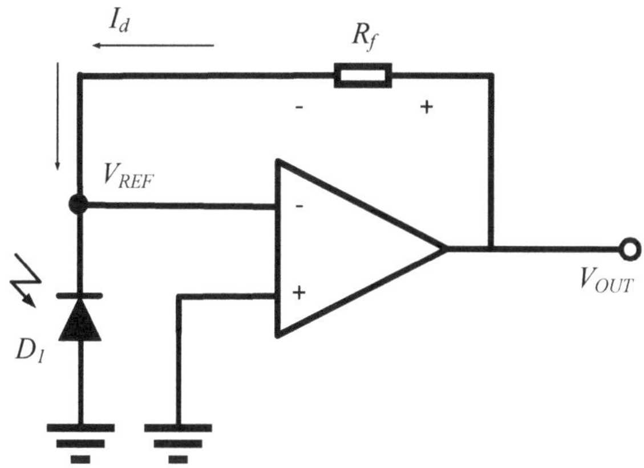

Fig. 4.2 A simplified schematic representation of an inverting trans-impedance amplifier using an op-amp and feedback resistor to provide gain

the photodiode at zero volts. The diode current I_d flows from the output of the op-amp through the feedback resistor R_f and through the photodiode towards ground. The output voltage V_{OUT} is determined by

$$V_{OUT} = I_d \times R_f \tag{4.4}$$

which varies with the intensity of the incident light on the active area of the photodiode (Orozo 2014). The *I–V* relationship of the photodiode is similar to that of a regular diode; the only difference is that it shifts on the *y*-axis (amplitude of the current) based on the photon-intensity of the light. A photodiode will output a small amount of current even if no light is present. This is called the dark current. The dark current increases as the applied reverse voltage on the photodiode increases and is typically specified by manufacturers at a recommended reverse-biased voltage for optimal operation. A common specification of the op-amp, which must be minimized for trans-impedance amplifiers used for photodiode output generation, is the input offset voltage. The input offset voltage appears across the terminals of the photodiode and increases the dark current, which leads to an increased system offset error. An increased offset error in the system decreases the total dynamic range of the amplifier. In addition, the op-amp used in this configuration should have a low input leakage current, the amount of current that flows into the input of the op-amp. In op-amp amplifier derivations this current is usually assumed to be zero (negligible), but in practice a small current leakage will occur, which may result in measurement errors. Implementation of trans-impedance amplifiers therefore requires careful design and consideration of the op-amp specifications, as well as external factors such as the printed circuit board (PCB) RF-design to minimize any possibility of current leakage in the system.

Two considerations of an op-amp based trans-impedance amplifier for photodiode current conversion in the AC domain is its signal bandwidth, closed-loop bandwidth and its equivalent noise bandwidth. The closed-loop bandwidth of the trans-impedance amplifier depends on three primary parameters: the open-loop

Fig. 4.3 A simplified schematic representation of an inverting trans-impedance amplifier using an op-amp and feedback resistor to provide gain and compensation feedback capacitor

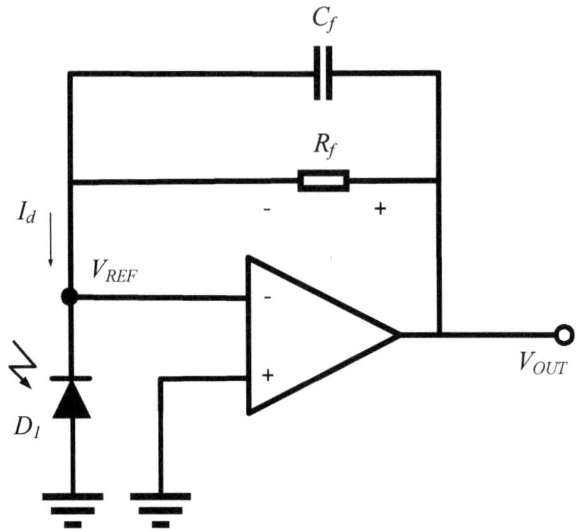

bandwidth, the gain (feedback) resistor and the total input capacitance to the amplifier inverting input. The total input capacitance depends on the capacitance of the photodiode, the common-mode and differential capacitance of the op-amp input terminal and any additional parasitic capacitance in the system, especially dominant in high-speed operation. The capacitance of the photodiode is process-specific and is determined by the manufacturing process, quality of materials and the manufacturing laboratory quality and cleanliness, although it can be compensated for electronically and post-manufacture by a feedback capacitor. Introducing a capacitor at the input of the op-amp can cause the system to become unstable and oscillate under normal operating conditions. Introducing a capacitor in the feedback loop (C_f) limits the closed-loop bandwidth, but it increases the stability of the circuit. The circuit configuration is adapted from Prokes (1996) and is shown in Fig. 4.3.

Given the schematic representation in Fig. 4.3, to determine a first-order approximation of the maximum possible closed-loop bandwidth that will result in a 45° phase margin[1] and a stable circuit, the following equation can be used:

$$f_{45} = \sqrt{\frac{f_u}{2\pi R_f(C_{IN} + C_M + C_D)}} \tag{4.5}$$

[1]The phase margin is the degree of freedom to choose between the amount of overshoot in the impulse response and the recovery speed. Recovery speed and overshoot are trade-offs in system design and reducing the phase margin decreases the overshoot at the expense of recovery time. The feedback capacitor is used as compensation to achieve an optimum phase margin.

where f_u is the unity-gain frequency of the amplifier, C_{IN} is the input capacitance, which includes the photodiode capacitance, feedback capacitance and other parasitic capacitances seen at the input terminal, C_M is the common-mode capacitance of the op-amp and C_D is the differential-mode capacitance of the op-amp. Generally, it is practical to assume that the capacitance of the photodiode dominates the total input capacitance of the system, especially at lower frequencies where parasitic capacitances are less dominant.

In terms of noise bandwidth, two dominant contributors to output noise are the input voltage noise of the amplifier and the Johnson (thermal/Nyquist) noise of the feedback resistor. Although the resistor noise is not amplified towards the output of the amplifier, increasing the resistor to increase the output voltage will inevitably also increase the Johnson noise seen at the output of the amplifier. The Johnson noise will, however, only increase proportionally with the square root of the resistor value and therefore it is preferred for high-gain applications rather to increase the size of the resistor as opposed to adding amplifiers in cascade, which will amplify the noise throughout the cascade to higher levels. Applications operated in photoconductive mode typically use two cascaded amplifiers, the first being the trans-impedance amplifier to convert the current to a voltage, with a second stage to amplify the voltage. This configuration can be designed to have high gain in the second stage without compromising the bandwidth of the first stage. The second dominant noise source is the noise of the amplifier itself. This noise level is the input-referred voltage noise amplified by a factor determined by the gain of the amplifier. This noise is not a constant value in the frequency spectrum, since it is dependent on variations in the feedback network and on the input capacitors.

If a compensation capacitor is used in the feedback loop, parallel to R_F as seen in Fig. 4.3, the closed-loop operating bandwidth of the system after gain peaking compensation (phase margin compensation) can be determined by

$$f_{3dB} = \frac{1}{2\pi R_f C_f} \tag{4.6}$$

where f_{3dB} is the resulting closed-loop operating bandwidth in Hz. The equivalent noise-bandwidth (ENBW) of a single pole system can be derived from the closed-loop bandwidth by

$$\text{ENBW} = \frac{\pi}{2} \times f_{3dB} \tag{4.7}$$

which can be used to determine the RMS noise due to the feedback resistor and the op-amp current noise. As mentioned, the Johnson noise of the feedback resistor will appear directly at the output of the amplifier, where the current noise will appear as an output voltage after passing through the feedback resistor. To determine the Johnson noise ($Noise_{R_f}$) and the current noise ($Noise_{current}$), the following equations are used:

$$Noise_{R_f} = \sqrt{4kT(\text{ENBW})R_f} \tag{4.8}$$

where k is Boltzmann's constant in J/K and T is the temperature in Kelvin and

$$Noise_{R_f} = I_N R_f \sqrt{(\text{ENBW})} \tag{4.9}$$

where I_N is the specified noise current density. The op-amp noise current is determined by R_f as

$$I_N\left[\frac{A_{rms}}{\sqrt{Hz}}\right] = \sqrt{\frac{4kT}{R_f}}. \tag{4.10}$$

To determine the output noise, the noise that is amplified by the trans-impedance amplifier by the noise gain from its input terminal is generally found by first considering the basic op-amp inverting amplifier with a feedback capacitor for stability compensation, as shown in Fig. 4.4.

The noise gain at the output node of the trans-impedance amplifier in Fig. 4.4 is determined by

$$Noise\ gain = 1 + \frac{Z_f}{Z_i} \tag{4.11}$$

where Z_f is the total impedance in the feedback loop considering the feedback resistor R_f and the feedback capacitor C_f and Z_i is the total input impedance of the

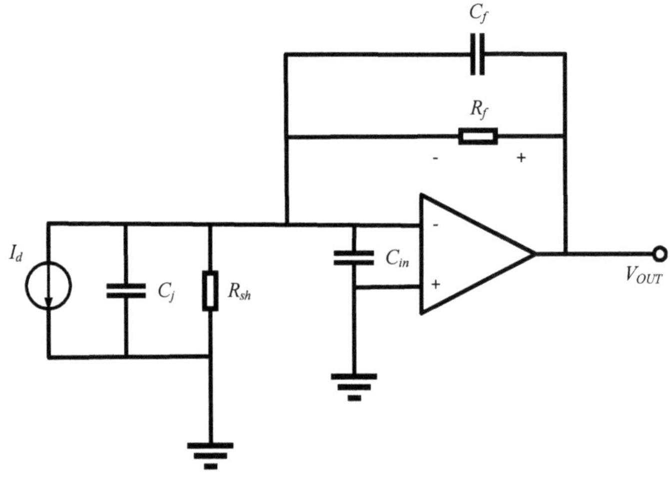

Fig. 4.4 An op-amp inverting amplifier with feedback resistor and input resistor used to determine the output noise amplitude

system considering the reactance of the photodiode shunt capacitance C_{sh} and shunt resistance R_{sh} in parallel with the input capacitance across the inverting and non-inverting inputs of the op-amp.

Derived from the discussion of the op-amp-based trans-impedance amplifier for receiving and amplifying an incoming light signal at the photodiode and adapted from Burr-Brown Corporation (2000), the following list highlights some design considerations when implementing photodiode amplifiers. These considerations can be used as general guidelines when implementing the trans-impedance amplifier and using a compensation capacitor in the feedback loop. The list considers the external circuitry and the optimum choices of these components to complement the overall operation of high sensitivity optoelectronic receivers.

4.4.1 Photodiode Capacitance

The system bandwidth can be improved and the additive noise contributions lowered by keeping the photodiode capacitance as low as possible. The photodiode junction capacitance and the amplifier feedback resistor form an additional noise-gain zero (feedback pole). In most cases the photodiode capacitance is determined by the manufacturing process and cannot be modified, but choosing a photodiode with the lowest capacitance that will suit the required application is recommended.

4.4.2 Photodiode Active Area

To decrease the photodiode junction capacitance during manufacturing, the photodiode active area should be as small as possible. This should again be a consideration when choosing a specific photodiode and bandwidth is a critical design specification. In addition, with a smaller active area, the shunt resistance of the photodiode equivalent model increases and this is advantageous, since it increases the signal-to-noise ratio of the integrated system as it lowers the noise generated by thermal carriers. This relationship between the thermal noise and the shunt resistance is given by

$$I_{jn} = \sqrt{\frac{4kT\Delta f}{R_{sh}}} \tag{4.12}$$

where Δf is the noise measurement bandwidth. This noise is the dominant current noise in photovoltaic (unbiased) mode. For applications that require higher optical gain, it is recommended to implement external optical gain by mirrors or lenses as opposed to choosing a photodiode with a larger active area. Capacitance is kept small by following this method and the advantage of adding essentially noiseless gain to the system through external optics optimizes overall system performance.

4.4.3 Large Feedback Resistor

Consider using a large feedback resistor for the amplifier as opposed to cascading amplifiers to achieve a higher gain, except when cascading a trans-impedance amplifier with low gain and a high voltage gain second stage amplifier. The larger feedback resistor introduces less thermal noise to the system compared to input-referred and amplified noise through a series of amplifiers. This method essentially increases the signal-to-noise ratio of the system by the root squared of the resistor value. A large feedback resistor does, however, decrease the response time of the circuit, since the time constant depends on the input capacitance of the photodiode and the value of the feedback resistor, $t = C_{in} \times R_F$.

4.4.4 Low Bias Current

To achieve high sensitivity for converting incident photons to electrical energy, use a low bias current for the op-amp. Op-amps are prone to present some value of current leakage through their input terminals, albeit a very small percentage of the bias current; if the bias current is increased, the voltage offset error will also increase. The input offset voltage error also varies with temperature and in most circuits, especially if operating in photovoltaic mode, the Johnson noise of the feedback resistor dominates the system noise, which will be amplified if presented at the op-amp input terminal.

4.4.5 High Photodiode Shunt Resistance

To reduce the gain of noise, voltage offsets and amplifier drift, the photodiode shunt resistance should ideally be as high as possible. The noise and offset voltages are amplified by a factor $1 + R_F/R_{sh}$, whereas if $R_{sh} \gg R_F$, the total gain simplifies to only 1 V/V. If the photodiode shunt resistance is comparable to or smaller than the feedback resistor, the input noise and offset voltages will introduce large and unwanted components at the amplifier output.

4.4.6 Photovoltaic Mode

If the application requires a high sensitivity circuit, it is advisable to operate the photodiode in photovoltaic mode, therefore with zero bias voltage across its terminals. Operating the photodiode at zero bias voltage effectively lowers the generated dark current leaked by the photodiode and dark current offset errors will effectively be reduced. The photodiode will only be able to operate at much lower frequencies in this mode to the advantage of higher sensitivity. Photovoltaic mode is recommended for low-frequency applications. If high-frequency operation is preferred, photoconductive mode is recommended.

4.4.7 Photoconductive Mode

If a fast response and high bandwidth are required, the photodiode should be operated in photoconductive mode, with a reverse bias voltage across its terminals. The photoconductive mode is generally the preferred mode of operating a photodiode, since fast response and high bandwidth are typically considered advantageous in photodiode applications. This mode reduces the photodiode junction capacitance by increasing the depletion region of the diode and also reduces the slow rise time diffusion tail, which can become a limiting factor in longer wavelength operation. This relationship between the junction capacitance C_j and the reverse-biased voltage is given by

$$C_j = \frac{\varepsilon_r \varepsilon_0 A}{W_d} \tag{4.13}$$

where ε_r is the dielectric constant of the material, ε_0 is the permittivity of free-space, A is the diffused area of the junction and W_d is the depletion region width. The depletion region width is determined by

$$W_d = \sqrt{2\varepsilon_r \varepsilon_0 \mu \rho (V_A + V_{bi})} \tag{4.14}$$

where μ is mobility, ρ is the resistivity of the material, V_A is the applied voltage and V_{bi} is the built-in voltage of the *pn*-junction. From (4.13) to (4.14) it can be seen that to increase the depletion region width, the applied reverse-bias can be increased, which will decrease the junction capacitance in (4.13), effectively increasing the response and bandwidth of the configuration. Photoconductive mode does increase the dark current and introduces 1/*f* noise to the system, but these are generally kept as low as possible for each device through advanced manufacturing techniques and high-quality epitaxial materials on wafer level.

4.4.8 External Shielding

External improvement to circuit and photodiode performance can be achieved by shielding the photodiode in a metal housing. The photodiode has high output impedance and therefore high sensitivity to spurious signals that may interfere with the operation of the circuit. Power supply noise is a common source of interference in many electronic circuits, especially high sensitivity circuits such as photodiode systems, and the power supply should be shielded effectively from crucial/sensitive devices.

4.4.9 Feedback Capacitor

Other common limitations and interferences (unplanned circuit operation) are unstable poles and zeroes that induce spontaneous oscillations. To avoid oscillations, several techniques exist and the phenomenon of oscillations is described in more detail in the following section. One method is to introduce a small capacitance across the feedback resistor to ensure loop stability, albeit at the expense of overall bandwidth.

Using bipolar transistors rather than MOS transistors in an amplifier generally depends on the application where it is used. Since for this discussion the application is a sensitive trans-impedance amplifier, which should ideally have high gain and low noise distortion, certain design parameters can be compared when using either bipolar or MOS transistors. Note that many of the inherent advantages of either transistor, for whichever parameter, can typically be improved upon by supporting circuitry and external optimization techniques. This essentially means that the advantages offered by the type of transistor do not mean that it must be used for the specific application; it depends on the advantages and trade-offs offered when considering the transistor type. Some of the parameters that are most commonly considered and compared when designing amplifiers, specifically trans-impedance amplifiers, include the power consumption, noise performance, voltage offset, input bias current, cost, chip area for comparable system performance, process deviations and matching of components, accuracy, efficiency and reverse-current sensitivity. These parameters will also be presented together with their mathematical derivations in subsequent sections to determine the dominant contributors of each parameter.

4.4.10 Power Consumption

CMOS transistors generally use less power compared to bipolar implementations, since CMOS transistors only draw current during the switching states, when the transistor changes its state from being on to being off, or vice versa. Bipolar transistors, however, draw current in their idle state and therefore increase the quiescent current consumption. The difference becomes less prevalent in high-frequency applications, where the time that the transistor is idle becomes very short. Therefore, as the bandwidth increases, the current consumption of CMOS transistors increases dramatically and they may even draw more current compared to their bipolar counterpart. In the case of trans-impedance amplifiers in opto-electronic receivers, the switching frequency is high and therefore the bipolar transistors with high cut-off frequencies, such as in BiCMOS processes, are advantageous. At low-frequency applications, CMOS transistors are still used, since the low quiescent currents are preferred in many applications, especially in mobile and battery-operated devices. The high input impedance and consequently low leakage current of CMOS transistors moreover count as an advantage in favor or CMOS when considering power consumption.

4.4.11 Noise Performance

The noise performance of CMOS and bipolar transistors is primarily dependent on process parameters and the physical construction of the transistors at wafer level. In a CMOS transistor the current flow is near the geometrical surface of the device and it is susceptible to possible defects on the surface of the semiconductor material. The susceptibility to defects on the surface of the material increases the low-frequency noise, such as flicker noise. Flicker noise is generally associated with irregularities in the conduction path. Bipolar transistors have a conduction path buried in the silicon and the irregularities at this level of the material are much less pronounced compared to those at the surface, since very little material processing is done at this level. At the surface, during wafer processing and semiconductor manufacturing, various etching and implantation processes occur, which could change and damage the surface of the material. Silicon is a well-defined and extremely consistent material to process and could potentially present a smaller probability of imperfections in the material, but in many other semiconductor materials the process itself can cause surface imperfections and irregularities. CMOS transistors also have lower transconductance compared to similarly sized bipolar transistors and this effectively results in higher broadband noise.

4.4.12 Input Offset Voltage (Transistor Matching)

The input offset voltage of an amplifier is the difference in voltage between the inverting and non-inverting input terminals and is primarily determined by the mismatches between the input transistors. Temperature also influences the input offset voltage, as a change in temperature may cause the offset voltage to drift, amplified by possible mismatches between the input transistors. Matching deals with statistical differences between pairs of identically designed active or passive components used in microelectronic circuit design. Mismatch is due to time-independent random fluctuations in physical magnitudes of on-chip devices. Types of mismatches that affect adjacent elements on a chip include process biases (shrinking and expanding of an actual component during photolithography, etching, diffusion and implantation), variations in etch rate, stress gradients and temperature gradients. Accuracy of analog devices such as op-amps demand extremely well-matched components to achieve the accuracy required in most applications. Temperature drift is also becoming more prevalent in many circuits where millions (up to billions) of transistors operate simultaneously and inevitably generate high volumes of thermal dissipation. For MOS transistors, the mismatch of drain current between adjacent elements may depend on the threshold voltage, body factor and current gain factor. The main source of error is usually the threshold voltage and the current gain factor because of the difficulty in controlling the diffusion process, oxide thickness and surface charge formation. Bipolar transistor mismatch sources include fluctuations in the base doping level, emitter junction area, densities of recombination in the emitter-base depletion region and area-to-periphery ratio variations. Base doping and recombination are process-specific, whereas the size of the emitter junction and the periphery can be controlled during the circuit design. Essentially, matching transistors can become an art form on its own and determining if CMOS or bipolar transistors offer better matching can become a complex task. In Wong it is proposed that *n*-MOSFET transistors offer better matching compared to *n*- and *p*-type bipolar and *p*-MOSFET transistors.

4.4.13 Input Bias Current

The input bias current in an op-amp must be as low as possible (ideally zero ampere), since the current flow is through external impedances and produces voltages that may add to system errors. The input bias current is therefore the amount of current that flows into the amplifier input terminals to bias the transistors. From the input offset voltage discussion, it is clear that the offset voltage is proportional to the input bias current too, hence the requirement to keep the bias current as small as possible. For bipolar transistors the input bias current is generally in the μA range, since these components are current-operated. CMOS transistors are voltage-operated and have very little current that appears at the gate of the transistor. CMOS input bias currents are generally in the pA order and are therefore inherently associated with smaller input bias current compared to bipolar transistors. Again, as the operation frequency increases to the high-GHz range, the input bias current for the bipolar and CMOS transistors become more comparable and bipolar transistors may even present smaller input currents at extremely high frequencies.

4.4.14 Transconductance

A method to characterize the gain performance of a transistor in an amplifier such as the trans-impedance amplifier is to determine its intrinsic gain, referred to as its transconductance. The transconductance of a CMOS transistor depends on the gate bias and can be quantified using the drain current (I_d) and gate-source voltage (V_{gs}) characteristic where

$$I_d = K_n\left(V_{gs} - V_t\right)^2 \tag{4.15}$$

where the equation for K_n is given as

$$K_n = \frac{\mu C_{ox}}{2}\frac{W}{L} \tag{4.16}$$

where μ is the mobility, C_{ox} is the oxide capacitance, W is the width and L is the length of the gate. It can be shown that the transconductance, g_m, of the MOS transistor is proportional to the ratio

$$g_m = 2K_n\left(V_{gs} - V_t\right) = \sqrt{2I_D\mu C_{ox}\frac{W}{L}} \tag{4.17}$$

where V_t is the threshold voltage. From (4.17) it can be seen that there is a square root relationship between the output current and the gain of the transistor. The intrinsic gain for bipolar devices is similarly equated by

$$g_m = \frac{qI_c}{kT} = \frac{I_c}{V_T} \tag{4.18}$$

where V_T is the thermal voltage, a constant value if the temperature is kept constant. From (4.18) it can be seen that the transconductance of a bipolar transistor is proportional to the output collector current, not in a square root relation. This brief analysis shows that bipolar transistors inherently have larger gain compared to MOS devices. Similar to the distinct advantages of MOS devices when considering the input bias current and voltage offsets, gain optimization can reduce the difference between the final gain at the amplifier output, but bipolar transistors have better gain characteristics by design. Pawlikiewicz and Hess (2006) state that at the peak operating current, the bipolar transistor achieves roughly three times the transconductance and hence three times the drive capability of its MOS counterpart. The bipolar transistor therefore always has a higher gain when biased at the same current level as a MOS transistor.

4.4.15 f_T and f_{max}

Process miniaturization by lateral and vertical scaling in CMOS and BiCMOS processes leads to lower parasitic capacitances, which increase the cut-off frequency f_T and maximum operating frequency f_{max}. BiCMOS has an additional degree of freedom in its ability to manipulate the doping concentration of the graded base. Furthermore, the reduction in the epi-layer thickness increases the collector transit time and further increases f_T. The relationship of BiCMOS f_T and its transconductance is given by

$$f_T = \frac{g_m}{2\pi(C_\pi + C_\mu)} \tag{4.19}$$

where C_π is the base-emitter capacitance and C_μ is the collector-base junction capacitance. The CMOS transistor f_T is given by

$$f_T = \frac{g_m}{2\pi(C_{gs} + C_{gd} + C_{gb})} \tag{4.20}$$

where C_{gs} is the intrinsic gate-source capacitance, C_{gd} is the date-drain capacitance and C_{gb} is the gate-bulk (substrate) capacitance of the MOS transistor. As seen in (4.19) and (4.20), the cut-off frequency is largely related to the transconductance of the transistor and the intrinsic capacitances, which are process-related. If it is assumed that C_{gs} is much greater than (C_{gd} + C_{gb}), it is shown in Gray et al. (2001) that the f_T of the MOS device is

$$f_T = 1.5\frac{\mu_n}{2\pi L^2}(V_{gs} - V_t) \tag{4.21}$$

where μ_n is the electron mobility and L is length of the transistor gate. Similarly, the f_T of the bipolar transistor can be given in another form, where

$$f_T = 2\frac{\mu_n}{2\pi W_B^2}V_T \tag{4.22}$$

if it is assumed that the forward transit time τ_F is much greater than the term (C_{je} + C_μ)/g_m and using the Einstein relationship $D_n/\mu_n = kT/q = V_T$ where D_n is the channel length. The thermal voltage $V_T = 26$ mV for bipolar transistors is constant but the f_T of a MOS transistor can be increased by increasing ($V_{GS} - V_t$). The base width of the bipolar transistor W_B is a vertical dimension. It is determined by diffusions or ion implantation depth and can normally be made considerably smaller than the channel length L of a MOS transistor. The channel length of a MOS transistor predominantly depends on the surface geometry and the accuracy and resolution of the photolithographic processes. For these reasons, bipolar transistors typically have a higher f_T compared to MOS transistors with similar processing limitations.

Process miniaturization also decreases the intrinsic resistances of transistors and therefore the f_{max} figure of merit. The f_{max} of the BiCMOS transistor is related to the f_T by

$$f_{max} = \sqrt{\frac{f_T}{(8\pi C_{bc} R_b)}} \tag{4.23}$$

where R_b is the intrinsic base resistance. For the MOS transistor, this relationship between f_T and f_{max} is

$$f_{max} = \frac{1}{2} \frac{f_T}{\sqrt{2\pi f_T C_{gd} R_g + \frac{R_g}{r_o}}} \tag{4.24}$$

where R_g is the gate-resistance and r_o is the output resistance of the MOS transistor. In (4.24) the relationship of $f_T/\sqrt{f_T}$ introduces an intrinsic limit on the maximum oscillation frequency of the MOS transistor. This is shown in Fig. 4.5.

From Fig. 4.5 it can be seen that the effect of the $f_T/\sqrt{f_T}$ relationship places an inherent limit on the increasing f_{max} of MOS transistors and in general the f_{max} of bipolar transistors can achieve higher values. This is advantageous in high-speed trans-impedance amplifier applications and the choice of bipolar transistors in this respect is usually a preferred alternative.

An average of the most commonly used FOM is presented in Table 4.1. These weighted factors are compared between bipolar (BiMCOS) and MOS transistors for implementation in trans-impedance amplifiers.

From Table 4.1, the weighted average of the CMOS versus bipolar technologies for trans-impedance amplifiers indicates that bipolar technology is superior for these applications. Note that the average derived in Table 4.1 is not based on

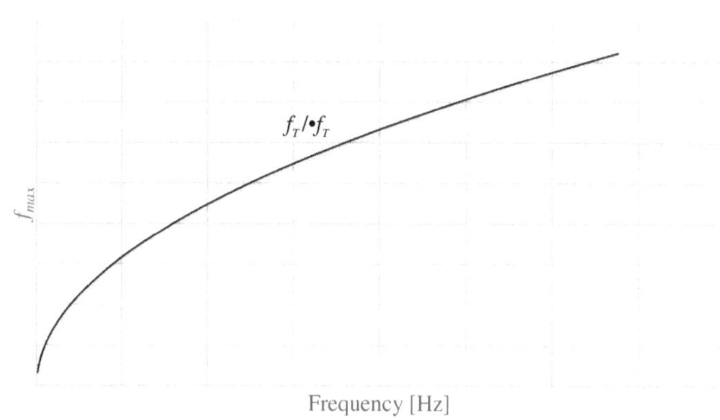

Fig. 4.5 Intrinsic limitation of the maximum oscillation frequency of MOS transistors due to the $f_T/\sqrt{f_T}$ relationship

Table 4.1 Figures of merit when considering bipolar or CMOS transistors specifically for an op-amp-based trans-impedance amplifier; weighting factors are awarded 1 for the lowest contribution and 5 for the highest contribution

#	Parameter	Bipolar	CMOS
1	Accuracy	3	3
2	Input voltage offset	2	5
3	Noise (high-frequency)	4	2
4	Noise (low-frequency)	4	3
5	Process degrees of freedom	5	4
6	Chip area	5	2
7	Voltage capability	3	4
8	Robust at high power/current	4	3
9	Current consumption (high-frequency)	5	3
10	Current consumption (low-frequency)	2	5
11	Reverse-current sensitivity	3	4
12	Cut-off and maximum operating frequency	5	3
13	Current gain linearity	5	2
14	Transconductance (gain)	5	3
15	Input impedance	2	4
	Average	3.8	3.3

scientific models, but rather gives a general indication of the advantages of either technology and should not be used as a deciding FOM without considering all tradeoffs for the specific application. Another non-technical consideration when choosing the technology for an amplifier (or any application involving tradeoffs between MOS and bipolar technology) is its economic benefits, as given in Pawlikiewicz and Hess (2006).

4.4.16 Economic Considerations

For each technology improvement and device scaling—i.e. node miniaturization—CMOS gate lengths have shrunk and gate-oxide thickness and power supply voltages have decreased, all leading to higher performance of devices, especially regarding the maximum operating frequencies. In addition, these advances have increased the density of components on chip and have lowered the power consumption when considering equal numbers of transistors for an application. Unfortunately, passive components such as inductors, capacitors and resistors have not been able to scale proportionally to the active devices and still take up significant space when used in CMOS designs. The speed advantages of BiCMOS processes are significantly superior to CMOS processes. A general observation is that BiCMOS and CMOS performance is approximately equal if two scaled generations are compared (for instance a 180 nm CMOS versus a 600 nm BiCMOS process) in terms only of speed and not physical size. For a given node, CMOS typically requires fewer masking steps compared to a BiCMOS process and this decreases the overall cost of the CMOS process. The significance of the lower cost from fewer masking steps decreases with device scaling, as CMOS process scaling increases the number of masks to manufacture the smaller devices and BiCMOS quickly becomes comparable in terms of process cost.

The considerations discussed for trans-impedance amplifiers are commonly used to implement low-noise and high-gain amplifiers in optoelectronic systems. An important consideration during amplifier implementation is the stability of the amplifier and its probability of oscillating rather than amplifying the signal.

4.5 Oscillations in Trans-Impedance Amplifiers

To identify the circumstances where and why an amplifier circuit could oscillate, consider the equivalent circuit of the photodiode in Fig. 4.6.

According to Fig. 4.6, I_d is the current flowing through the ideal photodiode, C_j is the junction capacitance of the photodiode, R_{sh} is its shunt resistance and R_s its series resistance. In photodiode approximations, it is generally fair to consider the series resistance R_s to be close to zero Ohm and simplified as a short-circuit and for the shunt resistance to be infinitely large, simplified to an open circuit. Considering these simplifications, the equivalent circuit of the photodiode, which will benefit in determining the nature of possible oscillations, is given in Fig. 4.7.

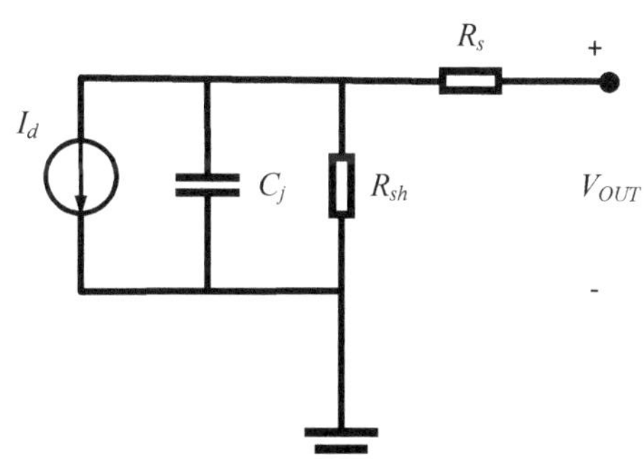

Fig. 4.6 Simplified equivalent circuit of a photodiode representing its series and shunt resistance and the junction capacitance

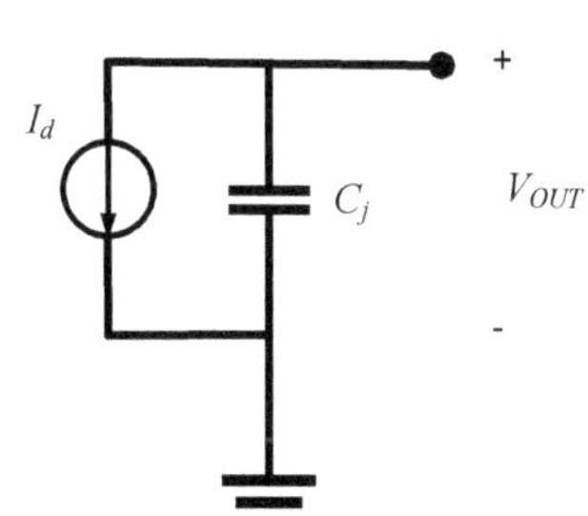

Fig. 4.7 Simplified equivalent circuit of a photodiode where the series resistance is assumed to be zero and the shunt resistance infinity

The only component that therefore has a substantial effect on the frequency transfer function and cannot be assumed negligible is the junction capacitance C_j of the photodiode. To investigate the possible sources of oscillation, first consider the frequency-dependent open-loop voltage gain response ($A_{OL(j\omega)}$) of the amplifier. Its single-pole response is determined by the output voltage and input voltage relationship

$$A_{OL(j\omega)} = 20\log\frac{V_{OUT(j\omega)}}{V_{IN(j\omega)}} \tag{4.25}$$

which translates mathematically to

$$A_{OL(j\omega)} = \frac{A_{OL}}{1 + j\left(\frac{\omega}{\omega_{PD}}\right)} \tag{4.26}$$

where A_{OL} is the DC open-loop gain, ω_{PD} is the dominant pole frequency in radians per second and ω is the operating frequency, also in radians per second. The feedback network adds another pole in the frequency response, comprising the feedback resistor R_F and the total input capacitance C_{in}, which is the parallel combination of the amplifier input capacitance and the photodiode junction capacitance. Mathematically, the frequency-dependent feedback factor $\beta(j\omega)$ is given by

$$\beta(j\omega) = \frac{X_{C_{in}}}{Z_F + X_{C_{in}}} \tag{4.27}$$

where X_{Cin} and Z_F represent the total impedance of the input capacitance reactance and the resistance of the feedback resistor, respectively, essentially forming a resistive-divider network. By simplifying (4.27) the feedback factor can be represented in a form that is practical to plot on the same frequency-versus-amplitude Bode plot similar to (4.26), and results in

$$\beta(j\omega) = \frac{1}{1 + j\omega R_F C_{in}} \tag{4.28}$$

where its reciprocal, $1/\beta(j\omega)$, represents the response curve of the feedback network. The open-loop gain and the reciprocal of the feedback factor are plotted on a frequency versus amplitude Bode plot, as shown in Fig. 4.8.

According to Fig. 4.8, the open-loop gain response A_{OL} of the op-amp is constant from DC up to the dominant pole corner frequency, where it decreases by 20 dB per decade until it reaches the second pole corner frequency (Graeme 1993). For the reciprocal of the feedback factor $1/\beta$, the frequency response is a flat unity-gain curve at low frequencies where the resistive feedback dominates and it rises at 20 dB per decade from the corner frequency f_F. The Barkhausen stability

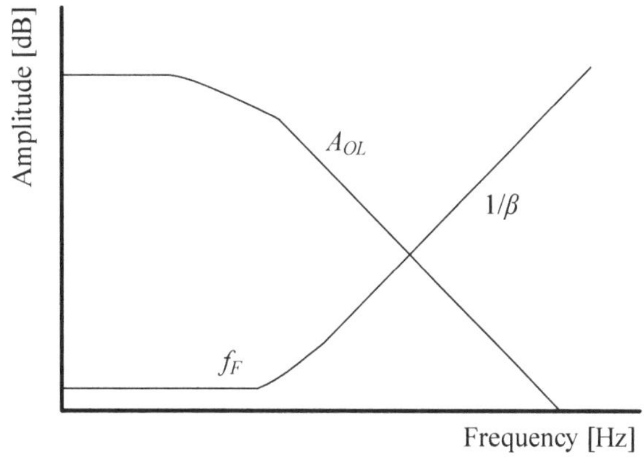

Fig. 4.8 The open-loop gain and the reciprocal of the feedback factor for a simple trans-impedance amplifier

criterion states that if the closed-loop trans-impedance amplifier circuit gain does not have a sufficient phase margin for $A_{OL}\beta \geq 1$, or put another way, $A_{OL} \geq 1/\beta$, the oscillation can occur in the circuit and compromise performance (Graeme 1993). This means that the intersection of the two curves denotes a critical intercept point for stability analysis. "*At lower frequencies the phase shift in the feedback signal is* 180° *due to the inverting nature of the negative feedback. As the frequency increases into the* −20 *dB per decade slope region of* A_{OL}, *the dominant pole of the op amp can add up to* 90° *of phase shift to the amplifier. Similarly, the pole introduced by the feedback network can add another* 90° *of phase shift, thus producing a phase shift of about* 360° *at* $A_{OL}\beta = 1$. *If the phase shift is* 360°, *self-sustaining oscillations will result. If the phase shift is close to* 360°, *heavy ringing is observed*" (Bhat 2012).

A commonly used method to provide compensation in the circuit and guarantee sufficient phase margin is a bypass capacitor placed in parallel with the feedback resistor. Mathematically, this can be represented by replacing the feedback resistor term in (4.26) by the parallel combination of the resistor and the added capacitor, such that

$$\beta(j\omega) = \frac{X_{C_{in}}}{Z_F \| X_{C_F} + X_{C_{in}}} \tag{4.29}$$

where X_{CF} is the frequency-dependent reactance of the feedback capacitor C_F. Simplifying (4.29) leads to the following equation for the feedback factor,

$$\beta(j\omega) = \frac{1 + j\omega R_F C_F}{1 + j\omega R_F (C_{in} + C_F)} \tag{4.30}$$

which accounts for the added phase-compensation capacitor. The feedback capacitor, C_F, therefore introduces an additional zero in the feedback factor $\beta(j\omega)$. This zero effectively compensates for the phase shift initially presented by the feedback network. The addition of this capacitor will also reduce the available bandwidth in the circuit and it is therefore recommended to determine the minimum value of the feedback capacitor, which will serve as phase compensation and thus eliminate the possibility of oscillation but also not significantly decrease the system bandwidth. To achieve this, determining the intercept frequency f_i of the open-loop gain and the feedback factor, therefore where $A_{OL}\beta = 1$, by solving for

$$f_i = \frac{1}{2\pi R_F C_F} \tag{4.31}$$

which has two unknowns at this point, the intercept frequency f_i and the feedback capacitor C_F. Essentially through geometric manipulation presented in Bhat (2012), it can be shown that the second equation for the intercept frequency is

$$f_i = \sqrt{\frac{\mathrm{GBWP}}{2\pi R_F (C_F + C_{in})}} \tag{4.32}$$

where GBWP is the unity gain-bandwidth product of the amplifier (usually taken as 60 % of the value on the op-amp datasheet to account for variations over process corners) (Bhat 2012). By substituting (4.31) and (4.32), C_F can be found as

$$C_F = \frac{1}{4\pi R_F \mathrm{GBWP}}\left(1 + \sqrt{1 + 8\pi R_F C_{in} \mathrm{GBWP}}\right) \tag{4.33}$$

which therefore represents the minimum value of C_F required to compensate for the phase margin without unnecessarily limiting the system bandwidth. The loop gain of the amplifier is the difference between the open-loop gain (A_{OL}) and the noise gain. The loop gain is an indication of the amount of negative feedback that can be applied to the amplifier. A functional block diagram showing the relationships of the open-loop gain and feedback factor is given in Fig. 4.9.

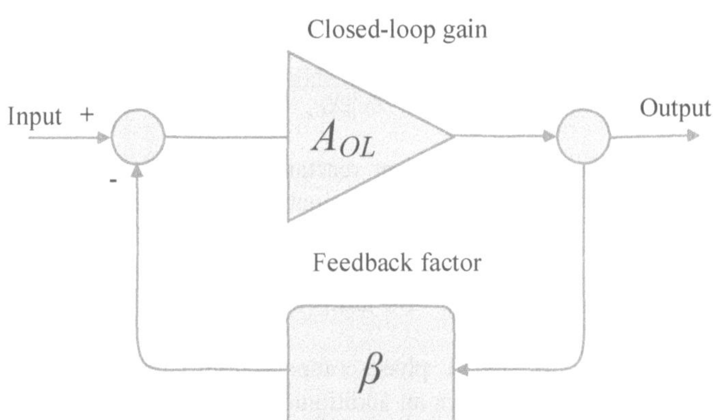

Fig. 4.9 The open-loop gain and the noise gain for a single-pole trans-impedance amplifier Bode plot

Figure 4.10 represents loop gain, open-loop gain, noise gain and closed-loop gain of an amplifier in a generalized Bode plot (Ramus 2009).

From Fig. 4.10, a logarithmic frequency versus noise in dB Bode plot, it is seen that the gain of an op-amp generally has a flat band gain at DC (and very low frequencies), which is determined primarily by the feedback resistor. This gain is the open-loop gain of the amplifier, the combined closed-loop gain and loop gain where the open-loop gain starts dropping at the open-loop breakpoint and follows a −20 dB per decade slope from this point onwards. As the gain passes through the closed-loop breakpoint frequency f_{CL}, the closed-loop gain can be determined at this point and the loop gain is the difference between the noise gain of these two points. The transfer function intersects the vertical axis at unity gain (0 dB) and this point determines the gain-bandwidth product of the amplifier. The noise in a trans-impedance amplifier, as briefly defined in (4.8)–(4.11), can either be amplified by the system, or appear directly at the output of the amplifier. The bandwidth and accuracy of the trans-impedance amplifier degenerate with excess noise at the input and output of the system. The following section discusses the dominant noise sources in the trans-impedance amplifier used to convert the current generated by a photodiode to an output voltage. The noise in the system also depends on the mode of operation of the photodiode, either used in photovoltaic (zero bias) or photoconductive (negative bias); this is also highlighted in the following section.

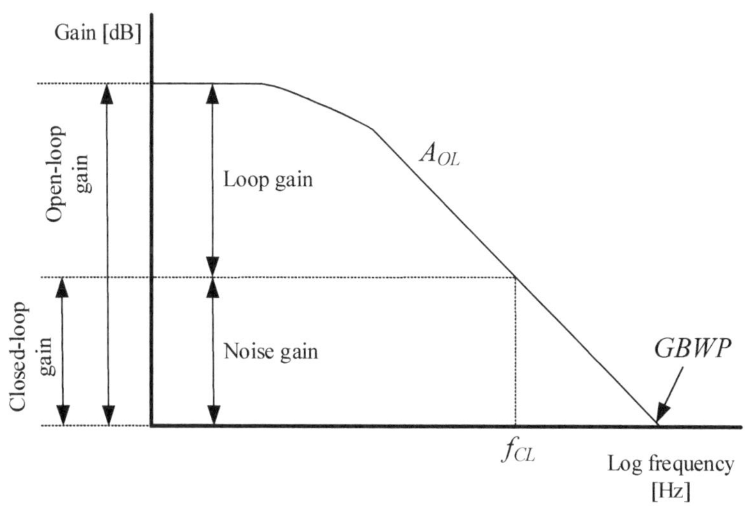

Fig. 4.10 The loop gain, open-loop gain, noise gain and closed-loop gain represented on the same Bode plot

4.6 Noise in Trans-Impedance Amplifiers

The accuracy of the input signal and the minimum acceptable input signal in a photo-receiver system are dependent on the noise within the system. Stringent limits on the detection of photons are placed on many military-based optoelectronic devices and analyzing and modeling the noise accurately helps to predict the operating limits of the system. Determining the particular effects and dominant source of noise in any system can be a cumbersome task, since many external factors contribute to noise and are likely to be random occurrences. This section focuses on the noise analysis of only the trans-impedance amplifier used in the optical receiver to amplify the incoming signal at the photodiode.

Noise is a arbitrary quantity and the average voltage noise of a source is nil; however, the average power of the noise is not zero and adds to the degradation of the output signal. In Morikuni et al. (1994) the traditional method of determining the optical receiver noise as originally presented by Personick (1973) is presented in a logical approach. Consider the trans-impedance amplifier equivalent circuit given in Fig. 4.6. The most common method of approaching a noise analysis is to refer all noise sources resulting from the preamplifier circuit to the receiver input, hence the term input-referred noise. The dominant types of noise in an amplifier are the Johnson noise given by

$$\langle i_n^2 \rangle = \frac{4kT}{R}\Delta f \tag{4.34}$$

where Δf is the frequency bandwidth where the noise is measured. The second dominant noise source is shot noise, which is a disturbance caused by a constant presence of random discrete events such as the flow of carriers across a potential barrier or the detection of single photons in a photodetector in low-light conditions. The equation for the shot noise is given by

$$\langle i_n^2 \rangle = 2qI\Delta f \tag{4.35}$$

where I is the current flow and q is the elementary electron charge. The third large contributor to amplifier noise is low-frequency flicker noise, or $1/f$ noise. The equation for determining the flicker noise is given by

$$\langle i_n^2 \rangle = \frac{k_d I^\alpha}{f^\beta}\Delta f \tag{4.36}$$

where k_d is a semiconductor device constant and can vary even by orders of magnitude between devices on the same wafer. Typically, MOS transistors have a larger k_d compared to bipolar transistors. The variable I is again the current flowing through the device where its raised power, α, is the fabrication constant and ranges between approximately 0.5 and 2. Δf is the frequency bandwidth and β is a constant ranging between approximately 0.8 and 3. As mentioned, by convention, the noise

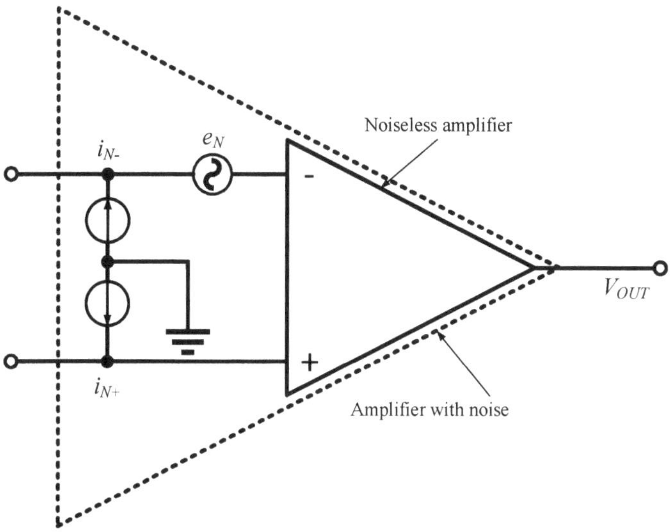

Fig. 4.11 Equivalent noise model of an op-amp with feedback resistor to determine the total noise at the output terminal

quantities are referred to the input of the amplifier and represent the amount of noise that would appear at the input and cause a resultant noise at the amplifier output, where further signal processing is required. To model the noise sources in an op-amp circuit, the voltage noise and current noise sources at the inputs of the op-amp can be represented by voltage and current sources as shown in Fig. 4.11, adapted from Burr-Brown (1994).

In Fig. 4.11 the noise currents can be expressed as voltages if the current flows through a resistor, typically applicable in the feedback circuit and if the photodiode is connected to the amplifier and operated in photoconductive mode. For the trans-impedance amplifier, the input-referred voltage noise (e_N) and the input-referred current noise (i_N) contribute to the noise behavior at the output in some way. The input-referred noise voltage consist of flicker noise at lower frequencies and thermal noise dominates at higher frequencies. The *1/f* corner frequency of an amplifier refers to the transition frequency where thermal noise starts dominating the flicker noise. The input-referred current noise at the inverting input of the op-amp is not amplified by the circuit, whereas the input-referred current noise at the non-inverting input and the input-referred voltage noise are amplified by the system gain. The system gain is dependent on the feedback resistor value and the feedback capacitor, if present. Table 4.2 identifies the type of gain that each noise source experiences towards the output of the amplifier.

From Table 4.2 it can be seen that the input voltage noise and the non-inverting input noise current are amplified by the amplifier gain. The non-inverting input resistor also adds noise to this terminal, which is amplified to the output, whereas the source resistor (generally 50 Ω) has a *negative* gain, as seen at the output of the

Table 4.2 Noise contributions at the output of the trans-impedance amplifier

Noise source	Noise type	Noise gain
Input voltage noise	e_N	$1 + R_F/R_G$
Inverting input noise current	$i_{N-} \times R_F$	1
Non-inverting input noise current	$i_{N+} \times R_1$	$1 + R_F/R_G$
Feedback resistor Johnson noise	$\surd(4kTR_F)$	1
Source resistor Johnson noise	$\surd(4kTR_G)$	$-R_F/R_G$
Non-inverting input resistor Johnson noise	$\surd(4kTR_1)$	$1 + R_F/R_G$

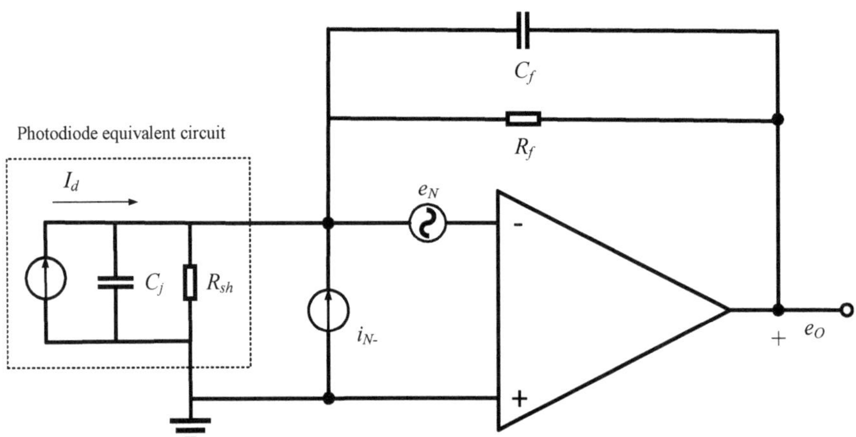

Fig. 4.12 Equivalent noise model of an op-amp with feedback resistor and feedback capacitor including the equivalent circuit for the photodiode at the amplifier input (adapted from Jiang and Yu 1998)

amplifier. The complete circuit of the trans-impedance amplifier as it is typically used in optoelectronic receivers, with the photodiode represented as its equivalent model, and all the contributing noise factors, is given in Fig. 4.12.

The output voltage noise e_O of the amplifier shown in Fig. 4.12 can therefore be determined by

$$e_O = e_n \left[\frac{A_{OL}}{1 + A_{OL}\beta} \right]. \tag{4.37}$$

Note that for a large loop gain ($A_{OL}\beta \gg 1$), (4.37) simplifies to

$$e_O = e_n \frac{1}{\beta}. \tag{4.38}$$

In lieu of the performance characteristics of the trans-impedance amplifier as discussed in this chapter, specifically based on the type of technology, CMOS or

BiCMOS, the following section lists examples of implementations of trans-impedance amplifiers in both technologies.

4.7 Performance Characteristics of Trans-Impedance Amplifiers

The list of implementations of trans-impedance amplifiers given in Table 4.3 highlights the most crucial characteristics when considering implementing CMOS or BiCMOS technologies. In EW and military operations such as laser rangefinders, radar applications, command and control and fiber-optic weapon sights, the choice of technology and performance limitations become crucial to ensure accurate, low-noise and high-speed applications. The parameters included in Table 4.3 are 3-dB bandwidth specified in GHz, trans-impedance in dBΩ, noise performance in pA/√Hz, input capacitance in pF and power consumption in mW. The technology type and technology node are also specified in Table 4.3 for comparative analysis.

Various implementations of trans-impedance amplifiers in CMOS and BiCMOS nodes are presented in Table 4.3. From these results, it can be noted that BiCMOS implementations generally present relatively low power consumption at high frequencies because the CMOS transistors use more power when fast switching is required. The BiCMOS trans-impedance amplifiers also present very low noise

Table 4.3 Technology implementations and performance characteristics of trans-impedance amplifiers

References	Bandwidth (GHz)	Z_T (dB Ω)	Noise (pA/√Hz)	C_{in} (pF)	Power (mW)	Technology
Scrodinger et al. (2002)	1.25	63.5	16	–	40	350 nm CMOS
Park et al. (2007)	2.1	64	32	2	50	180 nm CMOS
Jiao et al. (2013)	2.7	70	2.44	0.6	73	350 nm BiCMOS
Lu et al. (2010)	7	55	17.5	0.3	18.6	180 nm CMOS
Takashi et al. (2009)	25.8	70	3.2 μA_{rms}	–	74	65 nm CMOS
Li and Palermo (2013)	26	53	21.3	–	28.2	250 nm SiGe BiCMOS
Sedighi and Scheytt (2012)	28	71	18	–	57	250 nm BiMCOS
Kim and Buckwalter (2010)	29	50	51.8 (simulated)	–	45.7	130 nm CMOS
Jin and Hsu (2006)	30.5	51	34.3	–	60.1	180 nm CMOS
Knochenhauer et al. (2011)	34.2	65.2	25	–	80	130 nm SiGe BiCMOS

Units are specified at the top of the table, unless differently specified in the table characteristics, a major advantage in optoelectronic applications. Table 4.3 therefore defends the hypothesis that BiCMOS (specifically SiGe) is an increasingly important technology consideration for these types of applications and benefits not only military applications but also commercial devices.

4.8 Conclusion

The optoelectronic communications link as described in Chap. 3 typically uses a trans-impedance amplifier at the receiver end to convert and amplify variations in the photodetector voltage as incident photons fall on the detector. The inherent degradation of the transmitted signal through an optical medium is not discussed in detail in this book, as it falls outside the scope of the book. Focus is placed on the discussion of the electronic characteristics of the trans-impedance amplifier with emphasis on the transistors used to realize the trans-impedance amplifier. Op-amps are typically used to design trans-impedance amplifiers and the characteristics related to the semiconductor process are discussed, such as the noise considerations, possible oscillations in these circuits and how to compensate for them, as well as the gain of the circuit as a function of the transistors. The concluding sections of this chapter list trans-impedance amplifier implementations in CMOS and BiCMOS technologies and these comparative results point to SiGe BiCMOS being a superior alternative for current and future implementations.

References

Bhat, A. (2012). Stabilize your transimpedance amplifier. *Application Note 5129*. Maxim integrated.

Burns, H. N., Christodoulou, C. G., & Boreman, G. D. (1991). System design of a pulsed laser rangefinder. *Optical Engineering, 30*(3), 323–329.

Burr-Brown Corporation. (1994). Noise analysis of FET transimpedance amplifiers. *Application Bulletin*. Texas Instruments Incorporated.

Burr-Brown Corporation. (2000). Designing photodiode amplifier circuit with OPA128. *Application Bulletin*. Texas Instruments Incorporated.

Graeme, J. (1993). Op amp performance analysis. AB-045. *Application Bulletin*. Texas Instruments Incorporated.

Gray, P. R., Hurst, P. J., Lewis, S. H., & Meyer, R. G. (2001). *Analysis and design of analog integrated circuits* (4th ed.). John Wiley and Sons, Inc.

Jiang, H., & Yu, P. K. L. (1998). Equivalent circuit analysis of harmonic distortions in photodiode. *IEEE Photonics technology Letters, 10*(11), 1608–1610.

Jiao, Y., Li, W., Zhang, D., Wen, W., & Zhang, Z. (2013). A low noise transimpedance amplifier for 2.5 Gb/s optical communications. In *25th International Conference on Microelectronics* (pp. 1–4).

Jin, J. D., & Hsu, S. S. (2006). 40-Gb/s transimpedance amplifier in 0.18-μm CMOS technology. In *Proceedings of* the 32nd *Solid-State Circuits Conference* (pp. 520–523).

Kim, J., & Buckwalter, J. F. (2010). Bandwidth enhancement with low group-delay variation for a 40-Gb/s transimpedance amplifier. *IEEE Transactions on Circuits and Systems I: Regular Papers, 57*(8), 1964–1972.

Knochenhauer, C., Sedighi, B., Ellinger, F. (2011). 40 Gbit/s Transimpedance Amplifier with High Linearity Range in 0.13 μm SiGe BiCMOS. *Electronics Letters.* 47(10), 1-2, May 2011.

Li, C., & Palermo, S. (2013). Low-power 26-GHz transformer-based regulated cascode SiGe BiCMOS transimpedance amplifier. *IEEE Journal of Solid-State Circuits, 48*(5), 1264–1275.

Lu, Z., Yeo, K. S., Lim, W. M., Do, M. A., Boon, C. C. (2010). Design of a CMOS Broadband Transimpedance Amplifier with Active Feedback. *IEEE Transactions on Very Large Scale Integration (VLSI) Systems.* 18(3), 461-472, 2010.

Morikuni, J. J., Dharchoudhury, A., Leblebici, Y., & Kang, S. M. (1994). Improvements to the standard theory of photoreceiver noise. *Journal of Lightwave Technology, 12*(4), 1174–1184.

Orozo, L. (2014). Programmable Gain Transimpedance Amplifiers Maximize Dynamic Range in Spectroscopy Systems. *Analog Dialogue, 47*(5), 2013.

Park, K., Oh, W. S., Choi, B. Y., Han, J. W., & Park, S. M. (2007). A 4-channel 12.5 GB/s common-gate transimpedance amplifier array for DVI/HDMI applications. *IEEE International Symposium on Circuits and Systems,* 2192–2195.

Pawlikiewicz, A. H., & Hess, D. (2006). Choosing RF CMOS or SiGe BiCMOS in mixed-signal design. *Mixed-Signal Design.* Retrieved from www.rfdesign.com.

Personick, S. D. (1973). Receiver design for digital fiber optic communication systems: Parts I and II. *Bell Systems Technology Journal, 52*(6), 843–886.

Prokes, A. (1996). The transimpedance amplifier noise optimization for the atmospheric optical link receiver. *Radioengineering, 5*(4), 1–4.

Ramus, X. (2009). Transimpedance considerations for high-speed amplifiers. *Application Report SBOA122.* Texas Instruments.

Schrodinger, K., Stimma, J., & Mauthe, M. (2002). A fully integrated CMOS receiver front-end for optic gigabit ethernet. *IEEE Journal of Solid State Circuits, 37*(7), 874–880.

Sedighi, B., & Scheytt, C. (2012). Low-power SiGe BiCMOS transimpedance amplifier for 25-GBaud optical links. *IEEE Transactions on Circuits and Systems-II: Express Briefs, 59*(8), 461–465.

Takashi, T., Fumio, Y., Hiroki, Y., Takuma, B., Masashi, K., Yong, L., et al. (2009). A 25-GB/s 2.8-mW/Gb/d low power CMOS optical receiver for 100-Gb/s ethernet solution. In *Proceedings of the* 35th European *Conference on Optical Communication (ECOC)* (pp. 1–2).

Chapter 5
Electronic Countermeasures and Directed Energy Weapons: Innovative Optoelectronics Versus Brute Force

5.1 Introduction

The relatively near future of EW will show significant advances in laser designators and new technologies in the offensive and defensive high-intensity laser pulse domain. Apart from the active research fields in various disciplines attaining laser and radar technology, the military laser systems market was estimated to be worth US $3.03 billion in 2015. The market share is estimated to reach US $4.63 billion by 2020 at a compound annual growth rate (CAGR) of 8.86 % from 2015 to 2020. The primary driving elements influencing the growth are the need for high-precision weapons and growing use of high-speed lasers in communication networks, as well as high-powered and short-pulsing penetrative lasers. Advances in the commercial sector are also offering new growth opportunities in the military as miniaturization and CMOS technology improvements are becoming more accessible through the commercial and military sectors. The US Army revealed plans in March 2016 to have laser weapons on the battlefield by 2023, able to shoot down artillery shells and missiles in flight. Lockheed Martin demonstrated a military laser in March 2016, named ATHENA, which is able to burn a hole through a military vehicle from more than 1 mile (1.6 km) away, directing a beam of 30 kW of energy downrange. The major advantages are that unlike traditional missiles and anti-aircraft weapons, these beams travel at the speed of light, are silent and a single round costs orders of magnitude less that firing traditional missiles. Lockheed Martin's chief technology officer said in a statement that "*Fiber-optic lasers are revolutionizing directed energy systems*" and that "*We [Lockheed Martin] are investing in every component of the system—from the optics and beam control to the laser itself—to drive size, weight and power efficiencies. This test represents the next step to providing lightweight and rugged laser weapon systems for military aircraft, helicopters, ships and trucks.*" These are important statements for several reasons, but in the case of this book, it highlights the fact that this market is in need of a dominant technology to achieve its goals regarding size, weight and power

W. Lambrechts and S. Sinha, *SiGe-based Re-engineering of Electronic Warfare Subsystems*, Signals and Communication Technology,
DOI 10.1007/978-3-319-47403-8_5

efficiency. A system is only as good as the sum of its parts; the optics, beam control and the laser itself must all be designed with these goals in mind. The Naval Research Laboratory (NRL) was established in 1923 to fulfill Thomas Edison's vision of a government research laboratory that could develop and translate scientific knowledge into effective military systems in times of war. The NRL Plasma Physics Division was established in 1966 to create X-ray simulators for testing the effects of nuclear weapons on materials and components of military hardware, to study the physics and effects of high-altitude nuclear explosions, and to perform nuclear fusion research (Melhorn 2014). A detailed and descriptive study on national security research in plasma physics and pulsed power is presented in Melhorn (2014) and reading this article is highly recommended to understand the development of these technologies on their way to a weaponized state.

In terms of optoelectronic communication, Chap. 3 of this book presents the optical transmitter with a technical review on the laser driver circuit. Chapter 4 focuses on the transimpedance amplifier in the optical receiver. Both these chapters highlight the requirements of high-speed, high-power capabilities, as well as low-noise, adaptable, modular and small footprint technologies—characteristics that make SiGe advantageous to use in these applications. This chapter describes the system design of a laser rangefinder to summarize the requirements for the supporting electronic equipment. The concepts presented for the laser rangefinder can be adapted for various other laser systems, such as laser designators, laser scanners, satellite laser ranging and laser altimetry. Also covered in this chapter is a relatively new and promising technology that enables terahertz radiation. Terahertz radiation is an important technology that allows devices to *see* through various materials, which has not been possible without this technology, and to identify various materials. Si/SiGe is one of the enabling technologies in this field and the limitations and progress of terahertz radiation are highlighted in this chapter. The following section presents the building blocks and environmental considerations of laser rangefinders.

5.2 Laser Rangefinders

In this section, laser rangefinders are discussed to highlight the common considerations and design and research procedures for low-energy laser transmissions for applications in military operations and EW. To present the relevant concepts of a laser rangefinder, the system block diagram is given in Fig. 5.1.

The system block diagram in Fig. 5.1 shows the general subsystems of a typical laser rangefinder. According to Fig. 5.1, an oscillator (typically a crystal oscillator for high stability) and system timer set the pulse frequency and generate a start pulse periodically. If this system timer is activated, it triggers a pulsed laser source and initializes the range counter. A collimated laser beam travels to the target through the transmitter objective lens and a portion of the laser beam is reflected by the target object. The receiver objective lens receives this partial signal, after which

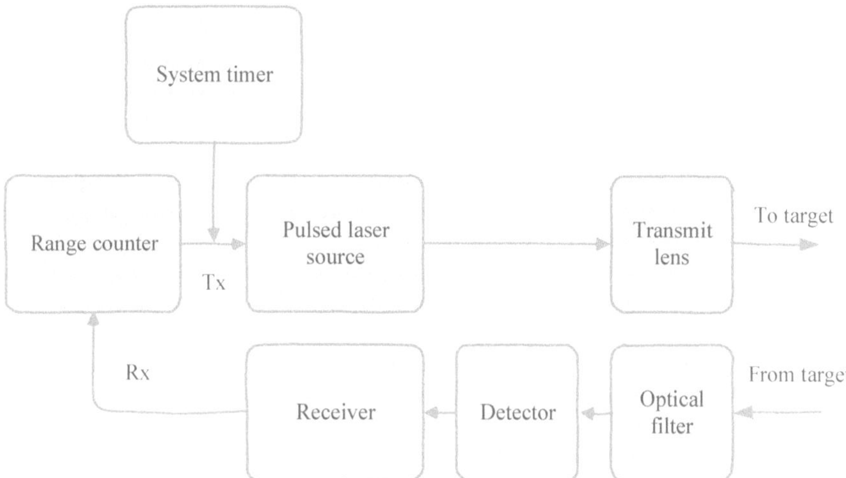

Fig. 5.1 The system block diagram of a typical laser rangefinder system

it is filtered in the required spectral band and focused onto the photodetector. The received signal is amplified and threshold-detected at the photodetector output signal. The range counter, a time-to-digital converter (TDC), is triggered to stop counting and the time between the transmitted and received signal is converted to distance. The clock frequency is not the same as the frequency of the light source and depends on the capabilities of the electronic IC. The frequency of the TDC clock signal is proportional to the accuracy of the measurement and the minimum measurable difference. SiGe, for example, can provide high clock frequencies in the GHz range without adding significant noise and interference to the system, allowing very small signals to be detected at the receiver. A further and more detailed discussion on TDCs is given in the following section.

5.2.1 Time-to-Digital Converter

The traditional method of implementing a TDC is to first convert the time interval into an analog voltage and then convert the voltage to a digital representation using an ADC. Essentially, the TDC performs the function of an accurate stopwatch and measures the time that elapses between a start pulse and (typically, multiple) stop pulses. A simplified block and signal diagram of a basic TDC is given in Fig. 5.2.

For the ADC in Fig. 5.2, the maximum time interval that can be measured and the maximum number of bits N of the ADC, i.e. the dynamic range (DR), are determined by a trade-off equation (Henzler 2010).

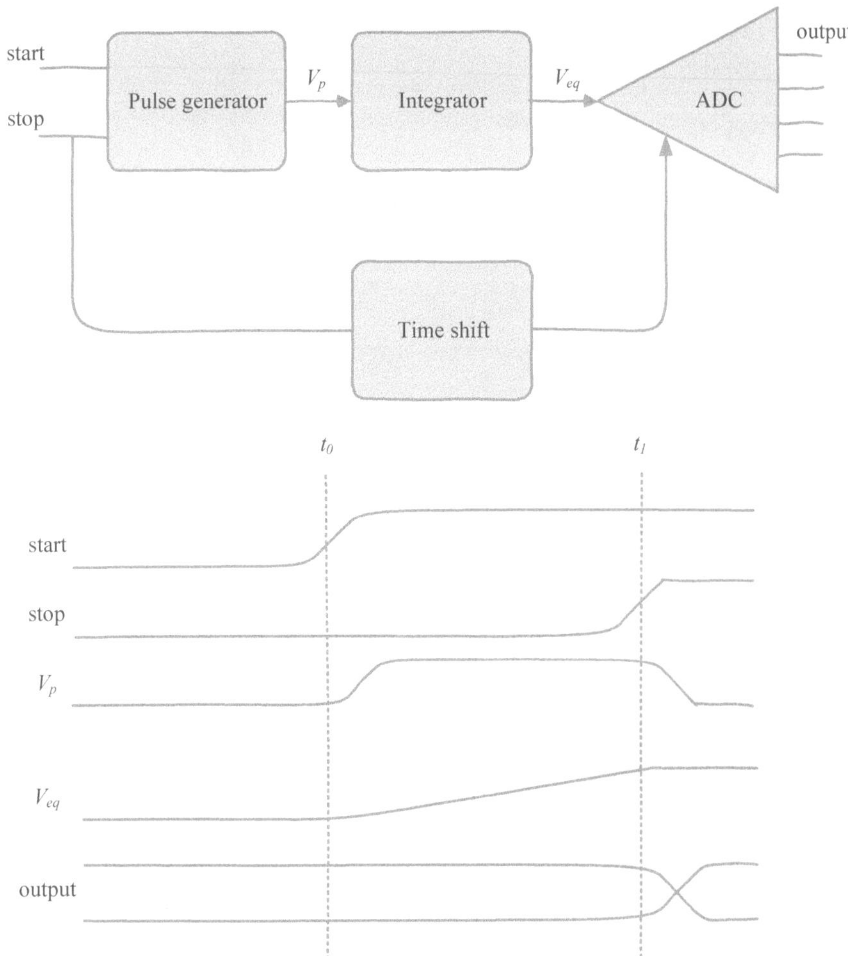

Fig. 5.2 The simplified block diagram of an analog time-to-digital converter (*top*) and the signal diagram showing its operation (*bottom*). Adapted from Henzler (2010)

$$DR = 2^N \times T_{LSB} \tag{5.1}$$

where T_{LSB} is the minimum time interval that can be resolved by the system. Assuming the dynamic range is kept constant, the minimum time interval and the maximum resolution of the ADC are tradeoffs in this equation. Advancements in CMOS and BiCMOS technologies improve upon the analog limitations of ADCs and typically improve upon the *DR* of a TDC. Additional improvements made to these circuits when technology advances are not considered include two-stage approaches where a coarse and a fine quantization are used or to integrate up and down periodically, effectively increasing the *DR* without compromising the minimum time interval. A coarse counter is based on a reference clock signal with a known and stable frequency f_0. If a start signal is detected, the counter starts counting each clock cycle either at the rise or the fall of the cycle. If a stop signal is received, the coarse counter stops counting and determines the time interval T between the received start and stop signals simply by

$$T = nT_0 \tag{5.2}$$

where T_0 is the period of one oscillation, determined by $1/f_0$, and n is the number of cycles counted. Time interpolation circuits or stretching methods are capable of measuring fractions of the clock resolution and can be implemented to determine the fine measurements. These circuits generally require two clock signals from slightly detuned oscillators with frequencies of f_1 and f_2. One such method is called the Vernier method. For the Vernier method, both frequencies start oscillating with the arrival of the start (or stop) signal. Once the leading edges of the oscillating signals overlap, the measurement stops and the number of periods of the oscillations leads the time interval T. This is mathematically described by

$$T = \frac{n_1 - 1}{f_1} - \frac{n_2 - 1}{f_2} \tag{5.3}$$

where n_1 and n_2 are the periods of the two oscillating signals. According to Fig. 5.1, the laser rangefinder operates on a similar principle as any optoelectronic communications system, the main difference being that the transmit beam is reflected from the target and received by the photodetector. This timing circuit is the critical component in determining an accurate range of the target by determining the time it takes for the signal to return to the receiver (housed in the same package as the transmitter). The distance R between the transmitted and the received signal is determined by

$$R = \frac{c\Delta T}{2} \tag{5.4}$$

where c is the speed of light and ΔT is the difference between the time of the signal transmission and the time the signal was received. It is fair to assume then that the target is ½ of the total distance R, since the signal traveled towards the target and a portion thereof back to the receiver. The small error due to the finite distance between the transmitter and receiver is generally negligible. According to (5.4), the speed of light is a relatively accurate constant (299,792,458 m s^{-1}), whereas ΔT is the only variable that can vary significantly based on the quality of the electronics, or in a practical sense, should be determined as accurately as possible to define the range of the target with high precision. Time-of-flight (TOF) measurements are often employed in laser rangefinders and sometimes in scanning laser radars. This technique can be achieved by pulsed or by CW TOF measurements. Both techniques are discussed in the following sections.

5.2.2 Pulsed Time-of-Flight

In pulsed TOF a transmitter transmits an optical pulse and measures the time until the reflected pulse is picked up at the receiver, situated close to the transmitting device. The ability to integrate the optical photodetector and the integrated active circuitry (transmitter, receiver, amplifiers and TDCs) on the same substrate is very advantageous in these situations, something that is possible when using CMOS or BiCMOS SiGe technology, for example. This technique requires high-quality telescopes, high beam quality and sensitive photodetectors to achieve ranges of several kilometers. The pulsed TOF method and the timing diagrams are practically shown in Fig. 5.3.

The pulsed TOF method as shown in Fig. 5.3 uses electrical charges Q_1 and Q_2, which accumulate in the two out-of-phase windows C_1 and C_2 to determine the distance between the transmitter and the target R by an equation, adapted from (5.4), given by

$$R = \frac{c\Delta T}{2}\left(\frac{Q_2}{Q_1 + Q_2}\right) \tag{5.5}$$

where ΔT is again the pulsed time that the target is illuminated. A typical circuit used to generate a pulsed signal by using a single transistor avalanche pulse method is shown in Fig. 5.4.

In Fig. 5.4, the single transistor avalanche pulsing circuit is shown. The circuit comprises of a capacitor, an avalanche transistor, a laser, a set of resistors and a set of inductors (Kilpelä 2004). These inductive components conduct high slew rates of

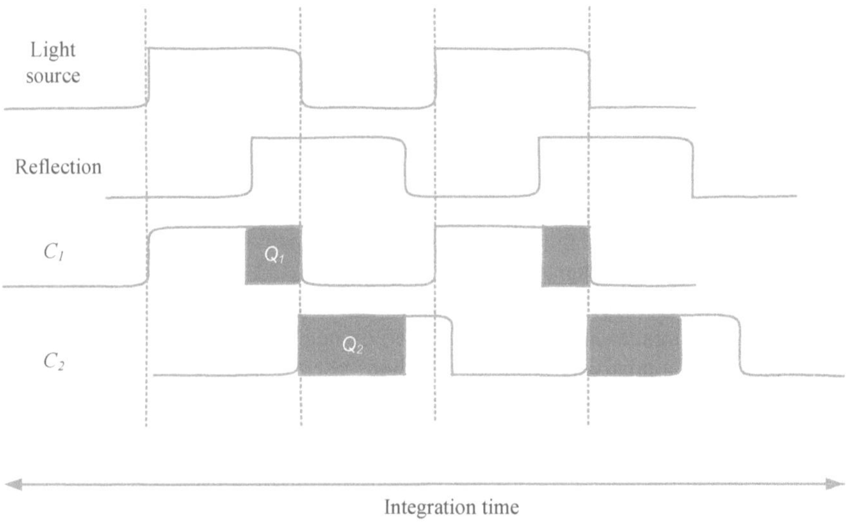

Fig. 5.3 Pulsed TOF showing the pulse duration of the light source, the reflected signal profile and two out-of-phase windows C_1 and C_2

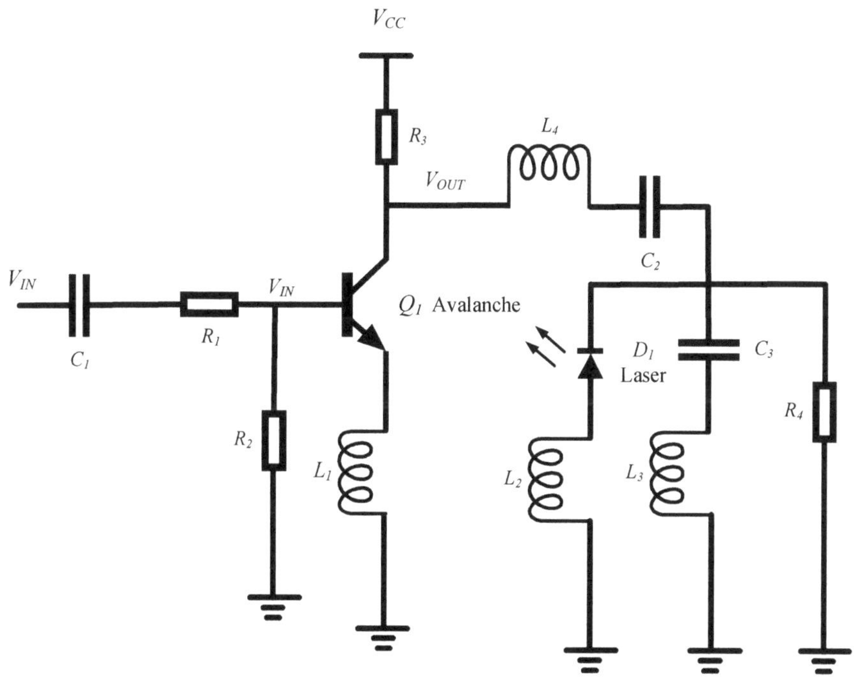

Fig. 5.4 The single transistor avalanche pulse circuit diagram

the current passing through the transistor. Capacitor C_2 is charged through resistor R_3 between the pulses and is discharged rapidly with the avalanche transistor through the laser device (Kilpelä 2004). To increase the pulsing frequency, the charging time of C_2 can be reduced by lowering the value of R_3 or by replacing it with an additional transistor switch. R_4 enables C_2 to charge and allows an increase in the peak current if C_3 is chosen at the correct resonant frequency. Typically, the load resistance is much larger than the collector resistor R_3 and therefore R_3 mainly determines the current supplied to charge C_2. The avalanche transistor Q_1 is triggered with a small current pulse applied to its base. In this configuration the transistor is biased such that the collector-emitter voltage (E_c) is close to the breakdown voltage of the transistor. An important consideration for Q_1 is its ability to sustain high supply voltages, as these circuits generally have voltage supplies in the 100s of volts, and are generally BiCMOS (bipolar) transistors; SiGe avalanche transistors present ideal characteristics to generate short pulses with high peak voltage. To increase the peak value of the output voltage further, circuit enhancements based on the Marx bank pulse circuit can be employed (Guo et al. 2013; Huang et al. 2014; Zhang et al. 2011). High-voltage nano- and sub-nanosecond pulses are used in a variety of applications, including ground-penetrating radar, wall-penetrating radar, biomedical applications, high-speed photography, ultra-wideband communications and high-intensity laser technology. The Marx bank pulse circuit is essentially a cascade of avalanche transistor circuits, as shown in Fig. 5.5.

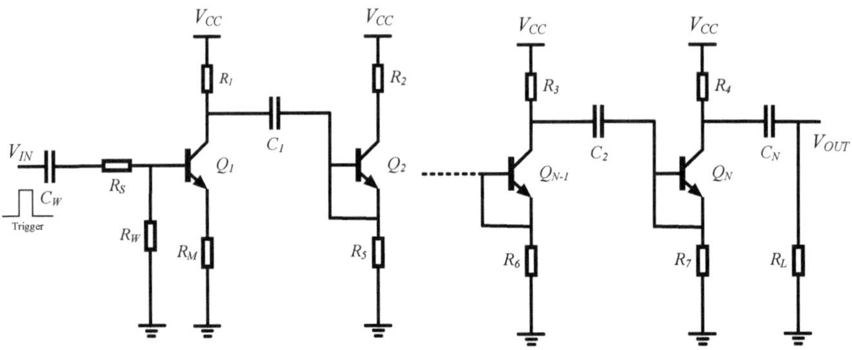

Fig. 5.5 The Marx bank pule transistor avalanche circuit diagram

The Marx bank pulse avalanche circuit as shown in Fig. 5.5 relies on a cascade of avalanche transistor circuit to achieve a high peak voltage over a very short pulse time. The ideal peak-out voltage is N times the supply voltage $V_{CC.}$ In practice though, the peak output voltage V_{peak} is scaled by the resistive divider formed between R_L and R_M. This relationship is described by

$$V_{peak} = NV_{CC}\left(\frac{R_L}{R_L + R_M}\right) \tag{5.6}$$

where the resistive divider ratio not only determines the output peak voltage, but also has an influence on the stability of the circuit. In Fig. 5.5, before the trigger signal is applied to the base of Q_1, all the avalanche transistors are in cut-off mode and the supply voltage charge is stored in the capacitors C_N. If a positive trigger signal (assuming *npn* bipolar transistors) is applied to the base of Q_1, then Q_1 will enter its avalanche mode and the charge of both C_1 and C_2 will be available at the collector of Q_2. An avalanche voltage of twice the value of V_{CC} forces Q_2 into instant breakdown. This process iterates through all the stages of the circuits with the charge on each consecutive avalanche transistor larger than the previous one by one V_{CC} factor. Peak voltages of up to 5 kV with 3 ns rising edges have been achieved using this method (Inokuchi et al. 2009). The theory and operation of the avalanche transistor and the benefits of using this component in laser transmissions are discussed in the following section.

5.2.3 Avalanche Transistor

An avalanche transistor is a bipolar transistor specifically designed for operation in the avalanche breakdown region and was first presented in September 1955 by Miller and Ebers (1955) as a follow-up paper on the design compromises and limitations between high-frequency and high-voltage germanium transistors described in March 1955 (Ebers and Miller 1955). At the time, it was already known that reverse biased Si and Ge junctions break down as a result of a multiplicative process, which is analogous to multiplicative breakdown in gases. Initial predictions of the possible applications of avalanche transistors included high-speed switching and transmission applications, which are inevitably exactly what the technology is used for in laser radiation circuits. In the avalanche breakdown region, the collector current and collector-to-emitter voltage is biased beyond the collector-to-emitter breakdown voltage. If the transistor is operated in the avalanche breakdown region, the device is also said to be in avalanche-mode operation. The key characteristic of avalanche-mode operation of a transistor is its ability to switch very high currents with lower than nanosecond transition times. If the avalanche effect occurs, the relationship between the collector current i_c and the emitter current $i_{.e.}$ is given by (Wu and Tian 2010)

$$i_c = \alpha M i_e \tag{5.7}$$

where α is the typical current amplification factor at low voltages and M is the avalanche multiplication factor (or Moll factor—as an acknowledgement by Miller and Ebers to J.L. Moll, who participated in many discussions leading to the invention of the avalanche transistor), which represents the current multiplication extent in the avalanche area. Essentially, the description of M is that if a single current carrier enters the transistor collector space-charge region, then M current carriers will exit the space-charge region and avalanche multiplication is perceived. The multiplication factor M is given as

$$M = \frac{1}{1 - \left(\frac{V_{CB}}{BV_{CBO}}\right)^n} \tag{5.8}$$

where BV_{CBO} is the collector-base breakdown voltage with the emitter open, n depends on the doping profile during the semiconductor manufacturing process and V_{CB} is the collector-base applied voltage. In addition, in avalanche breakdown mode, a negative resistance appears across the collector and emitter junction. In order to determine the collector-current during avalanche breakdown, the equation

$$I_C \approx \frac{I_{CBO} + \alpha I_B}{1 - \alpha - \left(\frac{V_{CB}}{BV_{CBO}}\right)^n} \tag{5.9}$$

is used, where I_{CBO} is the collector-base reverse leakage current and I_B is the base current. Avalanche transistor-based Marx circuits have been widely used and studied to generate high-voltage, nanosecond, high-repetitive frequency and portable pulses for many fields; these circuits are based on the avalanche transistor. In Oakley et al. (2015) an investigation of the steady-state safe operating conditions for large-signal SiGe HBT circuits is given. The study calculates capacitive currents with the intrinsic transistor; avalanche-inducing currents through the bipolar transistor junction are isolated and compared with DC instability points established through simulation and measurement. Oakley et al. (2015) therefore provide microwave circuit designers with the framework necessary to better understand the full-voltage-swing potential of SiGe HBT technology and the resultant hot carrier damage under RF operation. Chapter 6 of this book will expand on these theories to provide researchers and designers with a summarized look at SiGe HBTs in avalanche operation. CW TOF can also be used to drive a light source and this technique is discussed below.

5.2.4 Continuous-Wave Time-of-Flight

The light source can also be a CW source of typically sinusoid or square wave form. For a CW TOF method, four samples are taken for each measurement, with each sample phase-stepped by 90°. This method is shown in Fig. 5.6.

According to Fig. 5.6, the phase angle φ between illumination and reflection is determined by

$$\varphi = \tan^{-1}\left(\frac{Q_3 - Q_4}{Q_1 - Q_2}\right) \tag{5.10}$$

and the distance R is determined by

$$R = \frac{\varphi c}{4\pi f} \tag{5.11}$$

where f is the frequency of the CW signal. The measured pixel intensity A can then be calculated using

$$A = \frac{\sqrt{(Q_1 - Q_2)^2 + (Q_3 - Q_4)^2}}{2} \tag{5.12}$$

and the offset B is determined by

$$B = \frac{Q_1 + Q_2 + Q_3 + Q_4}{4} \tag{5.13}$$

where the reflected amplitude A and the offset B have an impact on the depth measurement accuracy σ, which can be approximated by

$$\sigma = \frac{c}{4\sqrt{2}\pi f}\frac{\sqrt{A + B}}{c_d A} \tag{5.14}$$

Fig. 5.6 CW TOF showing the pulse duration of the light source, the reflected signal profile and four out-of-phase windows C_1 to C_4

where c_d is the modulation contrast used to describe how well the TOF sensor separates and collects incident photoelectrons. The modulation contrast can attenuate at high frequencies as a function of the silicon substrate of the sensor. Because of the cyclic nature of the CW signal, which depends on the phase shift in the signal, aliasing occurs and is defined as the ambiguity distance d_{amb}, given by

$$d_{amb} = \frac{c}{2f} \tag{5.15}$$

which is also used to determine the maximum distance of the laser rangefinder based on the operating frequency. In order to increase the maximum measurable distance, the operational frequency can be decreased; however, this also decreases the accuracy of the measurement. Additional techniques such as adding multiple frequencies to the modulation scheme can be used to increase the accuracy while keeping the maximum distance constant (or increase the distance and keep the accuracy constant). On the receiver side, the single-shot measurement precision (the allowed statistical jitter of the measurement) can be approximated by

$$\sigma_R = \frac{0.35c}{2(BW)(SNR)} \tag{5.16}$$

where BW is the bandwidth of the device and SNR is the peak signal voltage-to-RMS noise ratio (Palojärvi et al. 2005). Light is essentially an EM signal with a specific frequency, undergoing the same diffraction, reflection and attenuation as RF and microwave signals. The characteristics of light, since it is the source of the transmitted signal in laser devices, is discussed in the following section.

5.2.5 The Frequency of Light

The oscillation frequency of light ranges from 2×10^8 Hz (200 GHz) for low-frequency microwave signals, between 4×10^{14} and 8×10^{14} Hz for visible light, and upwards of 10^{16} to 10^{18} Hz for UV and X-rays. A complete spectrum of EM radiation with the visible portion of light highlighted is shown in Fig. 5.7.

The significance of Fig. 5.7 becomes apparent when reconsidering (5.1) and observing that if it is possible to count the number of oscillations of light accurately in order to determine the distance R, the electronic equipment needed to distinguish between two consecutive oscillations of light should at least be able to operate (switch/count) in

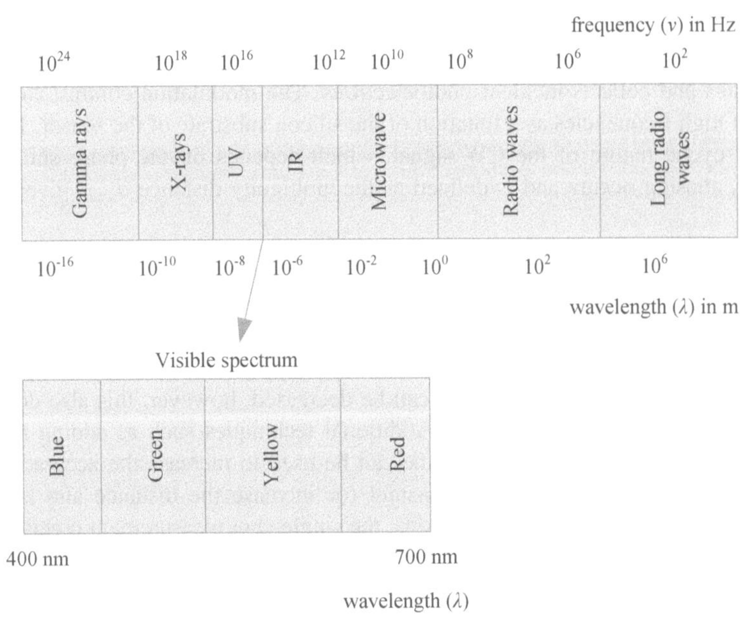

Fig. 5.7 The spectrum of EM radiation from long radio waves to gamma-rays

$$t = \frac{1}{f} \tag{5.17}$$

seconds, which approximates a 400 THz signal (red light) to $1/(4 \times 10^{14}$ Hz) = 2.5 ps (having the slowest oscillation frequency in the visible spectrum). The distance travelled during one oscillation of this signal is equal to its wavelength, being 700 nm. Effectively, this would therefore be the minimum measurable distance when using red light as a source and would require high pulse repetition frequencies, high accuracy and high resolution electronic equipment. Such high resolution distance measurements are generally not required in EW equipment such as rangefinders and designators, and would be required in other industrial applications such as semiconductor manufacturing; however, the principles remain the same. For a military laser rangefinder for example, the environmental conditions typically require sub-meter accuracy but resilient and noise-resistant signals. Chapter 4 presented the requirements of the trans-impedance amplifier in the optoelectronic receiver, which receives its variation in voltage from the photodetector. The signal received by the photodetector should also be considered in terms of its ability to convert incident phonons into electrical energy. Since the light source of laser systems are typically a distance R away from the photodetector (or $2R$ in the case of rangefinders that receive a signal back from the target), the typical geometry of such a system is considered in the following discussion.

5.2.6 Radiative Principles

In order to determine the total power incident on the photodetector that is available to be converted to electrical energy, the background radiance or power should first be determined. Consider a typical scenario of a laser source transmitting a light signal towards a photodetector, as shown in Fig. 5.8 (Burns et al. 1991). This scenario does not include the scattering and diffraction effects of the environment or a system where a portion of the signal is reflected back to the receiver and aims only to identify and present the principles of the radiated light.

From Fig. 5.8, the background flux incident on the detector can be determined, assuming a Lambertian[1] target and neglecting diffraction effects, by

$$P_B = L_\lambda A_s \cos\theta_s \Omega_D \Delta_\lambda T_R e^{(-\sigma R)} \tag{5.18}$$

[1]The apparent brightness of a Lambertian surface is the same regardless of the viewing angle. A Lambertian surface defines a perfectly diffusely reflecting surface with isotropic (uniform in all directions) luminance.

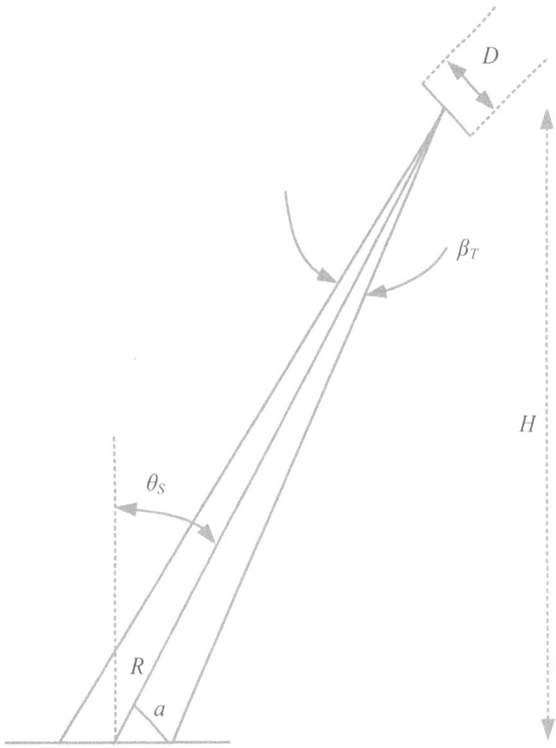

Fig. 5.8 Optical geometry of the background radiation towards a photodetector

where P_B is the background flux on the detector in watts, L_λ is the solar spectral radiance in W m^{-2} μm^{-1} sr^{-1}, A_s is the area of the detector active window in m^2, θ_s is the angle between the target surface normal to the line joining the target and receiver center points, Ω_D is the solid angle subtended by the laser receiver aperture in steradian, Δ_λ is the receiver spectral filter bandpass in μm, T_R is the transmission through the receiver optics, σ is the atmospheric extinction coefficient in m^{-1} and R is the slant range to the target in m. Mathematically it is convenient to represent the spectral transmittance $\tau(R)$ for a homogeneous medium by

$$\tau(R) = \mathrm{e}^{-\sigma R} \tag{5.19}$$

where R is the path length/slant range as above in m. If it is assumed that the radiation is isotropic, spatially homogeneous (uniform composition of space), unpolarized, and incoherent, enclosed in a cavity to balance the energy in the cavity wall and the EM waves, then by integrating over all wavelengths of light, it is possible to determine the thermal radiation as the total radiant exitance from the blackbody at temperature T in Kelvin through

$$M_e(T) = \sigma_e T^4 \tag{5.20}$$

where σ_e is the Stefan-Boltzmann constant of $\sigma_e = 5.67 \times 10^{-8}$ in units W $(m^2\ K^4)^{-1}$ and assuming that the environment is at 0 K with no incident flux. Exitance is the areal density flux on the source surface area, with flux flowing outward from the surface with no regard for angular density. The exitance leaving the surface can be caused by reflected light, transmitted light, emitted light or any combination thereof. Burns et al. (1991) show that the variables in (5.18) are defined by the set of equations given below. The area of the photodetector active window A_S, which is able to convert incident phonons to electrical energy, is given by

$$A_s = \frac{\pi a^2}{\cos \theta_s} \tag{5.21}$$

where

$$a = \frac{\beta_R}{2} R \tag{5.22}$$

and β_R is the receiver field of view in radians. From (5.21) and (5.22), the photodetector area can be rewritten as

$$A_s = \frac{\pi \beta_R^2 R^2}{4 \cos \theta_s} \tag{5.23}$$

with units in m^2. The receiver solid angle[2] Ω_D is determined by

$$\Omega_D = \frac{\pi D_R^2}{4R^2} \tag{5.24}$$

where D_R is the receiver clear aperture diameter in m. The reflected solar spectral radiance L_λ is calculated by

$$L_\lambda = \frac{E_\lambda \rho_B}{\pi} \tag{5.25}$$

where E_λ is the solar irradiance in W m^{-2} µm^{-1} and ρ_B is the background or target reflectance. The signal power from the transmitting laser can be determined by the set of equations below, given in further detail in Burns et al. (1991). Assuming that

[2]The solid angle is the two-dimensional angle in three-dimensional space that an object subtends at a point—a measure of how large the object appears to an observer based on the distance and viewing angle.

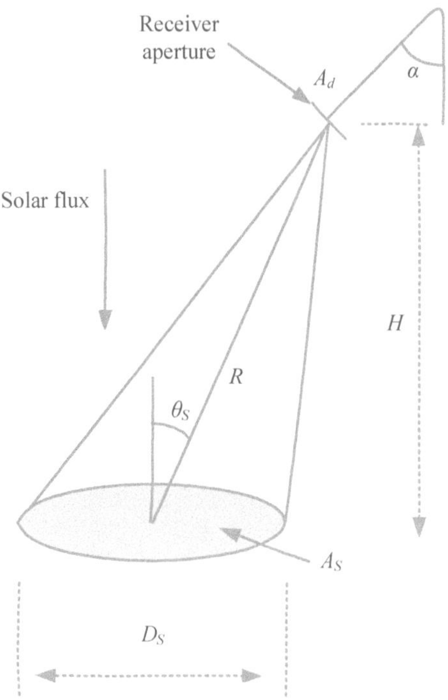

Fig. 5.9 Optical geometry of a laser transmission from an optical source towards a photodetector

the laser beam fills the target object, the geometry of the transmission beam is shown in Fig. 5.9.

In Fig. 5.9, if a Lambertian target and background are assumed, the irradiance of the laser beam at the target object is given by

$$E_T = \frac{P_T e^{-\sigma R}}{A_T} \tag{5.26}$$

where A_T is the area of the laser beam/footprint on the target in m^2. P_T is the total transmitted power in watts and is given by

$$P_T = P_L T_T \eta \tag{5.27}$$

where η is the collection efficiency of the transmit lens, P_L is the peak power of the laser and T_T is the transmission optical path transmission. Burns et al. (1991) show that the irradiance of the laser beam at the target can be equated to

$$E_T = \frac{4 P_L T_T \eta \cos\theta_s e^{-\sigma R}}{\pi B_T^2 R^2} \tag{5.28}$$

and the target exitance is given by

$$M_T = E_T \rho_T \tag{5.29}$$

where ρ_T is the target reflectance. Considering that $L_T = M_T/\pi$, the target radiance L_T is therefore given by

$$L_T = \frac{4\rho_T P_L T_T \eta \cos\theta_s e^{-\sigma R}}{\pi^2 B_T^2 R^2} \tag{5.30}$$

and the peak signal power P_S incident on the detector is given by

$$P_S = L_T A_T \cos\theta_s \Omega_D T_F T_R e^{(-\sigma R)} \tag{5.31}$$

where T_F is the receiver spectral filter transmission. Since the solid angle Ω_D is determined by the receiver clear aperture D_R, the incident signal power is rewritten as

$$P_S = \frac{P_T \rho_T \cos\theta_s \pi D_R^2 T_F T_R e^{(-\sigma R)}}{4R^2}. \tag{5.32}$$

The maximum range achievable with a laser rangefinder is strongly dependent on the reflectivity of the target. Datasheets of typical laser rangefinders generally specify the range performance for a diffusely reflecting Lambertian target of a specific reflectivity, generally 0.8 (80 %). A correction factor is then applied to the maximum range based on the variation of reflectivity of the target material. Table 5.1 lists some common diffusely reflecting target material reflectivity figures.

In Li et al. (2009) the maximum measurable distance for a pulsed laser rangefinder is given by the equation

$$R_{max} = \left[\frac{P_T A_R \sigma}{4\pi\theta^2 P_{s(\min)}} \tau_r \tau_t \tau_a\right]^{1/4} \tag{5.33}$$

where $P_{s(min)}$ is the minimum detectable power. The pulsed laser rangefinder noise probability density function follows a Gaussian distribution given by

$$P(x) = \frac{1}{\sqrt{2\pi}\sigma_n} e^{-\frac{x^2}{2\sigma_n^2}} \tag{5.34}$$

where σ_n^2 represents the output noise power. The directivity of a laser (or any other directed light) beam is the amount of diffraction that the directed energy undergoes as it travels through a medium. This directivity is a function of the wavelength λ of the source signal radiation and the power level P of the beam in watts such that the power density H of the beam at a target is given by

Table 5.1 Diffusely reflecting materials and surfaces

Material	Reflectivity (%)
Snow	0.8–0.9
Opaque white plastic	1.1 (110 %)
Reflecting foil	12.5 (1250 %)
Black rubber tire	0.02
Asphalt with pebbles	0.17
Opaque black plastic	0.17
Smooth concrete	0.24
Coniferous trees	0.3
Carbonate sand (wet)	0.41
Clear plastic	0.5
Carbonate sand (dry)	0.57
Deciduous trees	0.6
Limestone and clay	Up to 0.75
White masonry	0.85
Lumber (pine, clean, dry)	0.94
White paper	Up to 1.0 (100 %)

$$H = \frac{1.27P}{D^2 + \left(\frac{R\lambda}{D}\right)^2} \tag{5.35}$$

where R is the distance to the target and H is given in W/cm^2. This describes a diffraction-limited system and the maximum H is the theoretical best performance of the system. The average power density delivered to the target as calculated in (5.18) assumes no perturbing mechanisms such as atmospheric absorption, scattering or turbulence and uses ideal atmospheric conditions (possibly the most accurate results of this equation would be if used in space calculations).

SiGe has shown advantageous attributes to achieve terahertz radiation—with a variety of application in commercial, industrial and military space. The concept of generating terahertz radiation using SiGe structures is presented in the following section.

5.3 SiGe Quantum Cascade Lasers (Terahertz Radiation)

EM waves with frequencies between 300 GHz (1 mm wavelength) and 3 THz (100 μm wavelength), therefore between the optical and microwave spectrum, are classified as terahertz (far-infrared) signals. In this section, the term terahertz radiation will be used as opposed to THz radiation to distinguish between the technology and the frequency unit. These waves are non-ionized and therefore safe to be directed at humans and can pass through clothing, skin, paper, cardboard, plastic, wood, ceramics, fog and clouds. Terahertz waves cannot penetrate metal, concrete or water.

Terahertz radiation has proven difficult to implement on small scale while harnessing practical and high energies. Current terahertz radiation lasers can only generate a few milliwatts of radiation and require complex schemes for detection. Terahertz radiation is achievable using primarily artificial compounds such as Si/SiGe and GaAs and not by materials found *as-is* in nature. Generating EM waves at THz frequencies has potential in applications such as chemical analysis, security scanning, extremely high-speed communications and medical imaging. Terahertz spectroscopy, for example, provides a means of studying crystalline substances, which can be useful in identifying pharmaceutical substances or explosives. The ability to identify explosives rapidly and accurately is advantageous on the battlefield but also in airport security, since terrorist activity at airports and train stations has increased in recent years. Events such as the coordinated bombings in Brussels airport by the Islamic State of Iraq and the Levant on 22 March 2016, which killed 32 victims and injured another 300 people, can potentially be prevented if early detection is administered. Terahertz radiation allows explosives to be identified without physical contact, as currently being used in many airports globally. Scanning humans or animals with safe EM waves from strategic points such as entrances and exits would allow preventative identification of such risks. Another compelling range of applications of terahertz radiation is its use in space. Transceivers can be placed on satellites and measure atmospheric disturbances and climate change and also analyze interplanetary atmospheres. The lack of atmosphere in space has always been a big advantage for interplanetary communication, since EM signals are not attenuated by humidity, smoke, fog, pollution or rain in space. On earth, terahertz radiation can penetrate materials that block visible light and can potentially allow otherwise concealed objects to be identified and analyzed, which would be especially useful in military operations to give soldiers an opportunity to see obscured targets and assess the possible risks. Another sector where terahertz radiation of relatively high power (1–10 W primarily when implementing CW) has an advantageous impact is in infrared countermeasures to protect aircraft from incoming heat-seeking missiles.

There is a wide range of known materials that exhibit characteristic vibrational modes at terahertz frequencies. Using terahertz radiation, these materials can be analyzed and uniquely identified. Since it is currently mostly possible to design and build terahertz radiation lasers using physically large and costly equipment, research into miniaturizing these lasers has been receiving attention, especially since the early 2000s, several years after the invention of the quantum cascade laser (QCL) in 1994 at Bell Laboratories by Faist et al. (1994). The QCL seems to be the most promising device to generate terahertz radiation at energy levels high enough for practical use and by using devices that are relatively small, or at least mountable on aircraft or vehicles and possibly carried by soldiers on the battlefield. QCL structures are discussed in the following section.

5.3.1 QCL Structures

The basis of a QCL is stacking multiple layers of semiconductors on top of one another by growing each epitaxial layer through molecular beam epitaxy. Epitaxial layer growth improvements have led to newer and more efficient techniques such as metalorganic vapor phase epitaxy, which is currently being used as the primary growth method. This process of growing many layers is, however, very difficult to achieve, since the thickness of each layer must be sub-atomically precise and closely matched to the previous layer. The number of layers stacked on top of one another typically ranges between 1000 and 2000 depending on the application, but essentially a *the-more-the-better* approach is being followed. The operating principle of QCL light emission is fundamentally different from traditional laser sources in that it operates independently of the semiconductor bandgap (Jiang et al. 2016). The traditional laser principles are described in Chap. 2; the recombination of electrons and holes at the bottom of their respective conduction and valence bands produces light at a specific frequency, mathematically described by

$$E_2 - E_1 = \Delta E = hv \tag{5.36}$$

or

$$v = \frac{E_g}{h} \tag{5.37}$$

where E_g is the semiconductor bandgap and h is Planck's constant (given as 6.582×10^{-16} eV s). In a QCL there is only one type of charge carrier, electrons, and these electrons undergo quantum jumps between energy levels E_n and E_{n-1} to create a laser photon of specific frequency given by

$$v = \frac{E_n - E_{n-1}}{h} \tag{5.38}$$

where these energy levels do not exist naturally in the material active region. These energy levels are artificially constructed through the thin layering of the active region into quantum wells of thicknesses below the nanometer scale (Belkin and Capasso 2015). Using this technique in the mid-infrared spectrum has been adopted as one of the primary methods to create highly tunable (throughout the entire mid-infrared spectrum) emissions at low temperatures.

The relatively slow progress of terahertz QCLs is attributed to their inability to operate at room temperature (Belkin and Capasso 2015) and the low energy output. Recent advances in QCLs achieved by Belkin and Capasso (2015) have led to a new class of QCLs, which can operate at room temperature and have output energies in the range of 2 mW in the far-infrared spectrum. Conversely, in the mid-infrared spectrum, tunable devices with wavelengths between 3 μm and upwards of 20 μm and pulsed CW output power of 100 W have been achieved.

These achievements have shown major implications in the mid-infrared spectrum. Historically, the number of light source types that could generate light at these wavelengths was very limited and they only had a few nanometers of tuning capabilities as they depended solely on bandgap manipulation.

5.3.2 QCL Band Structure

SiGe intersubband QCLs have an emission wavelength of 30–100 μm, corresponding to the far-infrared spectrum. These devices have the unique benefit of monolithic optoelectronic integration on CMOS or BiCMOS silicon substrates, which makes them adaptable, modular and practically versatile. The advantages of SiGe intersubband lasers, especially their monolithic integration with active components, are driving the maturity of the technology and research into obtaining feasible results in terms of power output at room temperature. In the mid-infrared spectrum, InGaAs/InAlAs and GaAs/AlGaAs materials have been the dominant choice in achieving successful terahertz emissions, but without the advantages of combining the optoelectronic and active components on the same silicon substrate. In the far-infrared spectrum, however, increased free-carrier absorption loss, which increases waveguide attenuation, requires higher gain and more powerful pumping circuits. Again, operating these devices at low temperatures, typically around 20 K, makes them difficult to implement practically (Kelsall and Soref 2003). In the paper presented by Kelsall and Soref (2003), it is argued that device operation at 77 K (the temperature of liquid nitrogen) is achievable and can also present viable results. Compared to GaAs, SiGe has higher thermal conductivity. This is an especially important characteristic in active optoelectronic systems, since a change in temperature of the semiconductor leads to a shift in the emitted wavelength, which could be undetectable by the receiving photodetector if it falls outside the allowable tolerance. In Kelsall and Soref (2003) it is also stated that the indirect bandgap of SiGe, Si and Ge presents no noticeable disadvantage for QCL operation since the radiation is generated by intersubband transitions rather than interband recombination. A disadvantage of *p*-type SiGe QCL systems, according to Kelsall and Soref (2003), is that the hole masses are significantly larger compared to the III–V material electron masses. This means that the hole wave functions in SiGe heterostructures are relatively well-localized in individual quantum wells, but interwell coupling is weak; leading to lower device currents for comparable external bias of III–V devices.

To avoid the use of diffraction gratings to design and implement a QCL, which is a relatively inefficient technique, QCL design is mostly based on *p*-type SiGe heterostructures to realize true surface-emitting devices. For *p*-type Si/SiGe heterostructures the SiGe alloy layers represent the quantum wells and the Si layers act as the confining barriers. In strain-balanced structures the energy gaps

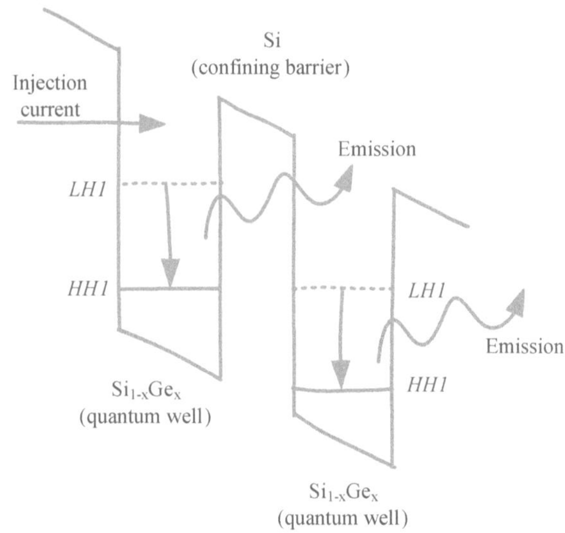

Fig. 5.10 Si/SiGe terahertz staircase design of energy emissions. Adapted from Kelsall and Soref (2003)

between different heavy hole subbands depend primarily on quantum well width, whereas the energy gaps between light hole and heavy hole subbands depend primarily on strain (therefore the alloy composition) and are relatively invariant with well width (Kelsall and Soref 2003). A design proposed by Friedman et al. (2001) and Soref et al. (1999) for QCL showing its energy levels is the quantum staircase design shown in Fig. 5.10.

The quantum staircase transitional diagram is based on the *LH1-HH1* inter-subband transitions. These transitions occur within each quantum well and carriers are transported between wells by non-radiative *HH1-LH1* transitions. To achieve radiative transitions through population inversion, the total interwell *HH1-LH1* transition rate should be faster than the total intrawell *LH1-HH1* transition rate. However, the interwell transition rates in QCLs are typically dominated by incoherent scattering processes rather than coherent tunneling (Kellsall and Soref 2003). Improvements on the quantum cascade laser design have been researched and proposed since its invention by Kolodzey et al. (2003), Bates et al. (2002), Grutzmacher and Mussler (2007), Lynch et al. (2006) and Paul et al. (2004) and are still an active research field today. Achieving high output power at room temperature operation remains challenging.

Weaponized lasers have been an attractive means of EW for countermeasures, offensive and defensive strategies. The advances in technologies and the attainable power output from lasers have made this a realizable possibility, which can come into being as battlefield-ready weapons in the near future. A description of the background of laser weapons and the critical subsystems is given in the following section.

5.4 Laser Weapons

This section purposely highlights publications and other works starting from the late 1970s up to the most recent works available. The presentation of the section in this manner gives an indication of the shift in opinion, potential, controversy and technological advances in lasers used as weapons. Historically, before lasers were able to produce the unprecedented levels of radiative flux that are possible today, the idea of lasers as weapons was somewhat taboo. Advances over the years show the change in opinion and the possibilities achieved to date and in a way how the perception of laser technology used as weapons has changed. Military lasers can be categorized as serving as either tactical or strategic devices. Low-energy pulsed lasers are generally classified as tactical devices and can be sold commercially and for research applications, whereas CW lasers of high intensity are often used for strategic military purposes or research (Stickley and Gerhardt 1985). In an article written by Stickley and Gerhardt (1985) more than 30 years ago, a comparative table of the primary differences between tactical and strategic lasers is presented. Many of these characteristics still hold true today and these parameters are presented in Table 5.2.

From Table 5.2, it is noticeable that tactical lasers, although also generating relatively high output powers, are generally not used to destroy their targets physically with high-intensity beams such as the laser application principles discussed for the laser rangefinder earlier in this chapter. Laser power levels that are likely to be employed in the future as standard optical power output for laser devices are 30, 60 and 125 kW optical power output (Gattozzi et al. 2015).

Table 5.2 Comparison of parameters for tactical and strategic systems

Characteristics	Tactical lasers	Strategic lasers
Primary targets	Missiles, aircraft, optics, electro-optics (communications)	Satellites, bombers, intercontinental ballistic missiles
Operating wavelength	Visible to infrared	Mid-infrared to X-ray
Target range	Kilometers	1000+ kilometers
Output power	Kilowatts	Megawatts
Mirror aperture	Centimeters	10+ meters
Pointing accuracy	Micro-radians	Nano-radians

Adapted from Stickley and Gerhardt (1985)

5.4.1 Tactical Lasers

Tactical lasers are used in applications such as (Stickley and Gerhardt 1985):

- Fire control and weapons guidance systems; this category includes solid-state laser rangefinders in military applications and the systems are used as standalone remote fire control systems. The types of lasers and materials (therefore wavelengths and efficiencies) are presented in more detail in Chap. 2 of this book.
- Laser surveillance radar such as forward-looking infrared (FLIR); the predominant research discipline in laser surveillance is achieving increased power output while maintaining low beam divergence, which can be achieved by, for example, phased arrays of diode lasers and waveguide lasers.
- Information transfer and communications, including data corruption or information-destroying offensive capabilities. The main advantages in information transfer and communications compared to traditional radio frequencies are the high-bandwidth capabilities, low susceptibility to EMI and secure channels with lower probability of signal interception. In EW, secure data transmissions between allied forces is considered a tactical advantage.
- Remote LOS meteorology; atmospheric sensing of aerosols and remote wind sensing, interplanetary atmospheric identification, pollution quantization and early detection of natural disasters are some of the applications of laser-based meteorology. Lasers are effective means of pulsing EM waves with large bandwidth through atmosphere-deprived channels.
- Chemical or biological agent detection; similar to meteorology, lasers can be used to identify chemical or biological constituents in the atmosphere, which is also an effective means of determining atmospheric pollution. An interesting adaption of this technology is identifying gun effluents to determine the location and distance of gunfire on the battlefield through analysis of the atmospheric particulate content.
- Data processing/holography; holographic communication is primarily based on laser principles, using femtosecond pulses through a spatial light modulator and a series of lenses, mirrors and Galvano scanners to direct the incoming laser beams precisely. Recent advances in this technology use safe-to-touch plasmas to generate the lasers, which allows for user interaction. The key to creating these virtual images is the short duration of the laser pulses. Longer pulses can be damaging to the surface they are directed at, but short consecutive pulses are generally harmless to the target materials. Considerable research activity in developing new ways to use optic devices for holographic communication and system pattern recognition continues.
- Laser material preparation; in the area of machining, one major application of lasers is the removal of material from another solid material such as in micromachining and drilling into ferrites and nonmetals without disturbing the magnetic or structural properties of the material. Laser welding and precision

annealing using lasers have the benefit of overcoming some physical mechanical limitations.

- Optical countermeasures; most importantly, optical countermeasures include optical jamming or destruction of optical sources/receivers with high-intensity optical light beams for EW purposes. Tunable lasers can offer additional advantages with the ability to shift frequency depending on the target optical source and can relatively easily be realized using alexandrite and chromium sources. Detailed material characteristics of laser materials and the advantages/disadvantages of each are presented in Chap. 2 in this book.

Characteristics and parameters of tactical lasers have been discussed in this book up to this point and the aim this section is to highlight the possibilities of strategic, therefore very high-power, lasers for military and EW applications in the field and in research.

5.4.2 Strategic Lasers

This topic has been controversial (Nahin 1977) and still is, but remains an active area of research and development although somewhat polemical, possibly because the extent of the damage and repercussions it may have is unknown. The gas dynamic laser was possibly the first developed strategic laser presented in the 1960s, since gas lasers have the inherent advantage of redirecting thermal effects due to the gas flow needed to generate a laser pulse. However, gas lasers are large, expensive and impractical devices to work with on the battlefield, but they laid the groundwork for advances in strategic lasers used as weapons and countermeasures. Using lasers for ballistic missile defense with counter-activity possible at the speed of light is possibly one of the most prominent advantages of strategic lasers. Underwater communication between submarines, for example, is another major advantage that drives the improvement of laser output and miniaturization of these devices (Stickley and Gerhardt 1985). There are three basic mechanisms for target damage (Nahin 1977) by a strategic laser weapon, namely:

- structural weakening and indirect mechanical damage as well as thermomechanical damage (generally due to extreme temperature increases in the target),
- shock waves and direct thermal damage (sudden increase in temperature of the target and rapid vaporizing and material explosively leaving the target at supersonic speeds), and
- X-ray radiation and ionization damage (emission of X-ray radiation from the target when shock waves are administered at a specific radiation wavelength).

HELs also require high-current recharging capabilities to recharge the capacitors required to supply large currents over short periods of time to the laser device.

5.4.3 Laser Recharging Unit

The recharge rate of the energy storage capacitor depends on the frequency of the laser, i.e. the time between successful high-energy pulses. The capacitor-charging source is a relatively complex and high-powered electronic device and typically comprises of a transformer followed by a voltage-rectifier bridge, a switching element in the primary windings of the transformer, a current-limiting element, a voltage sensor and control electronics (Koechner 2013). These subsystems are presented in block-diagram form in Fig. 5.11.

The DC voltage required for the energy storage in the capacitor is provided by the transformer and the rectifier. A semiconductor switch is added to the supporting circuitry to enable a variable DC voltage such that the output of the laser can also be varied. This control circuit includes either a solid-state relay, a triac, or a pair of inverse parallel thyristors/silicon-controlled rectifiers and is switched on at the start of the charging cycle and switched off if the predetermined voltage across the capacitor is reached. The rectifier diodes are protected by the current-limiting circuit, since the recharge path through the low-impedance transformer windings and the discharged capacitor is essentially a short-circuit. Laser charging units are preferably a constant-current circuit as opposed to varying rate or trickle charging and require the charging capacitors to be charged at the fastest rate possible. Assuming the charges on the capacitor plates are $\pm q$ and the voltage between the plates is given by V, then

$$C = \frac{q}{V} \tag{5.39}$$

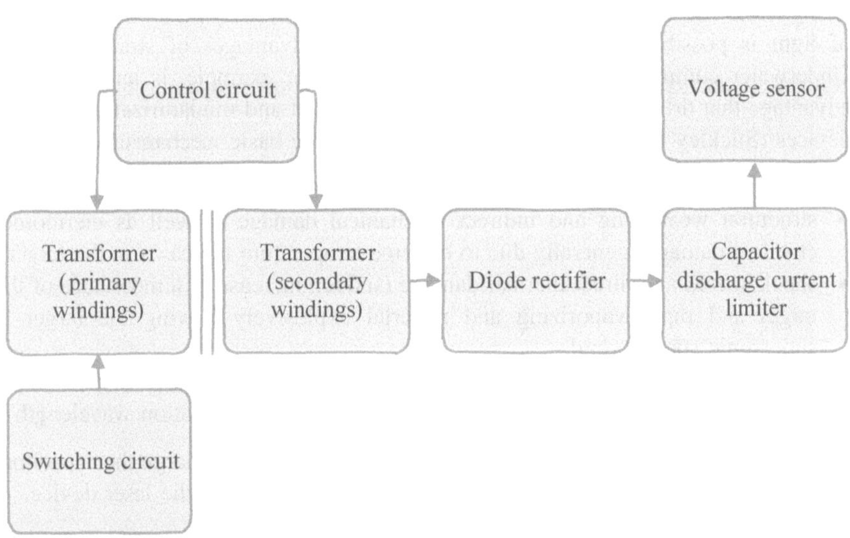

Fig. 5.11 Essential subsystems block diagram of a pulsed laser charging unit

where C is the capacitance of the capacitor. The voltage current relationship can be determined as

$$I(t) = C\frac{dV(t)}{dt} \tag{5.40}$$

where $dV(t)/dt$ is the rate-of-change of the voltage across the capacitance and $I(t)$ is the current at time t. To determine the required current needed to charge the capacitor to a specified voltage in a specified time (as is generally required in laser weapon systems where the repetition rate and output power are of concern) the charging current is defined by

$$I = \frac{CV}{t} \tag{5.41}$$

which should be limited by a current-limiting circuit. The easiest but also less efficient solution to limit the discharge current is using a resistive load. Improvements and replacement techniques to achieve this more efficiently are discussed below. The energy W stored in the capacitor can be determined by

$$W = \frac{1}{2}CV^2 \tag{5.42}$$

which is (under ideal circumstances) available to be transferred to the laser output device. In Koechner (2013) the most frequently used circuits for current limiting of the capacitor discharge current are presented. They are briefly highlighted here to emphasize the advantages and disadvantages of these circuit alternatives. In Fig. 5.12a–g the power supply charging circuits for pulsed lasers are shown.

In Fig. 5.12a the constant voltage source with resistance-limited charging is shown. This configuration dissipates an equal amount of power through the resistor as that stored in the capacitor, therefore the thermal losses and resistor-power requirements are high. These circuits could generally be tolerated if a long repetition rate is acceptable; however, in weapon systems, a low repetition rate is generally a primary specification. Removal of the discharge resistor from the output and insertion of an inductor in the primary windings are illustrated in Fig. 5.12b.

This technique introduces current limiting by inductive reactance during the initial phase of the charging when the secondary windings are shorted to ground through the uncharged capacitor. This inductance can also be integrated into the transformer as a primary leakage, as seen in Fig. 5.12c.

For HEL outputs with varied repetition rates and variation in the available energy from the source, it becomes more practical to apply optimization at the output/secondary windings of the transformer. This increases the available techniques and allows the transformer to operate at its specified rating without limiting its operation for relaxed requirements at the output. One such technique involves using resonant charging at the output, as shown in Fig. 5.12d.

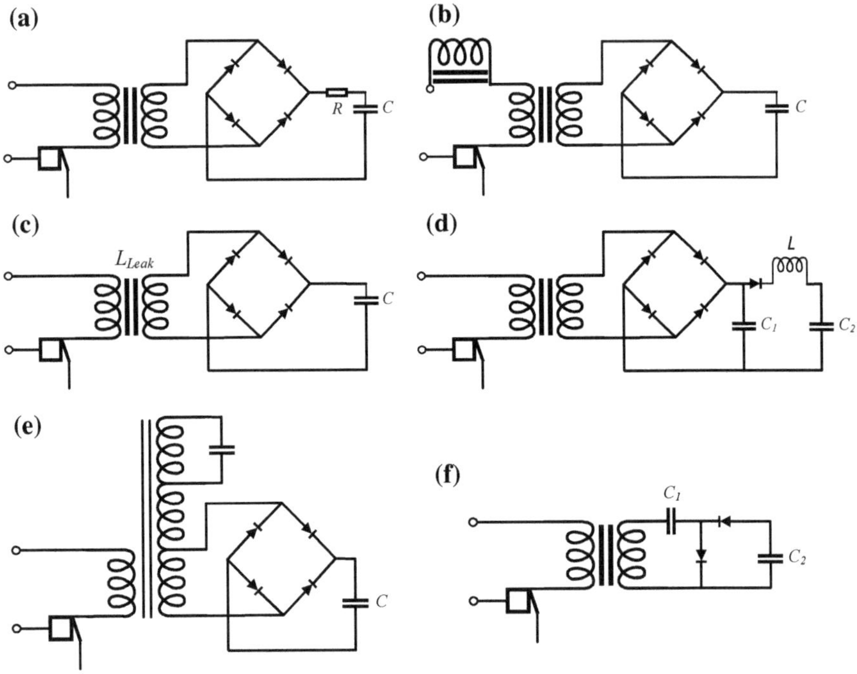

Fig. 5.12 Power supply charging circuits for pulsed lasers—**a** current-limiting resistor, **b** current-limiting with inductor, **c** high-leakage inductor, **d** resonant charging, **e** ferroresonant transformer, **f** ferroresonant transformer. Adapted from Koechner (2013)

In this circuit the current only flows from the designated DC source for the initial half-cycle of the resonant frequency and charge the capacitor to double the value of the source voltage. The peak current drawn from the power supply at this point depends on the inductor, resonant frequency of the *LC*-network and the voltage across the capacitor. A ferroresonant transformer circuit is shown in Fig. 5.12e.

In Fig. 5.12e the transformer characteristics are attained using a magnetic shunt and resonant tank circuit. The short-circuit current can be reduced by repeatedly transferring a small quantity of energy to the storage capacitor. In Fig. 5.12f, a voltage-doubling circuit is employed to limit the discharge current of the laser charging unit.

A small capacitor, C_1, transfers its charge to the primary charging capacitor during each cycle of the resonant frequency. These techniques to limit the discharge current and additional circuits used practically are discussed in further detail in Chap. 6, with specific focus on the enabling technologies of implementing such circuits as monolithic circuits and determining the viability of SiGe. This section only aims to identify the possible techniques that can be practically implemented.

5.4.4 Laser Target Material

The damage caused by a laser weapon is not only dependent on the energy and wavelength of the source, but also on the material of the target. For example, a painted aluminum target will experience up to three times the intensity of the damage of the laser beam compared to the same aluminum target, but unpainted (Callaham 1982). In Callaham (1982) a brief calculation result is presented, which concludes that in order to penetrate through a 3 mm aluminum plate, through a vacuum medium (such as space), a single laser pulse would require a 5 kJ/cm^2 energy flux. Conversely, semiconductor materials such as optical sensors, solar cells or certain active devices with active region close to the surface of the substrate would require only approximately 10 J/cm^2 of energy flux. The reason for quoting this from Callaham (1982) is to show that it does not necessarily require extremely high-powered laser beams to destroy an incoming target; it depends heavily on the target itself and the intended damage. Also, in 1982, the idea of generating 5 kJ/cm^2 laser beams was considered impossible; however, more than 30 years later in 2015, researchers in Osaka in Japan had already produced a laser beam (the Laser for Fast Ignition Experiment) of 2 petawatts for a 1 picosecond duration. This is equivalent to 1000 times the world's total electricity consumption. Of course the device is large (328 ft/100 m) and diffraction is limited by strategically placed glass panels so it is not a mobile device, but it shows that laser energy levels are reaching unprecedented levels and will continue to do so in the future. Japan is already planning to produce a 10 petawatt laser at the same institution. According to Callaham (1982) it is possible to illuminate and overheat, potentially destroying, a satellite at a 100 km altitude with an irradiance of 10 times solar irradiance (approximately 1.4 W/cm^2) with a 220 kW average power CO_2 laser, which can be mounted on a large aircraft or a land vehicle. This calculation gives a good idea of the (achievable) potential to destroy satellite communications during military operations by using lasers that are not overly complex or high-powered. Satellite communications have, since the issue of this article, become orders of magnitude more crucial for any country or military to convey information, communicate and use GPS navigation and is a major advantage in EW. Research and investigation into the behavior of composite materials subjected to laser illumination for ranges of power densities have been presented in studies such as the one by Fabrice et al. (2014). In Fabrice et al. (2014), the behavior is investigated for illumination sources varying between 100 and 2000 W/cm^2 specifically for 1.07 and 1.634 μm wavelength lasers on aeronautical composite materials to determine the damage and effects of the laser on the target. It is concluded in this brief study of Fabrice et al. (2014) that carbon epoxy materials are effective protection against laser weapon attacks although the mechanical properties of the composite materials rapidly decrease when subjected to continuous laser bombardment. Further damage assessment and combat effectiveness of anti-missile airborne laser weapons is presented in De-yun et al. (2015). This damage assessment divides the degree of laser damage on the target into interference, soft damage and hard damage and based on the accumulated data, divides the laser energy density into corresponding ranges for the different damage effects. The missile target is divided into four parts: the fuel tank, seeker, warhead and tail wing. It is important to distinguish the target in terms of its material composition and vulnerable parts when analyzing the effectiveness of a laser weapon. These types of studies have proven equally important as research into technology advances into laser energy output and efficiencies, since knowing the weaknesses of the target determines the design considerations and trade-offs of the source/weapon.

5.4.5 *Topical Research*

By 2003, Rheinmetall investigated solid-state MEL for anti-air defense application for distances of up to 7 km (Weise et al. 2003). The decrease in the physical size of lasers has led to this particular laser being a size that can be mounted on aerial vehicles as a secondary weapons system. The beam power of a MEL is in the range of 1–100 kW and the electric energy per pulse is in the kJ range. This allows such a weapon to destroy electronics and optoelectronics of targets up to 7 km away, in the earth's atmosphere. A MEL therefore aims to bridge the gap in anti-air defense operations between machine-guns and missiles in terms of firepower and cost-effectiveness. Adapted from Weise et al. (2003), the optical arrangement of a laser system is given in Fig. 5.13.

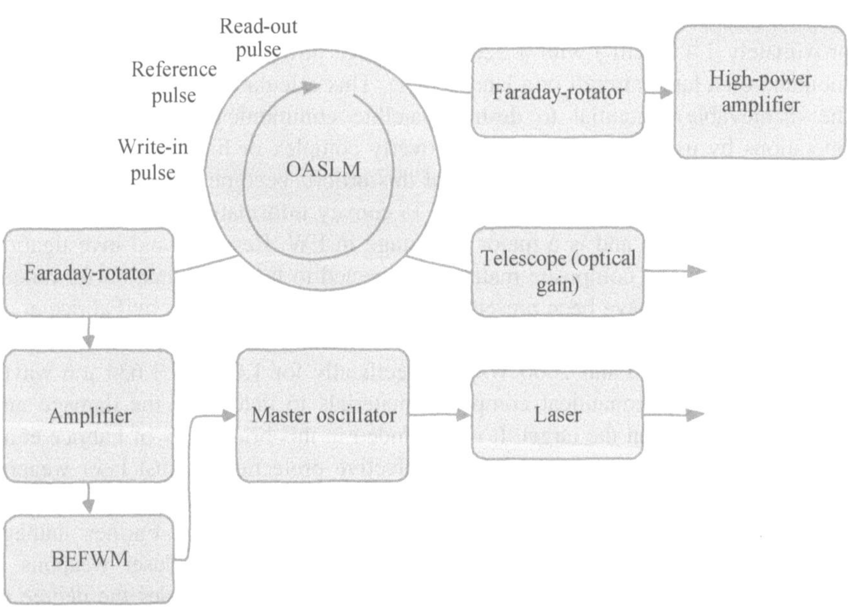

Fig. 5.13 Optical scheme of a laser system adapted from Weise et al. (2003). This scheme is reproduced here to show the integral dependability of the optical and the electronic systems, both susceptible to damage from laser pulses

The simplified optical scheme for a laser system as presented in Fig. 5.13 becomes exponentially more complicated when factoring in all the necessary components required to realize the subsystems presented in Fig. 5.11. In order to focus the beam energy on the optoelectronics of the target, the influence of the atmosphere must be compensated for. Internal amplification is required by converting the laser beam from a short pulse duration to a longer pulse within the beam system of the laser (Weise et al. 2003). When the fine target acquisition system works in continuous tracking mode, the laser pulse is initiated. The master oscillator generates the short pulse used for target illumination. This pulse is reflected by the target and received back by the telescope optics. The return pulse will therefore contain all the necessary information regarding the atmosphere turbulence and potential influence on the weapon-pulse. The pulse is then non-linearly phase-correlated by the Brillouin amplified four wave mixer device where the deformation of the laser pulse by the atmospheric turbulence is converted. The pulse is sent back again to the target where the atmosphere will reconstruct the original waveform and the pulse energy is focused on the small cross-section of the target optoelectronics. All the information of the pulse is stored within a hologram in the optical addressable light modulator and this information is transferred to the output pulse before final amplification by the high-powered solid-state laser and firing of the laser weapon to destroy the target. This technique of analyzing the atmosphere and preparing the high-powered laser for effective target acquisition and focused beam energy essentially summarizes the workings of laser weapons. In Weise et al. (2003) the main features and advantages of a MEL are summarized; these include permanent destruction of optoelectronics, extremely fast (not only the physical laser beam travelling at the speed of light but also the conditioning of the electronic circuit in preparation for the final weapon discharge), short-range effectiveness (although 7 km can already be considered practically long-range on the battlefield), non-lethal effectiveness when pointed at human targets, high precision and cost-effectiveness.

Various research opportunities for the subsystems and assemblies, as well as supporting circuitry of high-powered lasers, are exploited in a variety of disciplines. Examples of such research include developing models and prototypes of power systems and energy storage models for the integration of high-powered lasers on naval platforms. Although naval platforms such as ships and submarines are generally large enough to house the laser systems, the limitations of power generation and thermal management present several challenges during integration. Even on new ships, for example, the cost of the installation and thermal management of extreme proportions for weapons systems that may only be used sporadically prevents the technology from being commonplace in naval operations. In Gattozzi et al. (2015) the research progress of work conducted by the University of Texas Center for Electromechanics (UT-CEM) and the Naval Postgraduate School (NPS) in their collaboration to develop and simulate models of ship power systems to evaluate and guide the integration of pulsed laser loads into existing platforms is presented. The work focuses on the three major power generators that are likely to be considered for future laser implementations, these being battery-driven (both lead-acid and lithium-ion), flywheel power generation and charging of ultra-large capacitors. The collaborative program between UT-CEM and NPS explores all the combinations of storage technologies and laser output power (30, 60 and 125 kW) to develop best-case scenarios and a developmental framework.

In Brändle (2014) a HEL weapon demonstrator is designed with the focus on counter-rocket, artillery and mortar and counter-UAV scenarios. The pointing/tracking control design consists of four industrial fiber lasers that are geometrically coupled in one beam director telescope. The research is conducted on a system consisting of a Cassegrain beam director telescope mounted on an industrial robot for coarse pointing and target tracking.

Laser technology has become accessible on a commercial scale as well, as research by Soetedjo et al. (2014) shows. Soetedjo et al. (2014) use a Raspberry Pi® to detect a laser spot using the OpenCV[3] library and a web-based camera and shows the flexibility of commercial technology towards specialized laser equipment.

5.5 Conclusion

This chapter predominantly focuses on applications of laser in the lower energy output spectrum and uses the laser rangefinder as an example application to present and discuss the enabling principles. The characteristics that determine the quality, range and noise of the transmitted signal through a specific medium, such as a vacuum in space or through the earth's atmosphere, determines the requirements and specifications of the transmitter and the receiver in a laser system. The required power of the transmission, the optical properties of the detector and the minimum detectable signal at the receiver are all variables relevant to the application of the laser and the environment in which it is used. Lower energy laser applications, such as laser rangefinders, require specialized engineering practices and efficient and high-quality components to achieve long-range operation and low susceptibility to interference. SiGe lasers and the THz radiation that they can achieve have become a dominant research field since the invention of QCLs in 1994. QCLs can generate EM radiation in the terahertz range from relatively small semiconductor devices with thin stacked layers of Si/SiGe, which operate on the principle of quantum wells rather than the traditional bandgap emission principles.

In the HEL regime, weapons can be constructed if high enough energy can be supplied to lasers at specific wavelengths, depending on the target and the application of the weapon. The range of these weapons is also dependent on the atmosphere from which the laser emanates and the principles of scattering,

[3]OpenCV is available for academic and commercial use, free of charge, and is an open-source computer vision and machine learning software library that supports Microsoft® Windows®, Linux®, Google® Android® and Apple® Mac OS®.

diffraction and laser beam divergence limit the distance of its effectiveness. This chapter highlights the types of weapons, tactical or strategic, and provides methods, research opportunities and commonly used techniques to achieve high enough power to drive these lasers. This research field is still very active today and the US DoD aims to have battlefield-ready weapons already by 2023. A brief introduction on the spin-off research disciplines that form part of a high-energy system is also presented in this chapter.

References

Bates, R., Lynch, S. A., Paul, D. J., Ikonic, Z., Kelsall, R. W., Harrison, P., et al. (2002). THz electroluminescence from Si/SiGe quantum cascade heterostructures. *Summaries of Lasers and Electro-Optics, 1*, 629–630.

Belkin, M. A., & Capasso, F. (2015). New frontiers in quantum cascade lasers: High performance room temperature terahertz sources. *Physica Scripta, 90*(2015), 118002 (1–13, 2015).

Brändle, M. (2014). Pointing/tracking control design for a high energy laser system. In *2014 IEEE/ASME International Conference on Advanced Intelligent Mechatronics* (pp. 676–682).

Burns, H. N., Christodoulou, C. G., & Boreman, G. D. (1991). System design of a pulsed laser rangefinder. *Optical Electronics*, *30*(3), 323–328.

Callaham, M. B. (1982). Military: Laser weapons: A variety of targets might be neutralized by laser beams, but only if weapons employing them can be utilized with great finesse. *IEEE Spectrum*, *19*(3), 51–55.

De-yun, Z., Li-na, Z., Kai, Z., & Kun, Z. (2015). Damage assessment of airborne laser weapon to anti-missile. In *34th Chinese Control Conference (CCC)* (pp. 223–228).

Ebers, J. J., & Miller, S. L. (1955). Design of alloyed junction germanium transistors for high-speed switching. *The Bell System Technical Journal*, *34*(4), 761–781.

Fabrice, L., Alfred, E., Olivier, M., Vadim, A., & Gildas, L. (2014). Behavior of composite materials subjected to 1.07 μm laser irradiation in the range of 100 W/cm^2 to 2 kW/cm^2. In *2014 International Conference on Laser Optics* (pp. 1–1).

Faist, J., Capasso, F., Sivco, D. L., Sirtori, C., Hutchinson, A. L., & Cho, A. Y. (1994). Quantum cascade laser. *Science, 264*(5158), 553–556.

Friedman, L., Sun, G., & Soref, R. A. (2001). SiGe/Si THz laser based on transitions between inverted mass light-hole and heavy-hole subbands. *Applied Physics Letters, 78*, 401–403.

Gattozzi, A. L., Herbst, J. D., Hebner, R. E., Blau, J. A., Cohn, K. R., Colson, et al. (2015). Power system and energy storage models for laser integration on naval platforms. In *2015 IEEE Electric Ship Technologies Symposium* (pp. 173–180).

Grutzmacher, D., & Mussler, G. (2007). Si–Ge quantum well and cascade structures for optoelectronics. In *4th IEEE International Conference on Group IV Photonics* (pp. 1–3).

Guo, Y-L., Yan, N-N., Guo, S-H., & Zeng, G. (2013). 500 ps/1 kV pulse generator based on avalanche transistor marx circuit. In *2013 International Workshop on Microwave and Millimeter Wave Circuits and System Technology (MMWCST)* (pp. 296–299).

Henzler, S. (2010). *Time-to-digital converters* (Chapter 2). Berlin: Springer, ISBN 978–90-481-8627-3.

Huang, Z., Fu, Q., Chen, P., Yang, H., & Yang, X. (2014). High power pulse generator based on avalanche transistor marx circuit. In *2014 IEEE International Conference on Communication Problem-Solving* (pp. 315–317).

Inokuchi, M., Akiyama, M., & Sakugawa, T. (2009). Development of miniature marx generator using BJT. In *2009 IEEE Pulsed Power Conference* (pp. 57–60).

Jiang, Y., Vijayraghavan, K., Jung, S., Jiang, A., Kim, J. H., Demmerle, F., et al. (2016). Spectroscopic Study of terahertz generation in mid-infrared quantum cascade lasers. *Nature: Scientific Reports 6*. Article number 21169, Published online February 16, 2016.

Kelsall, R. W., & Soref, R. A. (2003). Silicon-germanium quantum-cascade lasers. *International Journal of High Speed Electronics and Systems, 13*(2), 547–573.

Kilpelä, A. (2004). *Pulsed time-of-flight laser range finder techniques for fast, high precision measurement applications*. Department of Electrical and Information Engineering, University of Oulu, thesis, October 2004.

Koechner, W. (2013). *Solid-state laser engineering*. Berlin: Springer, November 11, 2013.

Kolodzey, J., Adam, T. N., Troeger, R. T., Ray, S. K., Looney, G., Rosen, A., et al. (2003). The design and operation of terahertz sources based on silicon germanium alloys. *Silicon Monolithic Integrated Circuits in RF Systems, 1–5*, 2003.

Li, J., Zhao, B., Tang, L., & Zhao, X. (2009). Digital signal processing method and implementation for pulse laser rangefinder. In *The Ninth International Conference on Electronic Measurement and Instruments* (1-289-1-293).

Lynch, S. A., Paul, D. J., Townsend, P., Matmon, G., Suet, Z., Kelsall, R. W., et al. (2006). Toward silicon-based lasers for terahertz sources. *IEEE Journal of Selected Topics in Quantum Electronics, 12*(6), 1570–1578.

Mehlhorn, T. A. (2014). National security research in plasma physics and pulsed power: Past, present, and future. *IEEE Transactions on Plasma Science, 42*(5), 1088–1117.

Miller, S. L., & Ebers, J. J. (1955). Alloyed junction avalanche transistors. *The Bell System Technical Journal, 34*(5), 883–902.

Nahin, P. J. (1977). The laser BMD and other radiant energy weapons: Some thoughts. *IEEE Transactions on Aerospace and Electronic Systems, 13*(2), 96–107.

Oakley, M. A., Raghunathan, U. S., Wier, B. R., Chakraborty, P. S., & Cressler, J. D. (2015). Large-signal reliability analysis of sige hbt cascode driver amplifiers. *IEEE Transactions on Electron Devices, 62*(5), 1383–1389.

Palojärvi, P., Ruotsalainen, T., & Kostamovaara, J. (2005). A 250-MHz BiCMOS receiver channel with leading edge timing discriminator for a pulsed time-of-flight laser rangefinder. *IEEE Journal of Solid-State Circuits, 40*(6), 1341–1349.

Paul, D. J., Townsend, P., Lynch, S. A., Kelsall, R. W., Ikonic, Z., Harrison, P., et al. (2004). In search of a Si/SiGe THz quantum cascade laser. *Silicon Monolithic Integrated Circuits in RF Systems, 143–146*, 2004.

Soetedjo, A., Ashari, M. I., Mahmudi, A., & Nakhoda, Y. I. (2014). Rasperry Pi based laser spot detection. In *2014 International Conference on Electrical Engineering and Computer Science* (pp. 7–11).

Soref, R. A., Friedman, L., Sun, G., Noble, M. J., & Ram-Mohan, L. R. (1999). Intersubband Quantum-well terahertz lasers and detectors. In *Proceedings of SPIE* (Vol. *3795*, p. 515).

Stickley, C. M., & Gerhardt, I. D. (1985). The laser as tactical and strategic weapon: Laser technology has revolutionized conventional warfare and promises to transform dramatically the way battles are fought. *IEEE Potentials, 4*(2), 22–26.

Weise, T. H. G. G., Gowin, M., & Langhans, D. (2003). Solid state medium energy weapon laser. In *14th Pulsed Power Conference* (Vol. 1, pp. 245–248).

Wu, Q., & Tian, W. (2010). Design of electronic circuits of nanosecond impulser based on avalanche transistor. In *11th International Conference on Electronic Packaging Technology and High Density Packaging (ICEPT-HDP)* (pp. 774–777).

Zhang, X., Yang, X., Li, Z., & Yang, H. (2011). Subnanosecond pulsed power generator with avalanche transistor marx circuit. In *2011 International Conference on Computational Problem-Solving* (pp. 347–349).

Chapter 6
Frequency Response of Optoelectronic Receivers: The Motivation for Faster Transistors

6.1 Introduction

Optoelectronic receivers have several fundamental requirements that make them difficult to implement practically. The primary subsystems of the optoelectronic receiver are the photodetector and the transimpedance amplifier, which converts the optical signal received from a laser pulse, for example, into an electrical signal with a useable amplitude (typically in the high millivolts to low volts range). These systems can be realized either by hybrid or integrated approaches based on high-speed bipolar technologies forming what is known as optoelectronic integrated circuits (OEICs). Traditionally, these systems were primarily integrated using InP high-electron-mobility transistor (HEMT) technologies (Sharifi and Mohammadi 2007). Silicon photodiodes allow optical detection applications with spectral ranges between 400 and 900 nm and the operating principles of silicon photodiodes are well known. The hybrid and integrated approaches both have distinct advantages and disadvantages when implementing optoelectronic receivers. Hybrid receivers connect a photodiode fabricated on a separate substrate as the transimpedance amplifier. There are several ways of connecting the photodiode to the amplifier circuit; methods include wire bonding, flip-chip bonding or optical interconnects.

Hybrid integrations can be cost-effective, depending on the method and the capabilities within the institute, but there is a significant loss of sensitivity between the small-area photodiode and the supporting circuitry. Interconnect capacitances in high-frequency applications add an additional variable and must be accounted for in the design; ultimately increasing the complexity of the analogue design. Susceptibility to EMI also increases because of relatively long leads, compared to the operation wavelength, which could lead to unwanted circuit oscillations and parasitic inductances.

In integrated systems, the interconnects can be precisely controlled and specifically designed for, since the photodiode and the amplifier are situated on the same substrate, providing an inherent increase in data rate for a given optical power

W. Lambrechts and S. Sinha, *SiGe-based Re-engineering of Electronic Warfare Subsystems*, Signals and Communication Technology,
DOI 10.1007/978-3-319-47403-8_6

sensitivity. These systems are, however, more expensive to realize (flip-chip bonding versus traditional wirebonding for a hybrid approach, for example) and may introduce trade-offs between the operating speed of the transistor and the sensitivity of the photodiode (Sharifi and Mohammadi 2007).

BiCMOS technologies have an additional advantage over traditional CMOS technologies both for the photodetector and the active circuitry. BiCMOS substrates can be used, without any technological modifications, as PIN photodiodes. The buried *n*+-collector can be used as the cathode, the *n*-collector epitaxial layer can serve as the intrinsic layer and the base implant can serve as the anode (Zimmerman 1996). The transimpedance amplifier can take advantage of high-speed HBTs in its realization. CMOS technologies use the *pn*-junction available in the process as the anode and cathode of the photodiode. CMOS transistors inherently also have comparably slower response times than those of HBTs, even though CMOS presents low cost, low power consumption and small areas. PIN photodiodes are generally preferred over *pn*-junction photodiodes because of their higher bandwidth capabilities and high dynamic range resulting from the applied reverse bias, whereas *pn*-junction photodiodes do not operate with an applied reverse bias but typically perform better under low-light conditions. The photodetector and transimpedance amplifier have been described in Chaps. 4 and 5; this chapter focuses on the requirements emanating from the connection between the photodetector and the transimpedance amplifier. Careful consideration is required to optimize the frequency response of the system; these considerations include, for the

(i) Photodetector:

- Low junction capacitance for a high frequency response,
- low series resistance to minimize the response time, and
- high responsivity to reduce the necessary gain (and therefore decrease the chances of system oscillations);

(ii) Transimpedance amplifier:

- High-gain transistors to amplify weak inputs,
- high-frequency transistors to react to incoming pulses,
- low input capacitance and input impedance to maximize the frequency response, and
- insensitivity to electrical interference;

(iii) Interconnects:

- Short leads to limit capacitances to a minimum (integrated approach preferred over hybrid approach),
- sufficient power supply decoupling to filter interference, and
- efficient layout techniques to limit interconnect lengths for both integrated and hybrid approaches.

The drive towards generating sub-nanosecond laser pulses can be attributed to the use of the laser system. Essentially though, it concerns the fact that the information, the light pulse, travels at the speed of light and the electronic circuitry should be *ready* for the receiving pulse/information in a short time. The opto-electronic receiver components and the primary requirements for these components are shown in Fig. 6.1.

There are various subsystems in a laser transceiver (transmitter and receiver) that influence the overall speed of the system; all of these subsystems must be optimized if they are to have no negative impact on the performance of the entire system. At the transmitter, the laser driver must be able to generate pulses at high frequencies and the laser diode must be able to enter and exit population inversion at the same rate. On the receiver side, the photodetector must be able to receive and process consecutive incoming pulses at the same rate at which they are transmitted and the transimpedance amplifier must be able to amplify the current to a useable voltage, also at the same speed (or faster). Any additional supporting circuitry also needs to meet the demands in order not to decrease the measurement result acquisition time —this is where integrated systems as opposed to hybrid packaging become exceptionally effective and efficient. All these requirements place stringent design requirements on the system in terms of their frequency response. Several studies and reports have been presented on both CMOS and BiCMOS OEICs (John and Das 1994; Dutta et al. 1997; Brandl et al. 2014; Youn et al. 2015) and a common theme among these works is consideration of the system frequency response resulting from the individual subsystem. This chapter focuses on the subsystems of the receiver and specifically addresses the limitations and considerations to avoid any bottleneck decreasing overall system performance. The primary aim is to identify and isolate the intrinsic and external elements that influence the frequency response of the OEIC and/or hybrid approaches. By identifying factors on

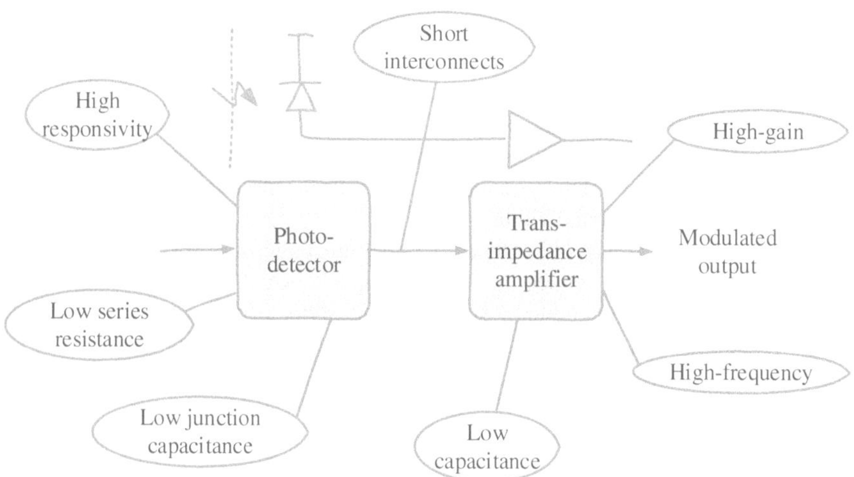

Fig. 6.1 A typical laser optoelectronic communication receiver showing the primary requirements for the connection between the photodetector and transimpedance amplifier

component level that influence system performance, improved strategies can be implemented when integrating these optoelectronic systems for space applications and when using them for military applications, which require high-performing, modular, small and low-noise components. The first component discussed in terms of its physical characteristics at fabrication level, which determine device bandwidth, is that of the photodetector.

6.2 Photodetector Bandwidth

If a fast response time and high bandwidth are required from a photodiode as opposed to only high sensitivity to incoming photons, it should be operated in photoconductive mode. The trade-off in photosensitivity is made up by the transimpedance amplifier gain, although this places additional design restrictions on stray capacitances of the amplifier. In photoconductive mode, a reverse bias is applied across the anode and the cathode of the photodiode to form a depletion region with specific width between the *p* and *n* type semiconductor materials and also the intrinsic layer in the case of PIN photodiodes. This mode reduces the photodiode junction capacitance by increasing the depletion region width of the diode (an inverse relationship) and also reduces the slow-rise-time diffusion tail, which can become a limiting factor in longer wavelength operation. The junction capacitance of the photodiode should be kept as low as possible to realize a high bandwidth and therefore fast response times to incoming light pulses. The junction capacitance C_J is defined in Chap. 4 and reiterated here. It follows that the junction capacitance is determined by

$$C_J = \frac{\varepsilon_r \varepsilon_0 A}{W_d} \tag{6.1}$$

where ε_r is the dielectric constant of the semiconductor material, ε_0 is the permittivity of free space, A is the diffused area of the junction (active area) and W_d is the depletion region width. This equation is used as the primary equation in determining the photodiode bandwidth.

The depletion region width and active area are therefore prospective variables that can be optimized to reduce the junction capacitance. The dielectric constant is determined by the semiconductor and the permittivity of free space is a constant value. Assuming that the active area of the photodiode is predetermined and kept constant, the remaining variable to consider is the depletion-region width, which is determined by

$$W_d = \sqrt{2\varepsilon_r \varepsilon_0 \mu \rho (V_A + V_{bi})} \tag{6.2}$$

where μ is the carrier mobility, ρ is the resistivity of the semiconductor material, V_A is the applied reverse-biased voltage and V_{bi} is the built-in voltage of the *pn*-

junction. Another form of (6.2), which highlights the effect of the donor doping concentration on the depletion region width, is given by

$$W_d = \sqrt{\frac{2\varepsilon_r\varepsilon_0(V_A + V_{bi})}{qN_D}} \tag{6.3}$$

where q is the elementary electron charge and N_D is the donor concentration of an ideally lightly doped n-type substrate. From (6.2) to (6.3) it can be seen that to increase the depletion region width externally and therefore decrease the junction capacitance without any modifications to the manufacturing process, the applied reverse-bias can be increased. If the required depletion region width is known, based on the responsivity requirements of the photodiode, then (6.3) can be rewritten to determine the required reverse-bias such that

$$V_A = \left(\frac{qN_D W_d^2}{2\varepsilon_r\varepsilon_0} - V_{bi}\right) \tag{6.4}$$

which is generally used to determine the system-level reverse-bias of the integrated photodiode. If, however, the value of the junction capacitance causes the amplifier circuit to oscillate, an increase in the reverse-bias voltage might not be adequate to overcome this unwanted operation by decreasing the junction capacitance and alternative techniques must be considered. Operating in photoconductive mode does, however, increase the dark current and introduces $1/f$ noise to the system, but these are generally kept as low as possible for each device through advanced manufacturing techniques and high-quality epitaxial materials on wafer level. If the depletion region width is large relative to the wafer thickness, additional capacitance is parasitically introduced in the form of a sidewall capacitance C_{SW}. The sidewall effect is represented in Fig. 6.2.

The depletion region perimeter only needs to be taken into account if the ratio of the perimeter length P multiplied by the depletion region width, therefore the surface area of the depletion region, exceeds 10 % of the parallel-plane area of the active area. If accounting for the perimeter and the parallel-plane area, the required reverse-biased voltage to achieve a depletion region width W_d changes to

$$V_A = \left(\frac{qN_D W_d^2}{2\varepsilon_r\varepsilon_0} - V_{bi}\right)\left(1 + \frac{P}{A}\frac{\pi}{4}W_d\right) \tag{6.5}$$

which illustrates the requirement of the reverse-biased voltage of the photodiode as seen in (6.4), as well as the effect of the sidewall capacitance. The right-hand side term in brackets in (6.5) is a positive, non-zero value and therefore the required reverse-biased voltage to achieve a specified depletion region width increases because of this sidewall effect. In addition, the wafer thickness and doping density affect the responsivity of the photodiode and its series resistance leads to a decrease in bandwidth. To determine the effects of the photodiode series resistance, the shunt

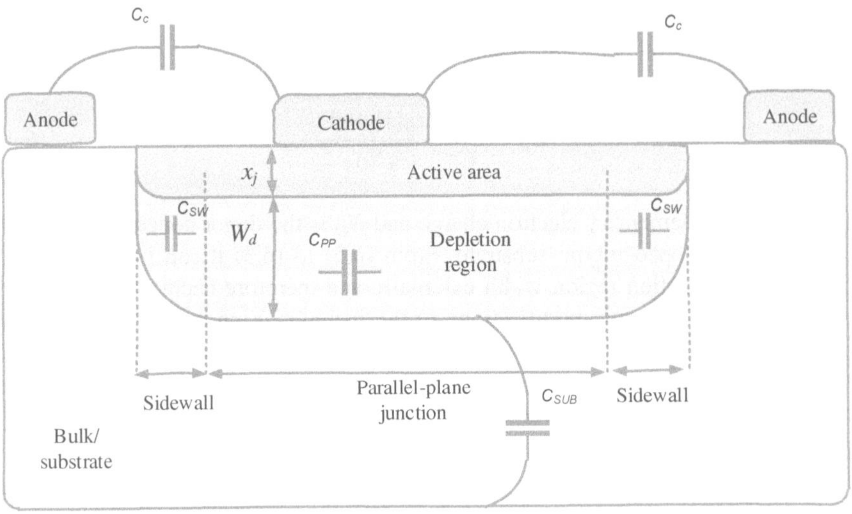

Fig. 6.2 Lateral layer structure of photodetector device showing the parasitic capacitances

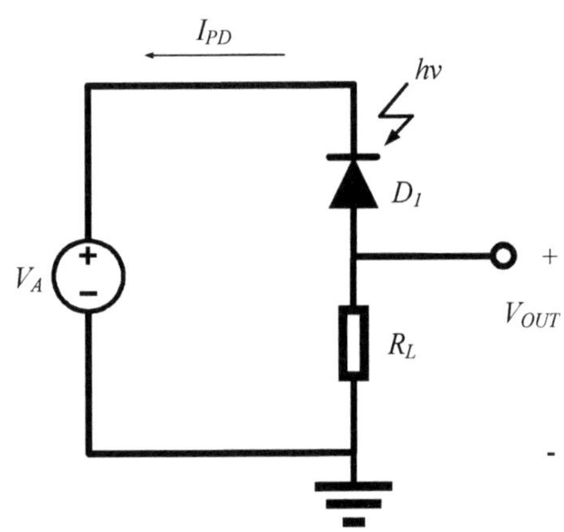

Fig. 6.3 Simplified photodiode reverse-biased representation when operated in photoconductive mode

resistance and the capacitance on the response and bandwidth of the device, a simplified circuit configuration to measure the photon-generated current in the photodiode is presented in Fig. 6.3.

As seen in Fig. 6.3, the reverse-biased voltage V_A is applied to the photodiode and the output is measured across the load resistor R_L, typically a 50 Ω load, and if light at a specified energy $E = hv$ is incident on the photodetector, a current I_{PD} is generated flowing through R_L. The output is then measured across the terminals of R_L. The circuit configuration as seen in Fig. 6.3 can be used to develop the equivalent circuit of the photodiode further, similar to the circuit configuration in Chap. 4 used to identify possible causes of oscillations in the transimpedance

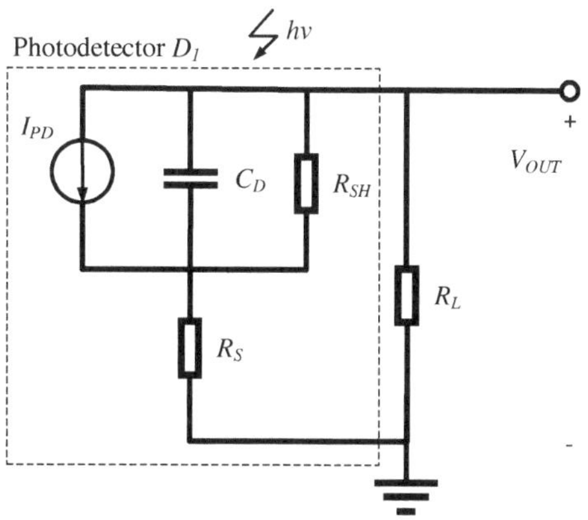

Fig. 6.4 Simplified equivalent circuit of a photodiode representing its series and shunt resistance and total capacitance

amplifier due to unaccounted parasitic effects at high frequencies. This configuration, which also shows the position of the 50 Ω load resistor, is used to quantify the bandwidth of the photodiode and is given in Fig. 6.4. The equivalent circuit of the photodetector is highlighted by the dashed-line box in Fig. 6.4.

In Fig. 6.4, I_{PD} is the current flowing through the ideal photodiode, C_D represents the total intrinsic capacitance of the photodiode, which includes the depletion region capacitance, sidewall effects, bonding pads, packaging and the external routing capacitances, R_{SH} is its shunt resistance (ideally a very high value in the high MΩ range and also called the device's R-zero or 'R_0') and R_S its series resistance (ideally in the low Ω range). To determine the total photodiode capacitance, the sidewall effect and parallel-plane capacitance C_{PP} should be taken into account such that

$$C_D = C_{SW} + C_{PP} \tag{6.6}$$

which is determined by

$$C_D = \frac{\varepsilon_r \varepsilon_0 A}{W_d} + \frac{2\pi\varepsilon_r\varepsilon_0 P}{4\ln\left(\frac{W_d}{x_j}\right)} \tag{6.7}$$

where W_d is constant if the device is fully depleted and equal to the junction depth subtracted from the wafer thickness. It is evident that if the reverse-biased voltage is increased and the depletion region increases, the sidewall capacitance and parallel-plane capacitance will also increase. The shunt resistance and series resistance of a photodiode are among the critical performance parameters and can be determined practically through *I-V* measurements. Figure 6.5 shows the typical measurement results of a photodiode *I-V* measurement.

As seen in Fig. 6.5, the current flowing through the photodiode when reverse-biased increases if photons are incident on the device, shown in Fig. 6.5 by the *Increasing light* arrow. In addition, the open-circuit voltage V_{OC} (knee-voltage)

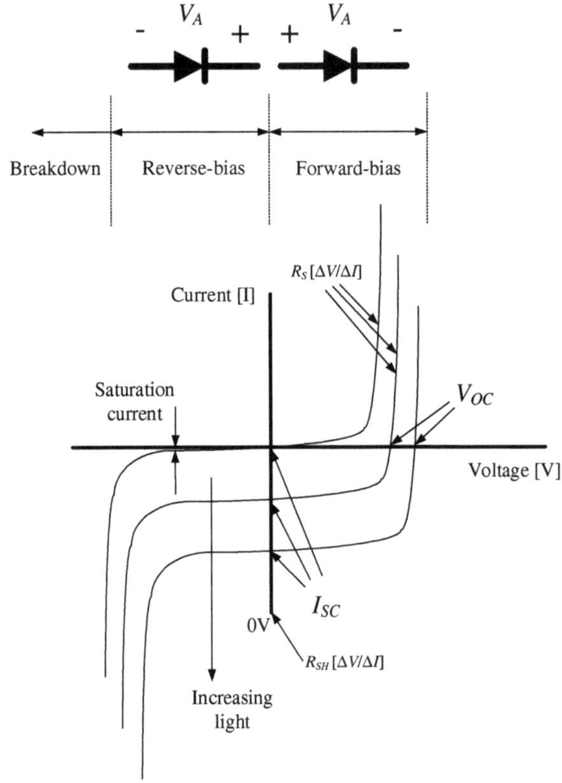

Fig. 6.5 Typical measurement setup and expected result of a photodiode *I-V* measurement to determine the series and shunt resistance

in forward-biased operation increases with an increase in incident photons. The series resistance of the photodiode can be calculated by the slope of the *I-V* curve where the device is forward-biased and conducting. This series resistance should be as low as possible to ensure that there is little resistance to current flow when the device is conducting. *R*-zero is the resistance of the photodiode when no bias voltage is applied to the circuit, therefore at the short-circuit current I_{SC} and typically determined around ± 10 mV. This resistance can be determined from the *I-V* characteristic as the slope around ±10 mV and mathematically defined as

$$R_0 = \frac{(V_0 + \Delta V) - (V_0 - \Delta V)}{I_1 - I_2} \tag{6.8}$$

where ΔV should be as small as possible, depending on the resolution of the measurement equipment, V_0 = 0 V and I_1 and I_2 are the currents measured at these two voltages. The photodiode current I_{PD} follows a typical diode characteristic and is determined mathematically by

$$I_{PD} = I_S\left(e^{\frac{qV_A}{nkT}} - 1\right) \tag{6.9}$$

where q is the electron charge, n is an ideality constant between 1 and 2, depending on the current flowing through the device, k is Boltzmann's constant and T is the operating temperature. I_S is the saturation current of a diode given by

$$I_S = \frac{qAD_n n_i^2}{N_A \omega_{anode}} + \frac{qAD_p n_i^2}{N_D \omega_{cathode}} \tag{6.10}$$

where A is the active area, D_n and D_p are the electron and hole-diffusion coefficients respectively, n_i is the intrinsic carrier concentration of the semiconductor, N_A and N_D are the electron and hole concentrations and ω_{anode} and $\omega_{cathode}$ are the positive p^+ and negative n^- depletion-region boundaries. Referring back to the equivalent circuit model of the photodiode shown in Fig. 6.4, the equivalent resistance of the photodiode is calculated as

$$R_{EQ} = \frac{R_{SH}(R_S + R_L)}{R_{SH} + R_S + R_L} \tag{6.11}$$

and since R_{SH} is generally large and R_S small, (6.11) can be simplified to

$$R_{EQ} = R_S + R_L \tag{6.12}$$

where the bulk of the substrate from the edge of the depletion region towards the thickness of the wafer T_W adds to the series resistance of the device. Only if the device is fully depleted will the substrate resistance be near-zero. The total series resistance of the device can therefore be determined by

$$R_S = R_L + \frac{\rho}{A}\left(T_W - W_d - x_j\right) \tag{6.13}$$

where x_j is the junction depth. The equivalent resistance of the photodiode and the effective total photodiode capacitance determines the RC time constant and therefore the response speed/rise time/bandwidth of the photodiode. The rise time of the device is the time it takes for the device to respond to in incoming photon-generated pulse. If the frequency of the incident light pulse is higher than the operating capabilities of the device, the system is unable to process incoming information efficiently and a slower incident pulse is recommended. The response speed t_R at which an output signal rises from 10 to 90 % of its peak value is determined by

$$t_R = 2.2C_D(R_L + R_S) \tag{6.14}$$

where R_S is generally much smaller compared to R_L Since R_L is an external factor and R_S can be minimized only by the physical dimensions, doping concentration and semiconductor characteristics of the photodiode during processing, the typical approach to minimize the response time of the device is to minimize the capacitance C_D. The capacitance is inversely proportional to the square root of the reverse-bias voltage and directly proportional to the active area A, such that

$$C_D \propto \frac{A}{\sqrt{V_A}} \tag{6.15}$$

which summarizes the internal and external parameters that can be varied to increase the response time (at the expense of responsivity and sensitivity). Another parameter that is often used to define the operating frequency capabilities of the photodiode is the cut-off frequency, related to the response time by

$$f_C = \frac{0.35}{t_R} \tag{6.16}$$

where f_C represents the cut-off frequency. The photodiode can also be cooled down to operate at high frequencies while depressing the negative impact on the responsivity. If the depletion region width is large, the total response time of the photodiode is determined by the rise time as well as the (typically longer) transit time t_T, which is the time a carrier takes to drift through the depletion region. The transit time is determined by

$$t_T = \frac{W_d^2}{\mu V_A} \tag{6.17}$$

and is therefore proportional to the square width of the depletion region and inversely proportional to the applied reverse-biased voltage and the carrier mobility. The total response time t_{total} is given as

$$t_{total} = \sqrt{t_R^2 + t_T^2} \tag{6.18}$$

and can be dominated by the transit time. These derivations therefore indicate that there are several external and internal factors of the photodiode that influence its frequency performance. During implementation where these photodiodes are used with supporting circuitry and additional subsystems such as the transimpedance amplifier, careful consideration of these parameters will ensure that the photodetector does not limit the overall bandwidth of the system. Since the transimpedance amplifier is typically the following stage in the optoelectronic receiver, whether integrated on the same substrate or an externally connected hybrid implementation, the output of the photodiode in terms of its capacitance and resistance influences the choice and performance requirements of the transimpedance amplifier. These bandwidth requirements and the crucial circuit elements are discussed in the following section.

6.3 Transimpedance Amplifier Bandwidth

Multiple sources of performance limitation on high-frequency amplifiers exist, which can be identified and analyzed when isolating the amplifier circuit to its individual components and their respective input and output connections. From an amplifier perspective, the main high-frequency gain performance limitations of the system are determined by internal capacitances of the active devices due to charge storage effects at and close to the *pn*-junctions and parasitic effects from external components, packaging, lead inductances and parasitic resistances. This section focuses on the internal capacitances in the bipolar transistor used as the primary active device to identify the performance characteristics of transimpedance amplifiers.

The dominant capacitances generated at the *pn*-junction of the bipolar transistor can essentially be separated into the junction capacitance related to the space-charge in the depletion region and the diffusion capacitance, also referred to as charge-storage capacitance. The junction capacitance involves the width of the depletion region resulting from the applied voltage and whether the junction is reverse-biased or forward-biased. The speed of the bipolar transistor (and also of the HBT) is a strong function of its base-transit time (Biswas et al. 2014), a function of the physical parameters of the transistor. In Biswas et al. (2014) it is presented that the base-transit time τ_B of an HBT is determined by

$$\tau_B = q\int_0^{W_B} \frac{n(x)}{J_n(x)}dx \tag{6.19}$$

where $n(x)$ is the minority carrier concentration, which can be found using transport and electric field calculations given in Yuan (1994), and $J_n(x)$ is the electron current density within the base region. To analyze the frequency response of the transistor, consider as an example the bipolar *npn*-transistor junctions shown in Fig. 6.6.

As shown in Fig. 6.6, the transistor junctions between the *n*-type and *p*-type region consist of inherent intrinsic capacitances due to the width of the depletion region and as a result of diffusion capacitance. The capacitance between the collector and the base of the transistor is denoted C_μ, whereas the junction capacitance between the base and the emitter and the diffusion capacitance in this same region are denoted C_{je} and C_{de} respectively. The total base-emitter capacitance is the sum of these two capacitances and is denoted C_π. It can be shown that the unity-gain bandwidth f_T of the bipolar transistor, which is defined as the frequency where the short-circuit current gain $h_{fe|sc}$ is unity, is given by

$$h_{fe|sc} = \frac{I_c}{I_b} = 1 \tag{6.20}$$

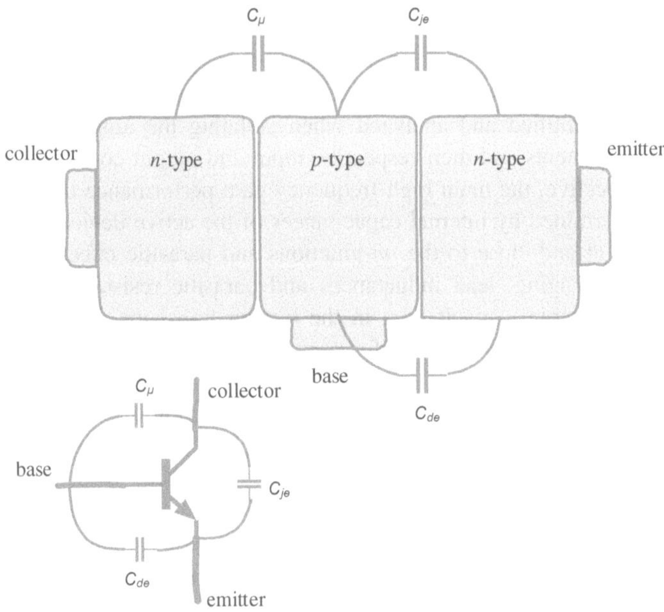

Fig. 6.6 Cross-section of a bipolar *npn*-transistor showing the junction capacitances between the *n*-type and *p*-type semiconductor materials used as the base, emitter and collector of the transistor. The circuit schematic and intrinsic capacitances are also shown

where I_c and I_b are the collector and base currents respectively. It is shown in Gray et al. (2001) that the current gain of the bipolar transistor is given by

$$h_{fe} = \frac{g_m r_\pi}{1 + j\omega r_\pi (C_\pi + C_\mu)} \tag{6.21}$$

where g_m is the transconductance of the transistor, r_π is the base-resistance and $\omega = 2\pi f$ is the operating frequency. Note in (6.21) that the base-emitter and base-collector intrinsic capacitances have a significant influence on the current gain. The unity-gain frequency, used to determine the GBP of the device, can therefore be calculated by

$$f_T = \frac{g_m}{2\pi (C_\pi + C_\mu)} \tag{6.22}$$

which shows the relationship between the frequency performance of the device and the transconductance, base-emitter and base-collector intrinsic capacitances of the transistor. If the unity-gain frequency of the transistor is specified on a datasheet or in process documents, as is often the case, it is possible to calculate the combination of (C_π + C_μ) given a specified bias current, since

$$g_m = \frac{qI_C}{kT} \tag{6.23}$$

where q is the elementary electron-charge, k is Boltzmann's constant and T is the temperature of operation in Kelvin. The combination of kT/q is also known as the thermal voltage, approximately 25 mV at room temperature (300 K). By using these results and the discussion on the intrinsic capacitances within the bipolar transistor, the frequency response limitations of the transimpedance amplifier can be reviewed. A transimpedance amplifier utilizes an op-amp with a feedback resistor between the output and the inverting input. An op-amp is essentially a differential amplifier with ideally infinite voltage gain, ideally infinite input impedance and ideally zero output impedance (Karanicolas et al. 1991). A differential amplifier is practically two symmetrical common-emitter amplifiers sharing a common emitter-resistor. Although there are differences in the input impedance and frequency response of the common-emitter amplifier and a differential amplifier, the inherent limitations of the transistor determine the overall performance of the amplifier.

The common-emitter amplifier stage is presented in Fig. 6.7, along with the required DC blocking capacitors, bypass capacitors, biasing resistors, gain-control resistor and output resistor.

Figure 6.7 represents the common-emitter amplifier configuration typically used as the enabling half-circuit in operational amplifiers that use a differential configuration of two common-emitter amplifiers sharing a common-emitter resistor. The input signal to be amplified is represented by the voltage source, V_{IN}. The impedance of the voltage source is represented by R_S (not to be confused with the series resistance of the photodetector) and is typically a 50 Ω resistive element within the source. R_B represents the parallel combination of the biasing resistors R_1 and R_2 ($R_B = R_1 \| R_2$) across the base terminal and the positive supply and across the base terminal and ground, respectively. Any DC offset in the input signal is removed by the blocking capacitor C_{C1} to ensure that the transistor T_1 is not biased by an unknown DC value that can vary depending on the input signal. The resistor R_C at

Fig. 6.7 Differential emitter-coupled amplifier with resistive biasing half-circuits

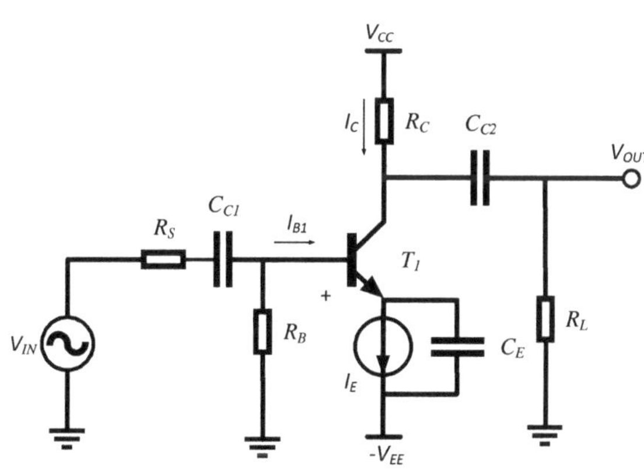

the collector terminal of the transistor determines the current flow through the transistor and therefore also determines the voltage gain A_V of the amplifier, since

$$A_V = \frac{V_{OUT}}{V_{IN}} = -g_m R_C \tag{6.24}$$

if no emitter resistor (emitter degeneration) is present and

$$A_V = -\frac{\beta R_C}{r_\pi + (\beta + 1) R_E} \tag{6.25}$$

if emitter degeneration is used (which is the case for differential amplifiers). Note from (6.25) that if $R_E = 0$, then the voltage gain becomes

$$A_V = -\frac{\beta R_C}{r_\pi} \tag{6.26}$$

where

$$g_m = \frac{\beta}{r_\pi} \tag{6.27}$$

and therefore (6.24) holds true. Another DC blocking capacitor is placed between the collector terminal and the output to remove the DC offset appearing as a result of the supply voltage. The output voltage V_{OUT} is therefore an inverted and amplified representation of the input signal centered at 0 V. The bypassing capacitor C_E ensures that any high-frequency components at the emitter of the transistor are shorted towards ground and do not appear across the base-emitter junction of the transistor. The current source I_E represents the current flowing between the emitter and ground, generated as a result of the resistance of R_C (and R_E if present).

The goal of this section is to analyze the frequency response of the common emitter amplifier to establish the effects of the transistor on the system. Comprehensive reference material for this analysis is found in Sedra and Smith (2004) and the adapted analysis applicable to this book is presented in this section. A typical frequency response of the common-emitter amplifier is shown in Fig. 6.8. This frequency response shows the voltage gain of the amplifier at low, medium and high frequencies and the analysis of each frequency band is described in the following sections, first with a brief analysis/summary of the overall response presented after Fig. 6.8.

As seen in Fig. 6.8, the common-emitter amplifier gain A_V in dB rises at frequencies up to f_L, called the low frequency of the common-emitter amplifier. This effect can also be described as the gain falling off as the frequency of operation lowers and approaches a DC value (0 Hz). The primary contributors to the lowering gain are the DC blocking capacitors and the bypass capacitors (C_{C1}, C_{C2} and C_E) in

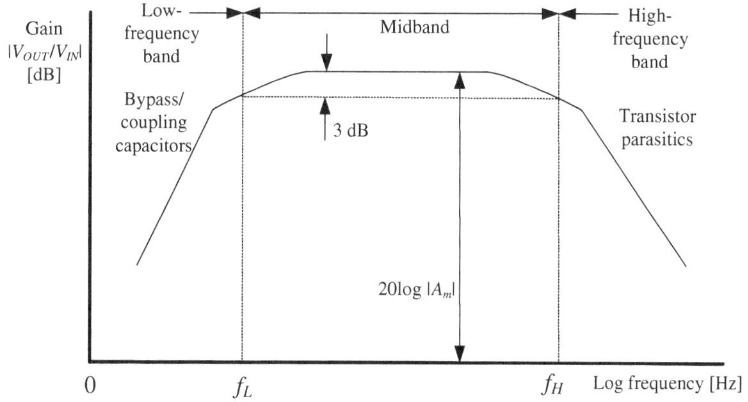

Fig. 6.8 Common-emitter configuration frequency response showing the low-frequency band, mid-band and high-frequency band effects

Fig. 6.7. This effect is therefore not attributed to the characteristics of the transistor T_1 and the choice of transistor would not make a significant variance in the low-frequency operation of the circuit.

For the mid-frequency band, the gain is ideally constant with increasing frequency, determined by the small-signal current gain β (or h_{fe}) of the transistor, its parasitic series resistance r_π, the collector resistor R_C and the emitter resistor R_E, as shown in (6.24) and in (6.25) if emitter degeneration is used. For this analysis though, R_E is initially omitted since mainly the transistor is evaluated for its frequency response effects on the overall system. DC blocking capacitors C_{C1} and C_{C2} are assumed to have very small impedances in this mid-frequency band and can essentially be replaced by short circuits. Recall that the complex impedance of a capacitor Z_C is given by

$$Z_C = \frac{1}{j\omega C} \tag{6.28}$$

where ω is the operating frequency in rad/s ($\omega = 2\pi f$) and C is the capacitance in farad. From (6.28) it can be seen that as the frequency of operation increases, the impedance (resistance to current flow) of the capacitor decreases. At a certain point, depending on the capacitance, the capacitor can be approximated as a short circuit. In addition, in the mid-frequency band of operation, the impedances of the intrinsic base-emitter capacitance C_π and that of the collector-base capacitance C_μ in Fig. 6.6 are large and are replaced by open circuits. These capacitances are very small (a few fF to pF) compared to the external DC blocking and bypassing capacitors (nF to μF). The placement and effects of these intrinsic capacitors are described in further detail in the mid-frequency and high-frequency operations of the common-emitter amplifier below.

For high-frequency operation, the blocking and bypassing capacitors C_{C1}, C_{C2} and C_E are still approximated at short circuits, but now the intrinsic base-emitter and collector-base capacitances cannot be approximated as open circuits and must be accounted for in the frequency analysis. The point at which the common-emitter gain starts falling off at high frequencies, therefore above f_H, is dependent on the transistor's intrinsic capacitances and resistances, as well as the Miller capacitance discussed below, and the choice of transistor size and technology heavily influences the frequency response of the circuit. This is where technologies such as BiCMOS HBTs processes using SiGe hetero junctions start to show significant dominance over purely CMOS technologies (Voinigescu et al. 2005) as well as over technologies such as GaAs. The low, medium and high-frequency operation of the common-emitter amplifier is described in further detail below to highlight the impact of the transistor on the overall performance of this circuit.

6.3.1 Low-Frequency Operation

It has been proposed in the above discussion that the DC blocking capacitors and the bypass capacitors are the primary contributors to the gain falling off as the operation frequency decreases and approaches DC. From Fig. 6.8 is appears as if the frequency response is affected by a single time constant, which determines the point at which the gain starts decreasing significantly. In actual fact, there is a more complex mechanism at work in this frequency band. The frequency response is affected by the cascaded product of all three high-pass time constant circuits resulting from C_{C1}, C_{C2} and C_E. An approximated solution of the low-frequency response is given by

$$A_V(\omega) \approx A_m\left(\frac{j\omega}{j\omega+\omega_{p1}}\right)\left(\frac{j\omega}{j\omega+\omega_{p2}}\right)\left(\frac{j\omega}{j\omega+\omega_{p3}}\right) \tag{6.29}$$

where A_m represents the mid-band voltage gain and ω_{p1}, ω_{p2} and ω_{p3} are the three poles resulting from the three individual capacitors. If the frequency response shown in Fig. 6.8 is zoomed in at the low-frequency point f_L, it is seen that this point is in fact not a single point but a more complex response of various poles decreasing the voltage gain. This zoomed-in effect is shown in Fig. 6.9.

From Fig. 6.9 it can be seen that the initial frequency where the gain starts falling off (moving from a high to a low frequency on the graph) is at f_{p2} (depending on the value in farad of the capacitor). At this point, towards lower frequencies, the gain decreases at 6 dB/octave up to the point where f_{p1} starts influencing the circuit, again depending on the value of the capacitor. From this point moving towards DC the gain falls off at 12 dB/octave where it finally reaches the frequency f_{p3}, where the third capacitor is decreasing the low-frequency gain at 18 dB/octave. From this

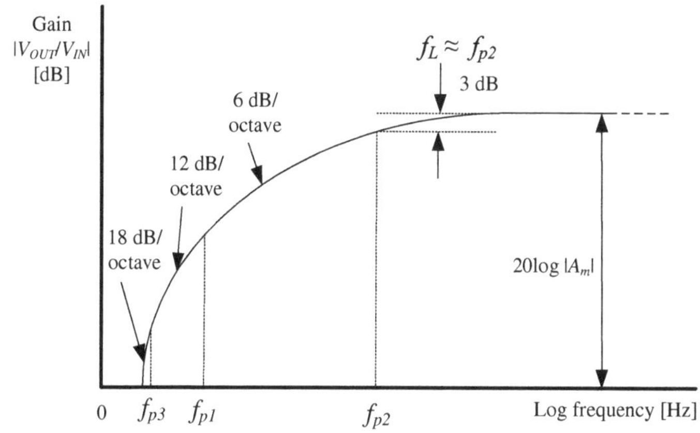

Fig. 6.9 Common-emitter configuration low-frequency response showing the effects of the blocking and bypass capacitors

analysis it should be noted that the choice of blocking and bypass capacitors are crucial for low-frequency gain control (high-pass filtering); however, if the circuit is only operated at higher frequencies, the choice of these capacitors becomes less significant for circuit performance. In order to determine an approximate solution for the low-frequency point f_L in Fig. 6.8, the cascaded configuration of blocking and bypass capacitors produces the following solution: a summation of the three pole-frequencies,

$$f_L \approx f_{p1} + f_{p2} + f_{p3} \tag{6.30}$$

which can be expanded mathematically and expressed by

$$f_L \approx \frac{1}{2\pi}\left(\frac{1}{C_{c1}R_{C1}} + \frac{1}{C_E R_E} + \frac{1}{C_{C2}R_{C2}}\right) \tag{6.31}$$

where R_{C1}, R_E and R_{C2} are the resistances seen by C_{C1}, C_E and C_{C2} respectively when the input signal is zero. From this result it is evident that the low-frequency voltage gain is mainly dependent on the external (larger) capacitors. The transistor limitations start becoming significant when moving out of the low-frequency domain towards higher frequencies (relative to the maximum capabilities of the transistor). The mid-band voltage gain A_m seen in (6.29) depends on various intrinsic characteristics of the transistor T_1 and is discussed in the following section. To introduce the mid-frequency response of the common-emitter amplifier, it is first necessary to consider the small-signal equivalent circuit of the bipolar transistor.

6.3.2 Bipolar Transistor Small-Signal Equivalent Circuit

To analyze the mid-frequency and high-frequency operation and performance of the bipolar transistor, it is recommended to use the small-signal equivalent circuit analysis to understand the underlying parasitic characteristics of the frequency operation. For this reason, the standard small-signal equivalent circuit is briefly presented in this section to serve as a reference for further discussions. Several assumptions regarding the parasitic components can be made for higher frequency operation and it is necessary to compare these simplified models with the standard equivalent circuit. A detailed description of the analysis to define each of these parasitic components is given by Gray et al. (2001). The small-signal equivalent circuit for the bipolar transistor adapted from Gray et al. (2001) is presented in Fig. 6.10.

Primarily, the technological process and manufacturing limitations—therefore using intrinsically non-ideal materials—determine and define the parasitic elements present in any semiconductor component. This is true not only for bipolar transistors but also for MOS transistors and diodes, as well as passive devices such as resistors, inductors and capacitors. In Fig. 6.10 the bipolar transistor small-signal equivalent circuit is shown. Refraining from presenting the precise origin of each of these parasitic elements, the parameters are listed here.

Significant base and collector resistances r_b and r_c appear in series with the base and collector contacts respectively. At high bias currents the resistance of the emitter lead, r_{ex}, also becomes a significant parasitic resistive element. Typical values of r_b, r_{ex}, and r_c are 50–500, 1–3 and 20–500 Ω respectively (Gray et al. 2001). The value of r_b varies significantly with the bias current. Since an increase in ΔV_{CE} results in a decrease in ΔI_B, this effect can be modelled by the inclusion of r_μ from the collector to the base of the model. The value of r_μ can be approximated by

$$r_\mu = \beta r_o \tag{6.32}$$

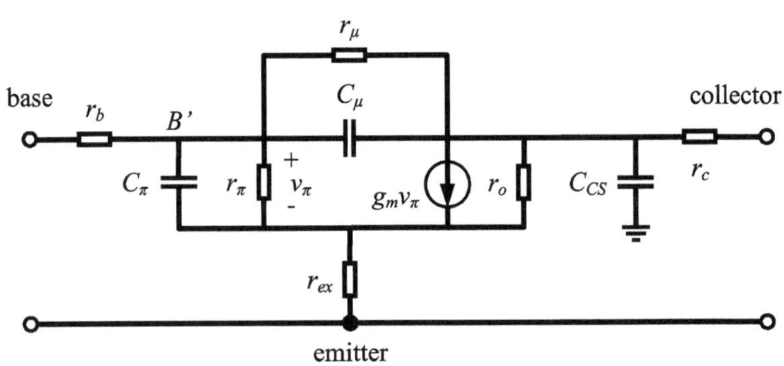

Fig. 6.10 Small-signal equivalent circuit for the bipolar transistor

where β is the small-signal transistor current gain (again, it can be represented by h_{fe} also) and r_o is its intrinsic output impedance due to the Early voltage and the collector-current.

Several voltage-dependent capacitances are also present in the bipolar transistor because of fabrication limitations. These parasitic capacitances include the base-emitter junction capacitance C_π and the base-collector junction capacitance C_μ. In addition, a parasitic capacitance between the collector and the substrate C_{cs} is also found in *npn* transistors, which is replaced by a base-to-substrate capacitance C_{bs} in *pnp* devices. Since the substrate is always connected to the most negative supply, all substrate connections are attached to AC ground in the small-signal equivalent circuit. The base-emitter capacitance contains the base-charging capacitance C_b and the emitter-base depletion layer capacitance C_{je} such that

$$C_\pi = C_b + C_{je} \tag{6.33}$$

where typical process values of C_{je} at zero bias are around 10 fF (Gray et al. 2001). The voltage-dependent current source with value $g_m V_\pi$ represents the internal gain of the transistor, where the voltage across the base-emitter junction is amplified by the transconductance, g_m.

With the small-signal equivalent model of the bipolar transistor defined, the mid and high frequency operation of the common-emitter amplifier is presented in the following sections.

6.3.3 *Mid-frequency Operation*

It is in the mid-band frequency response where the choice of transistor technology starts becoming a limiting factor of the circuit performance and this section discusses operation in this frequency band. In the mid-frequency spectrum, the blocking and bypassing capacitors have a very small electrical resistance to current flow, since these capacitances are relatively high. It is therefore feasible to replace these capacitors as short-circuits in the small-signal AC analysis. The intrinsic base-emitter and base-collector capacitor impedances are still very large in the mid-frequency spectrum and can be replaced by open circuits in the small-signal AC analysis. The resultant equivalent small-signal circuit model for the common-emitter amplifier operating at mid-band frequencies is given in Fig. 6.11.

In Fig. 6.11 the base-collector and base-emitter capacitances C_μ and C_π are replaced by open circuits, therefore not present in the analysis and obstructing the flow of charge through these junctions. The small-signal equivalent circuit in Fig. 6.11 includes the external components V_{IN}, R_S, R_B, R_C and R_L, the intrinsic elements r_x, r_π, r_o and the intrinsic transconductance amplification $g_m V_\pi$. The resistance r_x represents the base spreading resistance (which can also be denoted as r_{bb}), which is the bulk resistance between the base contact and the active region of

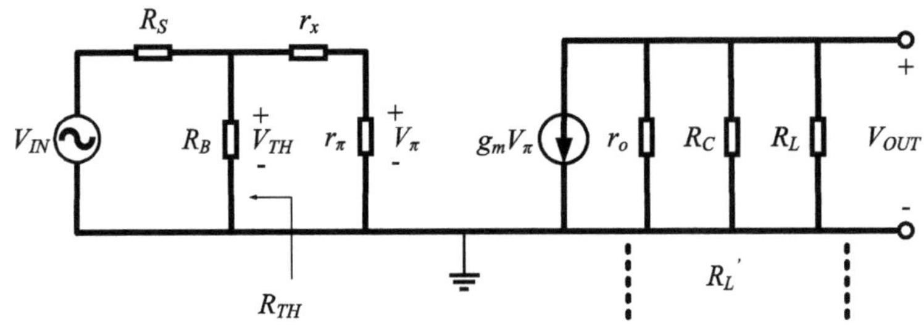

Fig. 6.11 Small-signal equivalent circuit for the common-emitter amplifier operating in the mid-band frequencies

the base under the emitter (Gray et al. 2001). The total output impedance $R_L^{'}$ of this circuit can be defined by the combination of r_o, R_C and R_L such that

$$R_L^{'} = r_o||R_C||R_L \tag{6.34}$$

and the output voltage V_{OUT} can thus be determined by the current flowing through the output impedance as

$$V_{OUT} = -g_m V_\pi R_L^{'} \tag{6.35}$$

where the negative sign indicates an inverted and amplified output signal when referenced to the input signal. By applying Thévenin's theorem for the intrinsic resistances of the equivalent circuit, the voltage V_π can be defined as

$$V_\pi = \frac{r_\pi}{r_\pi + r_x + R_{TH}} V_{TH} \tag{6.36}$$

where R_{TH} is the input resistance as seen from R_B and V_{TH} is the Thévenin equivalent voltage, which appears across R_B. Rewriting (6.36) by replacing R_{TH} and V_{TH}, it can be seen that

$$V_\pi = \frac{r_\pi}{r_\pi + r_x + R_B||R_S} \frac{R_B}{R_B + R_S} V_{IN} \tag{6.37}$$

which can be used to determine the mid-band voltage gain A_m. The mid-band voltage gain resulting from equating V_{OUT}/V_{IN} and using (6.34)–(6.37) is therefore given as

$$A_m = \frac{V_{OUT}}{V_{IN}} = \frac{-g_m r_\pi}{r_\pi + r_x + R_B||R_S} \frac{R_B}{R_B + R_S} (r_o||R_C||R_L) \tag{6.38}$$

which shows the relationships between the parasitic transistor elements and the external components in the common-emitter amplifier. Analyzing (6.38), it can be seen that the mid-band voltage gain is:

- increased with a higher transconductance of the bipolar transistor,
- increased with a higher output impedance of the circuit where R_C and R_L can be externally varied, and
- influenced by r_π, r_x, R_B and R_S in varying degrees. In order to determine the effect of these components on the mid-band voltage gain, consider Figs. 6.12 and 6.13, which vary r_π and r_x (the external source resistance R_S is held constant at 50 Ω and the external biasing equivalent resistor R_B is held constant at 1 kΩ).

The base-emitter resistance r_π is varied arbitrarily from 500 to 10,000 Ω and the resulting mid-band voltage gain is shown in Fig. 6.12.

From Fig. 6.12 it can be seen that with a varied base-emitter resistance r_π, the mid-band voltage gain increases (a decrease in the negative voltage gain) as the resistance of r_π increases. Also observed in Fig. 6.12 is that the mid-band voltage gain reaches an asymptote (limit) with increasing r_π, in this particular case at −16 V/V. A similar analysis can be done by varying the value of the base-spreading intrinsic resistance r_x, as shown in Fig. 6.13.

In Fig. 6.13 it is seen that an increasing base-spreading resistance r_x decreases the mid-band voltage gain (an increasing negative value) of the common-emitter amplifier. The decrease in the gain also slows down with higher r_x, but with very high values of r_x, the gain can theoretically reach 0 V/V (0 dB).

At high frequencies (relative to the transistor maximum operating frequency) a similar analysis of the intrinsic elements of the components, as well as the external components, can be done to determine the frequency response and the critical components that limit the performance of the common-emitter amplifier.

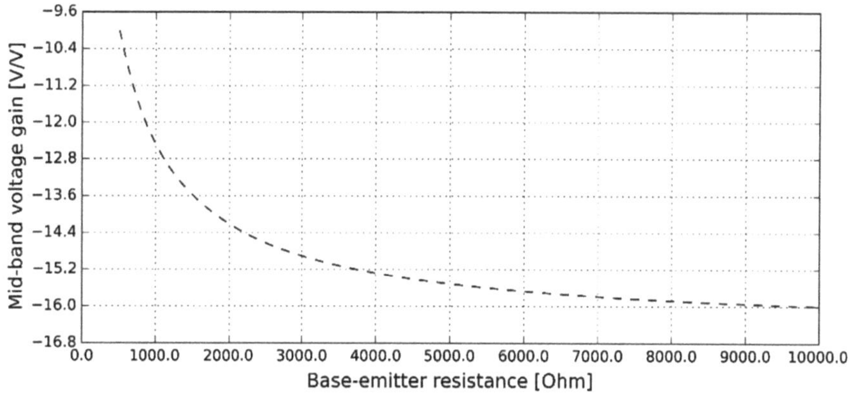

Fig. 6.12 The common-emitter mid-band voltage gain for varied base-emitter resistance r_π from 500 to 10,000 Ω

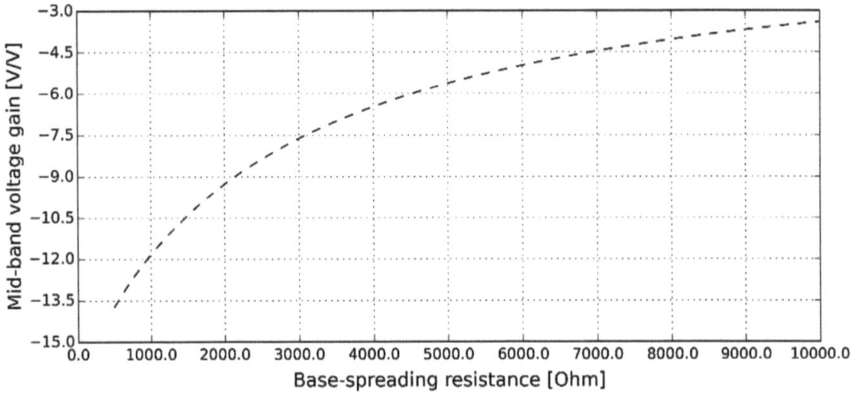

Fig. 6.13 The common-emitter mid-band voltage gain for varied base-spreading resistance r_x from 500 to 10,000 Ω

6.3.4 High-Frequency Operation

The crucial difference between mid-frequency operation and high-frequency operation is primarily the intrinsic capacitances C_π and C_μ, which cannot be assumed to be open circuits anymore and are included in the small-signal analysis at frequencies above f_H in Fig. 6.8. The small-signal equivalent circuit of the common-emitter amplifier using a bipolar transistor is shown in Fig. 6.14, similar to the standard small-signal equivalent circuit for the bipolar transistor shown in Fig. 6.10, but including the external components required to realize the common-emitter amplifier.

As seen in Fig. 6.14, the capacitances C_π and C_μ are now included across the base-emitter and base-collector junctions respectively. The input resistors R_S and R_B are again included, as well as the output resistors R_C and R_L, with all accompanying intrinsic resistors shown. To analyze this circuit, it is recommended to simplify the configuration by calculating the Thévenin equivalent circuit at the input and using the definition of $R_L^{'}$ as given in (6.34); this simplified circuit is shown in Fig. 6.15.

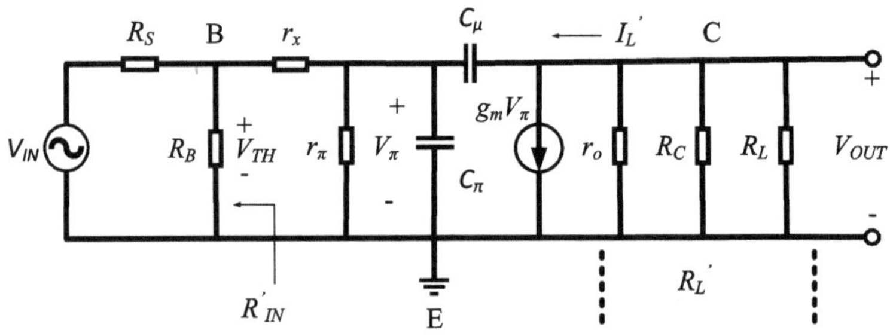

Fig. 6.14 Small-signal equivalent model for the high-frequency response of a common-emitter amplifier

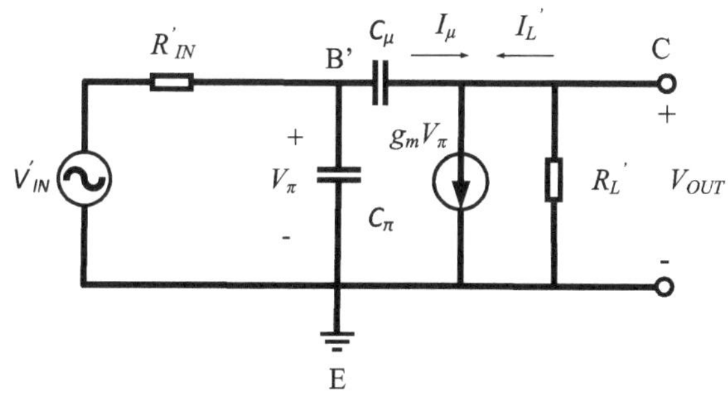

Fig. 6.15 Simplified Thévenin equivalent circuit of the small-signal equivalent model for the high-frequency response of a common-emitter amplifier

From Fig. 6.15 the equivalent input voltage when taking into account the Thévenin equivalent at the input and the equivalent output resistance $R_L^{'}$ can be defined by writing the equivalent input voltage $V_{IN}^{'}$ (similar to (6.37)) as

$$V_{IN}^{'} = \frac{r_\pi}{r_\pi + r_x + R_B||R_S} \frac{R_B}{R_B + R_S} V_{IN} \tag{6.39}$$

and by writing the equivalent input resistance $R_S^{'}$ as

$$R_S^{'} = r_\pi||[r_x + (R_B||R_S)] \tag{6.40}$$

which takes into account r_π and r_x. To analyze the frequency response, the current flow in Fig. 6.15 must also be defined as follows: following Kirchhoff's current loop at the output of Fig. 6.15, the current through the equivalent load resistance $I_L^{'}$, the current through the base-collector capacitor I_μ and through the voltage-dependent current amplifier $g_m V_\pi$ is given by

$$I_L^{'} + I_\mu = g_m V_\pi \tag{6.41}$$

which can be rewritten such that

$$I_L^{'} = g_m V_\pi - I_\mu \tag{6.42}$$

and for frequencies near f_H and higher, $I_\mu \ll g_m V_\pi$, and (6.42) can be simplified to

$$I_L^{'} \approx g_m V_\pi \tag{6.43}$$

and the output voltage V_{OUT} can therefore be defined as

$$V_{OUT} \approx I_L^{'} R_L^{'} \approx -g_m V_\pi R_L^{'} \tag{6.44}$$

where typically, if not used as a unity-gain amplifier (buffer), $g_m V_\pi \gg 1$. As seen in (6.44), the output voltage depends on the voltage V_π, which is in turn dependent on the intrinsic capacitance impedance as well as the source impedance. To define this

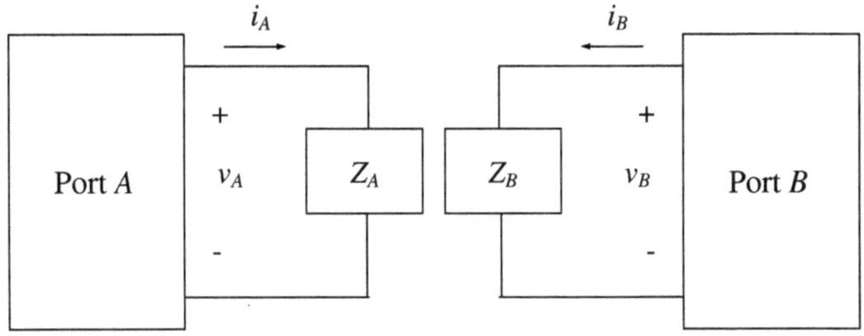

Fig. 6.16 The common-emitter amplifier equivalent circuit when using Miller's theorem

voltage, the process can be simplified by using Miller's theorem[1] and separating the input and output of the circuit as two individual ports, an equivalent representation of a single impedance Z_x (see Appendix 1) between an input and output port, as shown in Fig. 6.16.

The equivalent circuit in Fig. 6.16 represents the circuit topology by two ports, Port A at the input of the transistor, therefore across the base-emitter junction, and Port B at the output of the transistor, therefore across the collector-emitter junction. The combined impedance Z_A in Fig. 6.16 is determined by

$$Z_A = Z_x\left(1 - \frac{v_B}{v_A}\right)^{-1} \tag{6.45}$$

where the impedance Z_B in Fig. 6.16 is determined by

$$Z_B = Z_x\left(1 - \frac{v_A}{v_B}\right)^{-1} \tag{6.46}$$

and Z_x is the impedance linking the two-port network (Appendix 1). Similarly, it can be shown that the equivalent capacitances at ports A and B are defined by

$$C_A = C_x(1 - \frac{v_B}{v_A}) \tag{6.47}$$

and

$$C_B = C_x(1 - \frac{v_A}{v_B}) \tag{6.48}$$

[1]Refer to Appendix 1 for a discussion of Miller's theorem as typically used in multi-stage amplifier analysis.

where C_X is the capacitive element between the two-port network. Applying these terms, the capacitive elements C_A and C_B in (6.47) and (6.48) can be related to the intrinsic capacitances and the voltages in Fig. 6.15 by

$$C_A = C_\mu(1 - \frac{V_{OUT}}{V_\pi}) \tag{6.49}$$

and

$$C_B = C_\mu(1 - \frac{V_\pi}{V_{OUT}}) \tag{6.50}$$

and the equivalent circuit of Fig. 6.15 is redrawn in Fig. 6.17.

The circuit analysis presented in Fig. 6.17 leads to the required equations for the equivalent capacitances C_A and C_B to analyze the frequency response of the common-emitter amplifier using a bipolar resistor. This combination therefore leads to the circuit-related equations for C_A and C_B as

$$C_A = C_\mu(1 + g_m R_L^{'}) \tag{6.51}$$

and

$$C_B = C_\mu(1 + \frac{1}{g_m R_L^{'}}) \tag{6.52}$$

where if $g_m V_\pi \gg 1$ then $C_B \approx C_\mu$ and since the current flowing through the voltage-dependent current source is much larger compared to the current flowing through C_B, the capacitance C_B can be ignored for the final high-frequency small-signal equivalent circuit. The common-emitter amplifier is subsequently given in Fig. 6.18.

It can consequently be shown that the high-frequency cut-off f_H derived from the equivalent circuit in Fig. 6.18 through calculating the frequency-dependent output versus input voltage that

$$f_H = \frac{\omega_H}{2\pi} \tag{6.53}$$

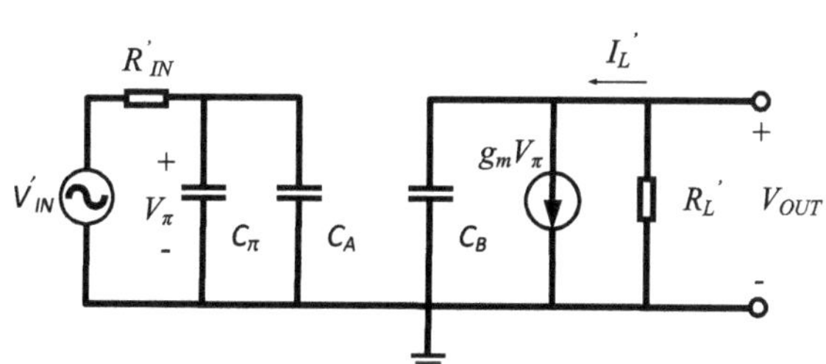

Fig. 6.17 The common-emitter amplifier equivalent circuit analyzed through Miller's theorem with equivalent capacitances C_A and C_B across the input and output terminals

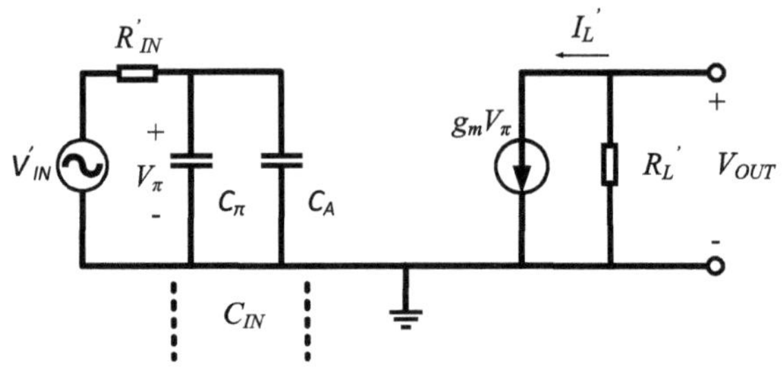

Fig. 6.18 Small-signal high-frequency common-emitter amplifier equivalent circuit using a bipolar transistor—Miller's theorem incorporated

where

$$\omega_H = \frac{1}{2\pi C_{IN} R'_{IN}} \tag{6.54}$$

and the input capacitance as seen in Fig. 6.18 is given by

$$C_{IN} = C_\pi + C_A \tag{6.55}$$

which is equivalent to

$$C_{IN} = C_\pi + C_\mu \left(1 + g_m R'_L\right) \tag{6.56}$$

which ultimately shows the proportional relationship between the high-frequency cut-off f_H and the intrinsic base-emitter and base-collector capacitances, with R'_L given in (6.34). Recall from (6.34) that the equivalent output impedance of the circuit is dependent on r_o, which in turn is dependent on the output current I_C through

$$r_o = \frac{V_A}{I_C} \tag{6.57}$$

where V_A is the constant value of the Early voltage and is dependent on the base-width of the transistor. To derive the high-frequency response, consider at the input, the voltage V_π in (6.44), which is derived as

$$V_\pi = \frac{Z_{C_{IN}}}{Z_{C_{IN}} + R'_S} V'_{IN}. \tag{6.58}$$

By substituting (6.58) into (6.44) and using (6.28) for the capacitor impedance, the output voltage can be described by

$$V_{OUT} = \frac{-g_m R'_L}{1 + j\omega C_{IN} R'_S} V'_{IN}, \tag{6.59}$$

then the output voltage divided by the equivalent input signal is given by

$$\frac{V_{OUT}}{V'_{IN}} = \frac{-g_m R'_L}{1 + j\omega C_{IN} R'_S} \tag{6.60}$$

and if (6.54) is substituted into (6.60), the result is

$$\frac{V_{OUT}}{V'_{IN}} = \frac{-g_m R'_L}{1 + j\frac{f}{f_H}} \tag{6.60}$$

where f_H is given in (6.53). To determine the overall transfer function of the high-frequency response, therefore V_{OUT}/V_{IN}, apply the theorem

$$\frac{V_{OUT}}{V_{IN}} = \left[\frac{V_{OUT}}{V'_{IN}}\right]\left[\frac{V'_{IN}}{V_{IN}}\right] \tag{6.61}$$

where the two terms in brackets in (6.61) can be replaced by (6.60) and (6.39) respectively, thus

$$\frac{V_{OUT}}{V_{IN}} = \left[\frac{-g_m R'_L}{1 + j\frac{f}{f_H}}\right] \cdot \left[\frac{r_\pi}{r_\pi + r_x + R_B||R_S} \frac{R_B}{R_B + R_S}\right] \tag{6.62}$$

and recognizing the term A_m as seen in (6.38), (6.62) is equivalent to

$$A_V = A_m \frac{1}{1 + j\frac{f}{f_H}} \tag{6.63}$$

where A_m is the mid-frequency voltage gain. This result therefore appears as the frequency response of a single-element low-pass filter, which is essentially what the high-frequency response represents. This response is shown in Fig. 6.19.

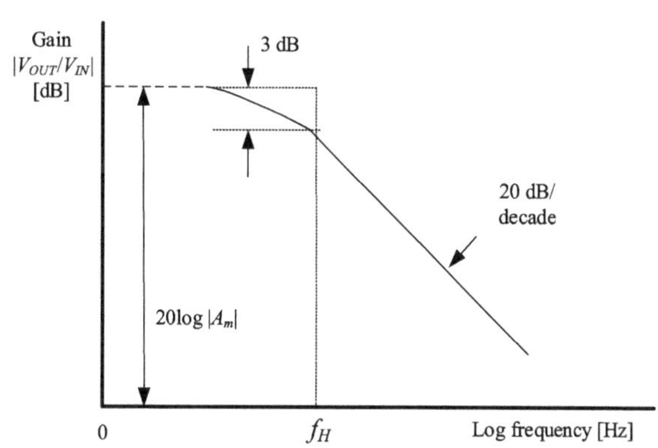

Fig. 6.19 High-frequency response of a common-emitter amplifier

In Fig. 6.19, the increase in operating frequency above that of the mid-frequencies shows a significant decrease in the gain of the amplifier from A_m. The −3 dB decrease in the overall gain is seen at the frequency f_H, after which the gain rolls off significantly at −6 dB/octave. All of these reductions in the gain-bandwidth product (GBP) and the frequency response for the amplifier are dependent on the technology used for the active devices. Including the best way to analyze many amplifier circuits, whether common-emitter, common-base or common-collector topologies and even from differential circuits forming the base of operational amplifiers used for transimpedance amplifiers, it all boils down to the inherent characteristics of the active device and the application in which it is used (low-current, high-gain or low-noise). The transimpedance amplifier is required to amplify extremely low incoming light signals in the nA range and convert these signals to voltages using 10,000–100,000 gain factors. These high-gain requirements significantly decrease the achievable bandwidth of the system and also place stringent requirements on minimum acceptable noise in the incoming signal. In addition, amplifier oscillations due to the high gain and added photodetector capacitance can be troublesome to overcome in practice. The following section provides a recommended setup environment when either measuring or using the photodetector and transimpedance amplifier in practice.

6.4 Detecting a Laser Pulse

Detecting an incoming laser pulse using a photodetector and transimpedance amplifier combination may become troublesome if the system is less than ideal. Stray inductance and capacitance, EMI, power supply noise and light interference can cause the measurements to be inaccurate or prevent measurement. Such a system relies on the detection of very small signals at high frequencies; any unwanted components in the closed system will have an effect on the detection. In Fig. 6.20 a recommended optoelectronic receiver setup is shown to minimize external effects interfering with the system. This setup is specifically for hybrid integration, which is often the case when the photodetector and transimpedance amplifier were not designed on the same substrate and possibly designed by discrete components from various manufacturers.

In Fig. 6.20, the typical optoelectronic receiver measurement or implementation setup is shown. This section highlights the critical components and interconnects in such a system to avoid unwanted interference in the system. Starting from the top left corner in Fig. 6.20, the photodetector biasing power supply is shown. This power supply should only be used to reverse-bias the photodetector. Depending on the photodetector used, the magnitude of the supply voltage can change significantly and should also be variable so that the width of the depletion region can be changed, effectively lowering the capacitance of the photodetector with higher bias voltage, as proposed in (3). Photodetectors such as InGaAs photodetectors, for example, typically with a small active area of a few hundred microns, require

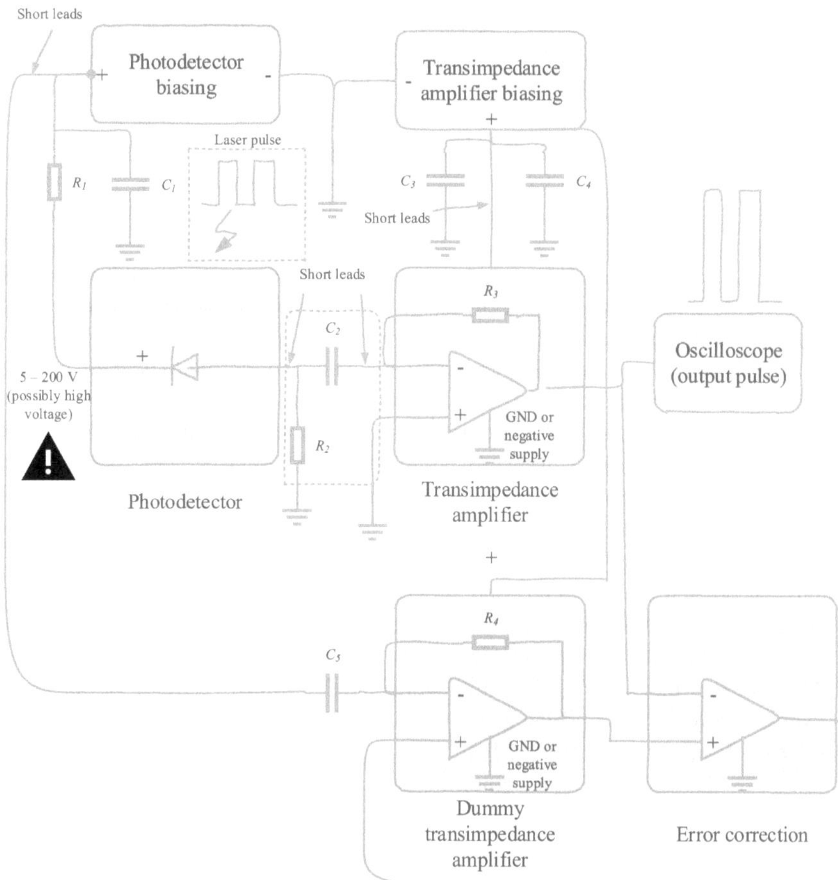

Fig. 6.20 Simplified practical setup to detect and amplify an incoming laser pulse

relatively small bias voltage of between 5 and 20 V. Si PIN photodetectors with large active areas in the millimeter range can require up to 200 V of reverse-bias; therefore, a high-voltage and variable power supply is recommended. To limit the current flow through this node and also reject any power-supply noise, resistor R_1 and capacitor C_2 should be added, as shown in Fig. 6.20. R_1 is typically in the range between 1 and 2 kΩ and C_1 could be a combination of capacitors in parallel to filter various frequencies of the unwanted disturbances. Recommended values range between 10 and 1000 nF for these parallel combination components.

The photodetector, as shown in Fig. 6.20, should be reverse-biased and the leakage current through the devices should be monitored. The laser pulse, therefore the incoming light, is directly pointed at this device. The surrounding area of the photodetector should ideally be kept in a dark enclosure with only the laser light visible to the detector. This will limit the amount of unwanted light triggering the device, which may be recorded. A simple technique to monitor if any current is flowing through the device in its dark state is to measure the voltages across both terminals referenced to ground. Ideally, the cathode of the photodetector should be approximately equal to the high-voltage bias and the anode should be at 0 V (meaning that no current is flowing through the device). Typically, a few hundred millivolts are recorded at the anode of the photodiode, since a small current will flow through the device, as its practical resistance is not infinity as would ideally be the case. Care should be taken when working with high voltages; although only small currents are flowing in these circuits, a short-circuit can cause high currents to flow to and from the power supply.

If using a hybrid approach, the interconnects between the photodetector and the transimpedance amplifier are one of the two crucial points where parasitic elements can introduce oscillations in the circuit. The other crucial point is at the supply rail for the op-amps, which is discussed below. The interconnect leads between the photodetector and the transimpedance amplifier should be kept as short as possible. A resistor R_2 at the output of the photodetector towards ground and a capacitive element C_2 between the two subsystems should be included to filter any unwanted signals introduced at this point and block DC signals from the inverting input terminal of the amplifier. The capacitive value is typically a few hundred pico-farads, whereas the resistive element is in the low-kΩ range.

The biasing terminals of the transimpedance amplifier should be properly AC-shorted towards ground using another capacitive combination of between 10 and 1000 nF in parallel. Any EMI or noise generated at these terminals will be amplified by the transimpedance amplifier and can cause the output of the circuit to oscillate. The transimpedance amplifier is typically biased with a positive and negative supply, $\pm V_{CC}$, but can also be referenced to ground depending on the required gain.

As shown in Fig. 6.20, a dummy transimpedance amplifier is typically introduced to the receiver and is not connected to the terminals of a photodetector. The output of this transimpedance amplifier can be used as an error-correction signal if fed into an offset-correction op-amp. This is usually only necessary in the final integration of the system and can be omitted during the research and development phase, testing or qualification of the system bandwidth. The primary output is measured and recorded at the output of the transimpedance amplifier connected to the photodetector. Again, short leads are recommended at this point, although this is less crucial because of the low output impedance of the op-amp. The oscilloscope should be connected in its AC-coupling mode and the trigger function used to capture the output signal. If the circuit components and parasitic elements do not introduce significant time delays in the system, the rise-time of the output signal should be similar to the rise-time of the original laser pulse. If the rise-time is slower when compared to the laser pulse, the photodetector and/or the transimpedance amplifier could be the limiting factor of the system bandwidth and should be investigated based on the approaches given in this chapter.

6.5 Conclusion

This chapter reviews the requirements, considerations and limitations of an optoelectronic receiver using a photodetector and TIA to detect a weak light pulse and amplify the response to a usable voltage. Optoelectronic receivers are known to operate under stringent requirements owing to the low intensity of the incident charges. The combination of the photodetector bandwidth and the bandwidth of the amplifier determines the overall system bandwidth. The ability of such a system to receive a high-frequency laser pulse depends on its bandwidth. Publications in this field rarely discuss these requirements of the combined system and typically focus on the individual subsystem requirements and performance. Implementing such a system, however, requires fundamental knowledge and practical experience to avoid interference in the system, which could detrimentally affect system performance or render the system unusable. The photodetector bandwidth and frequency response are discussed to identify the critical design and manufacturing characteristics that define its frequency response. This discussion is followed by the low, mid and high-frequency response of the TIA and the effects of the input capacitance, essentially determined by the capacitance of the photodetector, as well as the parasitic capacitances and resistances in the amplifier. TIAs typically make use of CMOS or bipolar transistors in their primary configuration and at high frequencies, the transistor response dominates the system response and limitations in the transistor relate directly to system performance. The comprehensive review on the frequency response of common-emitter amplifiers in the work of Sedra and Smith (2004) is adapted for the discussion regarding TIA in optoelectronic receivers, presented in this chapter. Using this analysis and the results obtained from the photodetector bandwidth discussion, a practical guide to implementing an optoelectronic receiver is also presented in this chapter. These guidelines, in accordance with the theoretical review on the subsystems, aim to provide the designer and researcher with a complete view of the critical elements in such a system.

Appendix 1—Miller's Theorem

Miller's theorem can generally be described that given any general linear network having one common terminal and two terminals whose voltage ratio, with respect to the common terminal. If an impedance Z is connected between the input and output terminals of a network that provides a voltage gain A_v, the equivalent circuit that gives the same effect can be drawn by removing Z and connecting impedance

$$Z_1 = \frac{Z}{1 - A_v} \tag{6.1.1}$$

across the input terminals and another impedance

$$Z_2 = \frac{Z}{1 - \frac{1}{A_v}} \tag{6.1.2}$$

which can also be written as

$$Z_2 = \frac{ZA_v}{A_v - 1} \tag{6.1.3}$$

across the output terminals. This transformation is shown in Fig. 6.21.

The analysis in Fig. 6.21 forms the basis of the frequency response analysis in this chapter. A similar technique is followed to represent the currents i_A and i_B flowing through the circuit. These currents for the original circuit are shown in Fig. 6.21a. By defining the factor α as

$$\alpha = \frac{i_B}{i_A} \tag{6.1.4}$$

the two conducting referred currents result in effective impedances of

$$Z_1 = Z(1 + \alpha) \tag{6.1.5}$$

and

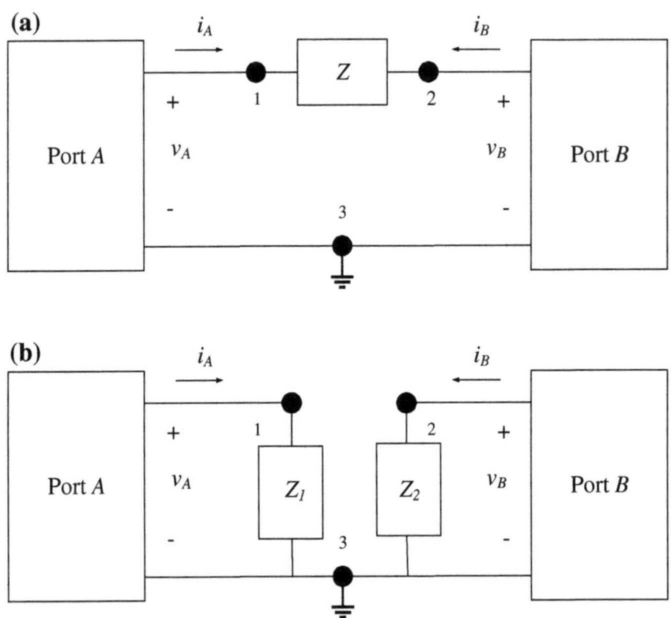

Fig. 6.21 Analysis for **a** Miller's theorem showing the **b** equivalent circuit

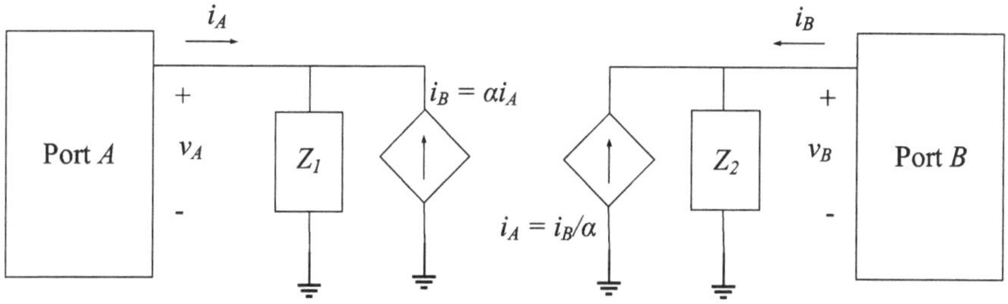

Fig. 6.22 Miller's theorem for currents flowing through input and output impedances showing the equivalent circuit

$$Z_2 = \frac{Z(1+\alpha)}{\alpha} \quad (6.1.6)$$

and the two-port network in Fig. 6.21b can be replaced by its equivalent, shown in Fig. 6.22.

The ideal current sources in Fig. 6.22 are therefore defined by using (6.1.4) to have equivalent currents of

$$i_B = \alpha i_A \quad (6.1.7)$$

and

$$i_A = \frac{i_B}{\alpha}. \quad (6.1.8)$$

Miller's theorem is frequently used in amplifier circuit analysis where a gain between the input and output is experienced.

References

Biswas, P., Mannan, R. A., Jahan, N., & Arafat, Y. (2014). Dependency of heterojunction transistors' speed on various physical parameters: A comparative study of SiGe & AlGaAs HBTs. In *International Conference on Electrical Engineering and Information & Communication Technology (ICEEICT) 2014,* (pp. 1–6). IEEE.

Brandl, P., Schidl, S., & Zimmermann, H. (2014). PIN photodiode optoelectronic integrated receiver used for 3-Gb/s free-space optical communication. *IEEE Journal of Selected Topics in Quantum Electronics, 20*(6), 391–400.

Dutta, N. K., Tu, K. Y., & Levine, B. F. (1997). Optoelectronic integrated receiver. *Electronics Letters, 33*(14), 1254–1255.

Gray, P. R., Hurst, P. J., Meyer, R. G., & Lewis, S. H. (2001). *Analysis and design of analog integrated circuits.* John Wiley & Sons.

John, E., & Das, M. B. (1994). Design and performance analysis of InP-Based high-speed and high-sensitivity optoelectronic integrated receivers. *IEEE Transactions on Electron Devices, 41*(2), 162–172.

Karanicolas, A. N., Kenneth, K. O., Wang, J. Y. A., Lee, H., & Reif, R. L. (1991). A high-frequency fully differential BiCMOS operational amplifier. *IEEE Journal of Solid-State Circuits, 26*(3), 203–208.

Sedra, A. S., & Smith, K. C. (2004). *Microelectronic circuits* (p. 2004). New York: Oxford University Press. Fifth Edition.

Sharifi, H., & Mohammadi, S. (2007). Heterogeneously Integrated 10 Gb/s CMOS optoelectronic receiver for long haul telecommunication. *IEEE Radio Frequency Integrated Circuits*, 515–518.

Voinigescu, S. P., Beerkens, R., Dickson, T. O., & Chalvatzis, T. (2005). Design methodology and applications of SiGe BiCMOS cascode opamps with up to 37-GHz unity gain bandwidth. In *IEEE Compound Semiconductor Integrated Circuit Symposium, CSIC'05*. (pp. 3). IEEE.

Youn, J., Lee, M., Park, K., Kim, W., & Choi, W. (2015). Low-power 850 nm optoelectronic integrated circuit receiver fabricated in 65 nm complementary metal-oxide semiconductor technology. *IET Circuits Devices and Systems, 9*(3), 221–226.

Yuan, J. S. (1994). Effect of base profile on the base transit time of the bipolar transistor for all levels of injection. *IEEE Transaction on Electron Devices., 41*, 212–216.

Zimmermann, H. (1996). Monolithic bipolar-, CMOS-, and BiCMOS-receiver OEICs. In *International Conference on Semiconductor 1996*, (Vol. 1, pp. 31–40). IEEE.

Chapter 7
SiGe for Radiation Hardening: Spearheading Electronic Warfare in Space

7.1 Introduction

Space electronic warfare (SEW) implies tactical advantages in electronic attacks, electronic protection and electronic support using satellites in low-earth orbit (LEO) for surveillance, communications or positioning, or denying the enemy these activities. EW space weapons are systems designed to project against attacks between earth and outer space or contained in space. Applications that fall under this category include antisatellite weapons, projectile platforms in space and ground-based lasers relying on orbiting mirrors in space to reflect energy to earth or into space. Preston et al. (2002) and DeBlois et al. (2005) categorize space weapons into four categories, namely DEWs delivering high-energy pulses at the speed of light, non-explosive kinetic-energy weapons colliding with their targets to inflict damage, conventional warheads delivered to or from space and microsatellites used for surveillance or even as kinetic-energy weapons to damage or destroy larger satellites. Modern satellite constellations are already prone to threats such as denial and deception of intelligence, jamming of communications for EW tactics, physical attacks on ground stations rendering the satellites unusable and anti-satellite weapons or even remote nuclear detonation. Modern society has grown more reliant on satellite communications, GPS and intelligence/information gathering and with this increased dependence, EW is also gaining traction in space and becoming a prevalent tactical advantage for militaries with space-based abilities. The United States DoD requested US $1.8 billion in its budget proposal in 2016 for funding of space launch activities and additional hundreds of millions of dollars (approaching a billion) for research into space technology and counter-space systems, with the detailed breakdown provided in Space Foundation (2016). The US Congress authorized an additional US $5 billion for the Pentagon space programs till 2020. This shows the significance of space research in the US alone and similar trends are being witnessed worldwide. The budget information provided in Space Foundation (2016) includes only the unclassified space programs, i.e. the technological

W. Lambrechts and S. Sinha, *SiGe-based Re-engineering of Electronic Warfare Subsystems*, Signals and Communication Technology,
DOI 10.1007/978-3-319-47403-8_7

concepts and research outputs that will definitely be applied for military implementation. Adapted from Preston et al. (2002), the advantages and disadvantages of space weapons are summarized below. The advantages include:

- Accessibility to targets that are difficult or impossible to reach by traditional earth-bound weapons and a global advantage to nations and militaries that have access to these weapons.
- A decreased period between instructions to fire a space weapon and the actual launch. Space weapons are *always* ready to be used, since the biggest challenge is to deploy the weapons in space; triggering them mostly depends on their location in orbit and authorization.
- The large distance between space weapons and for example their control room makes attacking the weapons or the control station difficult, compared to aircraft carrying weaponized payloads, where much shorter distances separate the weapon and the launching vehicle.
- There are extremely short time intervals between a space weapon being fired and hitting its target because of its high velocity and brief flight. Either in DEW or kinetic-energy weapons, defense against these attacks is difficult.

The primary disadvantages of space weapons, as adapted from Preston et al. (2002), are summarized as:

- Space weapons can be vulnerable to attack because of their static nature (static in terms of their location relative to the enemy attack). It is not possible simply to move the weapons out of harm's way and their relative position in space is fixed.
- Similar to their static nature, space weapons have a predictable orbit and this can be exploited by enemy weapons. In addition, if a stable-orbit weapon is destroyed, its debris could potentially damage nearby systems.
- Space weapons are expensive. The expenses of space weapons are not limited to the technology and materials used for the weapon, but include expensive payload logistics, logistics concerning launching a vehicle into space, maintenance (if possible) and licensing and application for these weapons that also drive up the overall cost.
- Adding to the cost of a single satellite or weapon in space, its effectiveness can only be increased if large numbers of these satellites or weapons are deployed to cover a larger area of earth-based or space-based targets.
- The use of terrestrial weapons is still illegal in most parts of the world and using such weapons offensively, especially weapons of mass destruction, has far-reaching legal consequences.

These advantages and disadvantages do not specifically justify the technology of space weapons or the medium through which the energy or particles move. Space has no atmosphere, therefore there is no need to consider weather changes, aerosols, particulate matter, dust or any other charged particles that attenuate the energy of the weapon; the $1/R^2$ relationship dominates the attenuation of the energy and therefore its range. Explosions in space are significantly different compared to explosions on earth. In space there is no accompanying shockwave during an explosion and an explosion is essentially only a cloud of expanding gas absorbed by nearby matter. For space weapons to be effective, kinetic impact or focused radiation is required. This is why DEW and lasers are at the forefront of space weapon research and technology. Lasers and directed-energy beams are likely to be used in space combat to blind surveillance equipment or incinerate satellite-integrated components. Another element of assault would be to hack or reprogram an opposing military's satellites and use them to send deceitful signals.

In this book both DEW and lasers in optoelectronic systems have been reviewed, primarily for use in earth-based applications. The environments where these devices are used on earth can be extreme, especially when considering military applications, but none of these environments compares to the harsh space atmosphere (approximately 200 °C day and night temperature variation on the moon, for example). The technology used to realize these types of weapons and guarantee their operation for long periods of time in harsh environments is an enormous challenge. It is not only high-powered technologies that generate high-energy directed electromagnetic waves that must be able to operate in space. The command and control, communications, radar and supporting circuits are equally crucial to the success of any space weapon or system. Each of these systems requires highly integrated (Gaucher et al. 2007 and Kuo et al. 2006), cost-effective, efficient and reliable modules.

The reliability of conventional microelectronic components, especially active devices such as transistors, decreases drastically when they are exposed to the harsh environment of space. Using standard off-the-shelf components, even military-specified components, is not a feasible option when designing and implementing electronic equipment for space applications, space weapons or satellites. The statistical probability of device failure due to extreme temperature or high radiation levels is large and the consequences of such a failure may be catastrophic. Commercial applications might permit more room for errors or random glitches; for military applications, however, the potential consequences outweigh the risks of using components that have not been approved for space use. One possible solution to decrease the risk of device failures, either SEUs or permanent damage and degradation, is to enclose the active circuitry in shielded boxes. This is not always practical for ideally modular designs, not an ideal solution and not necessarily possible in view of typical weight and size penalties. There are also demanding limitations to the materials that can be used for hermetic packaging for space application, typically silver-glass, gold-silicon eutectic and polyimide film or paste (Shah et al. 1998). Other techniques to reduce the statistical probability of failures are to employ special circuit design techniques to increase reliability through redundancy. Redundancy in space-based electronic systems is, however, a risky technique, since increasing the size and complexity of a system that can only be checked and maintained remotely and is constantly bombarded with radiation may paradoxically decrease its reliability. Apart from the discrete techniques available to radiation-harden electronic components for space applications, the ideal solution would be to use semiconductor components that are inherently resistant to extreme variations in temperature and high levels of radiation (Messenger 1969).

This chapter focuses on semiconductor transistor physics and its susceptibility to degraded performance or complete failure in space. The principal focus here is not to argue that BiCMOS HBT technology is better suited for applications where radiation hardening is required, compared to using CMOS technology, as this statement has been documented in sources such as Cressler (2007, 2013). The key arguments here are that BiCMOS HBTs manufactured in, for example, SiGe are vertical transport devices with less dependence on photolithography and the active region exists deep within the device; CMOS is a lateral transport device and more susceptible to degradation close to its surface (Yuan et al. 2009). SiGe has a two- to three-year generational advantage over CMOS (Cressler 2013) in terms of scaling and speed and it presents improved tolerance to incident radiation that may cause failures or upset events and operates in exceedingly wide temperature ranges. Ongoing research into SiGe HBT total dose damage due to protons, X-ray, gamma radiation, neutrons, cryogenic temperature shocks and varying dose rates has been conducted since the first paper on electrical performance degradation from radiation effects in n^+-Si/p^+-$Si_{1-x}Ge_x$ epitaxial diodes and n^+-Si/p^+-$Si_{1-x}Ge_x$/n-Si HBTs by fast neutrons and MeV electrons in 1994 (Ohyama et al. 1994). An interesting conclusion drawn by Ohyama et al. (1994) is that the degradation of electrical performance due to radiation decreases with germanium content; although requiring a modification of a standard BiCMOS process, this is still a valuable observation, as the germanium doping concentration also affects the dynamic response of the HBT. This chapter identifies the physical and performance aspects of BiCMOS and CMOS transistors when used in space applications and thematically applies these to EW and circuits and systems discussed in this book. The following section highlights some research works in the current body of knowledge, specifically aimed at investigating HBT performance when bombarded with radiation, as typically experienced for systems operating in space. These examples do not specifically highlight the physical mechanisms at work causing the radiation effects and discuss the radiation hardness results. The physical mechanisms of potential damage and single events in CMOS and BiCMOS technologies are discussed in subsequent sections.

7.2 Research on the Radiation Effects on Applied BiCMOS Circuits

In Thrivikraman et al. (2008) the irradiation results on high-frequency operated X-band (8–12 GHz) SiGe HBT and CMOS phase shifters for space-based phased-array antennas used in radar or wireless communication are reported. The research shows that when dealing with phase shifters, both CMOS and SiGe HBTs can be used for space applications without any specific radiation-hardening techniques. Significant in this paper is demonstrating that CMOS phase shifters using

NMOS devices are also TID tolerant (McLean and Oldham 1987), although the discrete devices are known to have substantial TID responses. Both CMOS and SiGe HBTs do present slight changes in DC leakage currents when exposed to 3 Mrad total dose exposure, somewhat more prominent in the CMOS circuit, but the reported RF performance shows no noteworthy degradation. The radiation response of phase shifters cannot be directly related to the radiation response of the individual active devices, as the transistors are not usually irradiated in diode-connected or pass-gate operating conditions and Thrivikraman et al. (2008) report on the radiation response of these circuits for their use in space applications. Also notable is that no changes were made to the commercially available CMOS and BiCMOS processes, which is an important consideration in cost-effectiveness when designing space-based electronic circuits.

Thrivikraman et al. (2007) evaluate the total dose radiation response of a low-power X-band LNA in SiGe, targeting high-altitude or space-based low-power density phased-array radar systems. In this work the LNA was radiated with space-relevant 63.3 meV protons with proton fluence as high as 5×10^{13} cm^{-3} and it presents the gain and NF response before and after a 6.7 Mrad dose proton radiation exposure. The paper concludes that almost no performance degradation of the post-radiated samples is recorded, which is consistent with the general understanding that SiGe HBTs are inherently total dose radiation tolerant, even by standard fabrication processes.

A radiation hardness evaluation on a SiGe BiCMOS technology with a laterally diffused MOSFET module for applications such as power switches in DC-DC converters conducted by Teply et al. (2011) shows little to no degradation in performance for radiation doses and fluence typically found in space; between 100 krad and 1 Mrad TID. This work concludes that such a technology is well suited to both MMICs and ASICs or space and high-energy physics applications; this is mostly attributed to the inherent radiation resistance of BiCMOS HBTs.

A publication presenting the performance impact of radiation exposure to SiGe HBTs in a 130 nm BiCMOS technology is seen in Rice et al. (2008). This research is, however, aimed at radiation exposure in the ATLAS detector of the LHC as opposed to space radiation, but the methodology and mechanisms of this research can be applied to space-based electronic circuits. The predicted target radiation level in this study is 100 Mrad of gamma radiation causing damage by ionization based on simulation studies. Results show that the degradation of the HBTs in this technology from displacement and ionization damage is within acceptable bounds for the expected radiation levels. The identified areas of concern are the dispersion of the final gains of the devices after irradiation and the temporary decrease in gain after short annealing times.

Adell et al. (2014) investigate the radiation robustness of a Wilkinson ADC within a cold-capable, analog sense ASIC manufactured in a 500 nm SiGe BiCMOS process. The ASIC is for a distributed motor controller architecture, which is a candidate for space applications for the next generation of Mars rovers. Its main function is to interface with various sensors to monitor the health of the rover such as its temperature, mechanical stress and pressure. The ADC in Adell et al. (2014) shows very good tolerance to cumulative radiation up to 300 krad but has shown some sensitivity to SEUs where the most sensitive blocks were identified, revealing the importance of the impedance network tied to the channel inputs to decrease SEU sensitivity. An important consideration in this work is its methodology to remove the current methods of enclosing the electronic circuits in shielded boxes to protect them from harmful radiation and highly variable temperatures. Adell et al. (2014) specifically follow this technique to decrease the size and weight constraints for interplanetary systems, a constraint that is applicable to most space missions.

In the study presented by England et al. (2014), SiGe BiCMOS transceivers are designed for basic compatibility with the RS-485 and ISO 11898[1] standards while operating over a temperature range between 90 and 390 K. The transceivers presented in England et al. (2014) were designed as part of a larger NASA project with the goal to demonstrate that SiGe BiCMOS technology was/is a viable candidate for implementing extreme environmental electronics, particularly targeting future missions to the moon and Mars. The system designed in this work was a SiGe remote electronics unit, which consisted of two ASICs, a remote sensor interface and a remote digital control. A total of 16 instrumentation channels of three different types and a multiplexing unit were used and the second ASIC was used as the outside communication port and controller for all gain, calibration and sensor stimulus settings. For the TID experiment, the hardware was irradiated at three levels: 200, 500 krad and 2 Mrad, with the samples removed and tested between each level. It is reported that both the receiver and the transmitter showed radiation hardness up to the 2 Mrad dose with measured values within 5 % of the acceptable range.

Jung et al. (2014) investigate the single-event effect sensitivity of three different commonly employed current mirror circuits implemented in C-SiGe HBT on silicon-on-insulator (SOI) technology. Current mirror circuits are chosen as the subject of this investigation, since these circuits can be considered essential and fundamental DC biasing blocks in ICs and virtually all RF circuits use current mirrors for their DC biasing. Current mirrors are also well defined and understood in the electronics discipline and effects from radiation can be conveniently isolated. England et al. (2014) found that basic *npn*-transistor current mirrors are recommended over cascaded implementations as the basic circuit shows a higher tolerance to SEU in terms of lower peak transient, shorter settling time and smaller area of impact across the HBTs. Utilizing *pnp* SiGe HBTs in the current mirrors can make the AC transients of the input and output currents induced by the ion strikes shorter and the peaks smaller. Their measurements showed that inverse-mode *pnp* current mirrors show excellent SEU tolerance at radiation levels up to eight times

[1]The RS-485 standard is published by the Telecommunications Industry Association and Electronic Industries Alliance (TIA/EIA) for digital communications networks specifically used over long distances and noisy environments. The International Organization for Standardization (ISO) 11898 standard specifies the physical and datalink layers of serial communication that supports distributed real-time control and multiplexing.

that of the expected radiation in such missions. Jung et al. (2014) attributed these results to the inherent SEU tolerance of SiGe HBTs on SOI with buried oxide and deep trench isolation, characteristics that traditional CMOS transistors do not exhibit.

Military systems operate over a wide range of environmental conditions, historically also considered in the vicinity of nuclear explosions (Messenger 1969), as well as in satellite communication and control systems. Radiation is inevitable in these circumstances and the effects of radiation on electronic equipment has been studied thoroughly, although its complete understanding frequently requires experimental results to verify theoretical assumptions. Two primary space radiation effects affect electronic equipment detrimentally. These effects are known as ionizing/non-ionizing dose and SEEs. These mechanisms are described in the following section and understanding them will complement the hypothesis that BiCMOS HBTs are more tolerant to radiation when compared to CMOS technologies. A brief overview on the X-band is given below to highlight its importance in space applications, as it is evident from the above works that the X-band plays a major role in military-space applications.

7.3 X-band Frequency Spectrum

The radio spectrum is a subdivision of the EM spectrum between 3 Hz and 3 THz, which is allocated to the transmission and reception of radio signals for various radio communication services and is strictly regulated to avoid congestion of bands leading to degradation of signals by noise and sub-band interference. The most obvious route towards more bandwidth for communication is to increase the frequency of operation of the equipment used, which is a large driver in researching and developing new technologies capable of operating at higher frequencies. The transistor is still the most prominent active device determining the capabilities of analog and digital circuits and an increase in transistor speed translates to systems operating at higher frequencies. In addition, the channel of communication determines the capabilities of the frequency spectrum where natural decay of EM signals due to clouds, aerosols, dust, PM, ionized particles, rain or fog and several other elements influence the integrity of EM signals. The X-band spectrum has been specifically allocated to the military for high-speed applications. The X-band offers several inherent advantages to support these challenging applications, summarized as:

- Superior for mobile applications, since double the amount of space between orbital slots for X-band satellites (4° as opposed to the traditional 2° spacing) leads to a smaller probability of adjacent satellite interference.
- The use of high power-density carriers is allowed, which is beneficial when using small antennas.

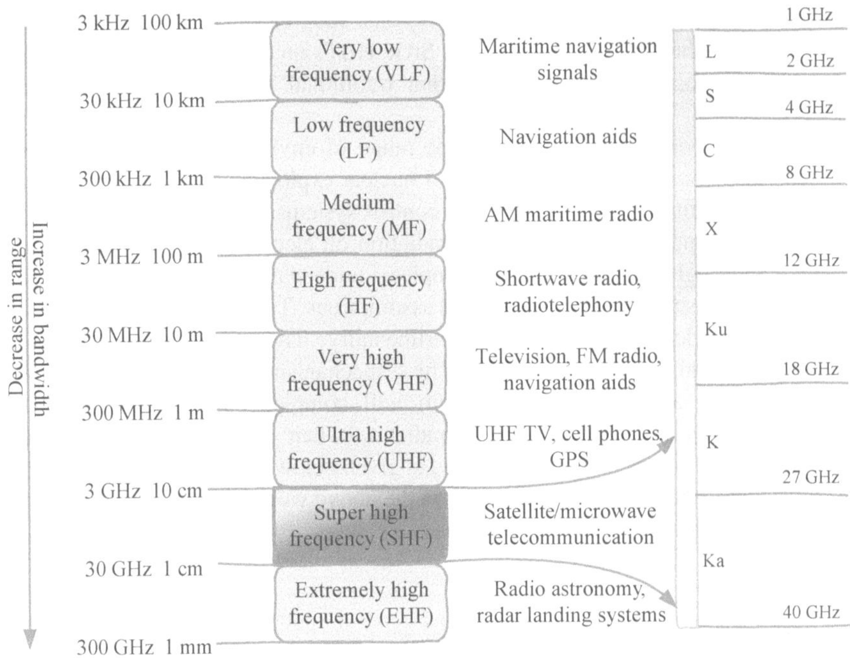

Fig. 7.1 The radio spectrum as a subset of the electromagnetic spectrum showing the satellite frequency convention as outlined by IEEE (2003)

- Spread-spectrum techniques are typically not required, translating to smaller bandwidth requirements in an already large spectrum.
- The X-band is robust in adverse conditions such as high rain, fog or snow. The link availability in these conditions is high.
- It has high efficiency of its link budget owing to the high power densities and since spread-spectrum is not a requirement.
- The users of this band are exclusively government entities such as the military. The number of users in this band is therefore significantly smaller compared to the commercial spectrum and there is a smaller probability of interference.

Figure 7.1 presents the standard frequency letter band nomenclature for frequencies between 3 kHz (100 km wavelength) up to 300 GHz (5 mm wavelength).

As seen in Fig. 7.1, the X-band, which falls under the SHF frequency range, is defined between 8 and 12 GHz, according to IEEE (2003). Figure 7.1 also presents the radar letter designation for frequencies between 1 and 40 GHz, where the L-, S-, C-, X-, Ku-, K- and Ka-bands are defined. These bands are important for commercial and military applications, satellite communication and television and radio broadcasting. Figure 7.2 shows the typical applications in each of these bands.

As seen in Fig. 7.2, the frequency allocations between 1 and 40 GHz are primarily concerned with satellite, radar and high-bandwidth television or radio

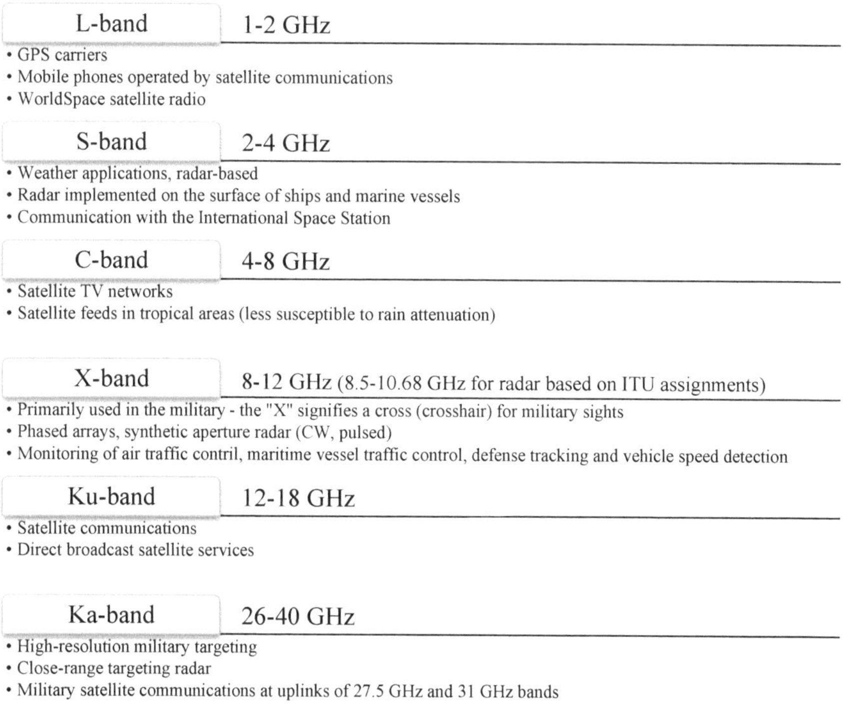

Fig. 7.2 Frequency band letters, ranges and applications within the 1–40 GHz spectrum

broadcasts. Several exclusively allocated military bands exist within these bands, but most prominent in modern radar and space communication is the X-band of frequencies between 8 and 12 GHz. IEEE (2003) also defines the frequencies between 8.5 and 10.68 GHz as specific frequency ranges for radar based on ITU assignments. For this reason, many research works for SiGe BiCMOS circuits and their tolerance to radiation are focused on the 8–12 GHz band.

A brief review of the potential radiation effects on electronic devices is presented in the following section.

7.4 Radiation Effects on Electronic Devices

The space radiation environment can directly or indirectly ionize atoms and disrupt the crystal structure of a semiconductor. In space, several types of radiation exist, including plasmas, particle radiation, neutral gas particles, UV and X-ray; space debris can also be regarded as a type of radiative element (the high-velocity physical debris bombarding a target) and these mechanisms are summarized by Label (2004) and adapted in Fig. 7.3.

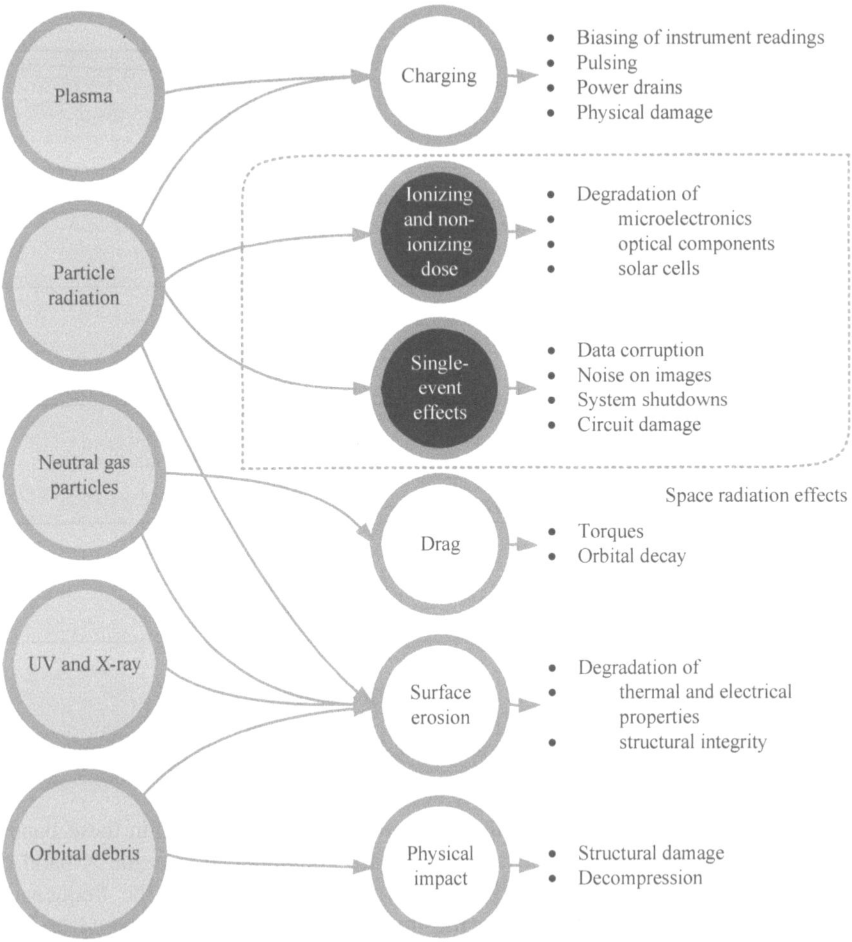

Fig. 7.3 The mechanisms present in space environments, which can degrade or destroy equipment, satellites and other systems orbiting in space. This summarized chart is adapted from Label (2004)

Figure 7.3 highlights the two primary mechanisms that may cause semiconductor materials in space to either degrade permanently from cumulative effects or radiation bombardment or cause random SEEs, which may lead to analog or digital signals being misconstrued temporarily. These mechanisms include ionizing/non-ionizing dose and SEEs. Figure 7.4 summarizes the two effects and the devices that are most prone to them.

In Fig. 7.4, the cumulative effects of space radiation in electronic components are gradual effects during the lifetime of the components while exposed to radiation. The tolerance of components exposed to cumulative effects is reached when the accumulated TID or particle fluence has reached its limit. Fluence is the number of

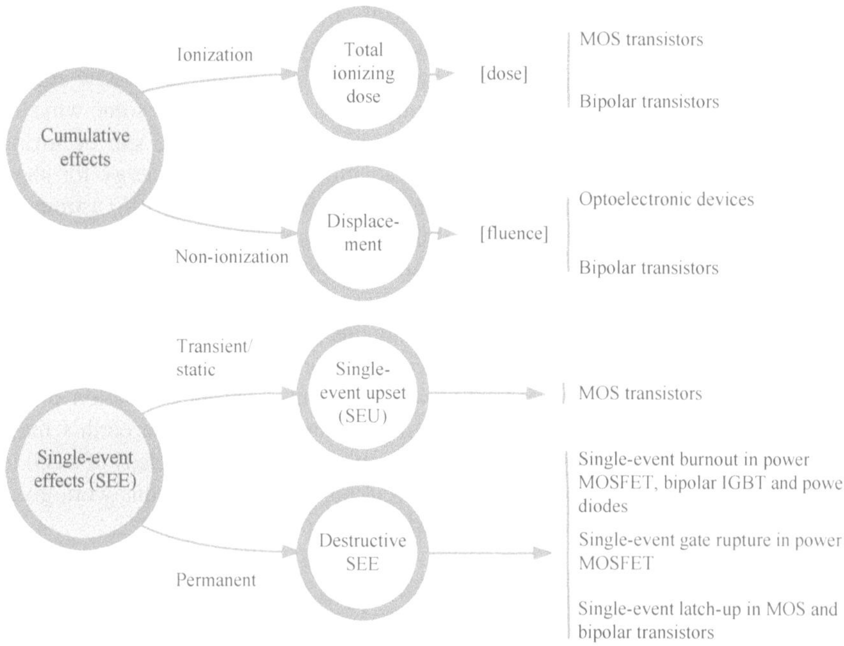

Fig. 7.4 Devices prone to space radiation cumulative effects from constant ion-bombarding and SEUs

particles crossing per unit area from either side of the semiconductor. Fluence can also be described as the flux integrated over time. In theory it is therefore possible to estimate the expected failure of such a device by determining the radiation intensity and the time of exposure to the radiation, assuming the radiation tolerance of the device is known and characterized. SEEs are a more statistically invariant and unpredictable mechanism, as they are triggered by the energy deposited by a single particle into an electronic device. The probability of SEEs is stated in relations to the cross-section of the exposed area (the importance of this measure will become clear during the discussion of CMOS and BiCMOS technology tolerance to SEE).

There are essentially two principle types of radiation in space that affect electronic equipment: galactic cosmic rays (GCR) and radiation emanating from the sun. The types of particles and their respective energies can vary widely, but protons and heavy ions are generally the most destructive. Electrons have a relatively short penetration depth and can be shielded from more easily. On earth, the magnetic field shields equipment from this radiation, whereas electronic equipment in space can only be shielded by using thick metal casings or semiconductor materials with a higher tolerance to radiation (the primary case for using BiCMOS HBTs in space). The major components of radiation in GCR are energetic protons and heavier ions, with approximately 2 % of the constituents being electrons and positrons. The source of GCR is unknown, but it is generated outside of our own solar system and this radiation is isotropic with a relatively constant fluence over time. The abundance of these ions decreases with an increase in atomic number; however, iron is abundant in GCR. The sun produces a constant flux of low-energy particles, mostly protons, and this constant flux is known as the solar wind or coronal mass ejections. Solar flares are magnetic storms from the sun occurring every 11 years on average, increasing the radiation density and energy for short periods. These storms pose a threat to electronic equipment (as well as to astronauts in space and biological entities on earth if the intensity of the magnetic storm is very large). Charged particles are confined in areas known as the Van Allen radiation belts. The inner Van Allen radiation belt is located between approximately 1.1 and 3.3. earth radii in the equatorial plane and contains primarily protons with energies exceeding 10 meV. The outer Van Allen belt contains mainly electrons with energies up to 10 meV. Curtis and Letaw (1989) presented an evaluation of the exposure of space travelers to galactic cosmic radiation outside the earth's magnetosphere through a sphere of 100 μm^2 (chosen to simulate the nucleus of a cell in the human body but adaptable for electronic circuit exposure area); this work gives a good guideline to expected radiation particles in space.

7.4.1 Total-Ionizing Dose

Defined under the category of cumulative effects in Fig. 7.4, TID is deposited by particles passing through semiconductor materials. TID is therefore a measurement of dose as energy from radiation in the form of ionization energy. The dose is always referred to by the absorbing material, x; common examples are Si, SiO_2 and GaAs. The International System of Units (SI) measurement unit of TID is the Gray (Gy(x)) unit, but the radiation absorbed dose (rad) unit is frequently used. The Gy (x) is defined as the absorption of one joule of radiation energy per kilogram of matter. This unit is mathematically described by

$$1\ \text{Gy}\,(x) = 1\frac{\text{J}}{\text{kg}} = 1\frac{\text{m}^2}{\text{s}^2} \tag{7.1}$$

where 1 Gy(x) is equal to 100 rad and the biological unit is the Sievert (Sv). The equivalent dose H_T of radiation is defined as

$$H_T = W_R D_{T,R} \tag{7.2}$$

where W_R is a weighting factor related to the type of energy particle and $D_{T,R}$ is the absorbed dose in material T due to radiation type R. The weighting factors W_R for photons and electrons (at all energies) are typically 1.5 for protons with energies higher than 2 meV and 20 for all alpha particles, fission fragments and heavy nuclei. The weighting factor of neutrons can be calculated as a continuous function of neutron energy E_n in MeV by

$$W_R = \begin{cases} 2.5 + 18.2e^{-\frac{[\ln(E_n)]^2}{6}}, & E_n < 1\,\text{MeV} \\ 5.0 + 17.0e^{-\frac{[\ln(2E_n)]^2}{6}}, & 1\,\text{MeV} \leq E_n \leq 50\,\text{MeV} \\ 2.5 + 3.25e^{-\frac{[\ln(0.04E_n)]^2}{6}}, & E_n > 50\,\text{MeV} \end{cases} \tag{7.3}$$

as described in ICRP (2007). The tissue-weighted sum of the equivalent doses is the effective dose E and is calculated by

$$E = W_T H_T \tag{7.4}$$

where W_T is another weighting factor, typically defined for biological tissue and used as a measure of risk (this equation is typically used to determine the risk of radiation treatment for cancer being absorbed by biological tissue of humans). These principles and mechanisms are applied to semiconductor materials to define and quantize the tolerance of the material to radiation. The mechanisms through which photons interact with a semiconductor material are the photoelectric effect, the Compton effect and pair production (Srour 1983). Charged particles interact with materials through Rutherford scattering or nuclear interactions (heavy particles) and neutrons interact with materials through mechanisms such as elastic scattering, inelastic scattering and transmutation effects. For each of these processes the primary results of the interaction between the bombarded proton and the target material is the creation of energetic secondary electrons.

Charged particles generate electron-hole pairs when passing through matter, which leads to ionization damage. Based on these interactions, the electron-hole pair density is proportional to the energy transferred to the target material (Oldham 1984). Stopping power or linear energy transfer (LET) is the particle energy loss per unit length and is a function of the particle mass and energy, as well as the target material density (Srour 1983). The unit for stopping power is MeV-cm^2/g and is typically measured against the particle energy for electrons and protons incident on the material, measured in MeV. Protons have a higher stopping power compared to electrons.

Electron and proton exposure, which results in cumulative effects, typically lead to oxide charge trapping and the creation of interface states at SiO_2 and Si interface due to chemical bonding changes at this interface. Oxide charge trapping can induce a shift in Early voltage V_{TH}, which leads to changes in the output current and added noise in the system. These are particularly significant in optoelectronic applications detecting very small signals and lead to an increase in leakage current for reverse-biased diodes. Interface states can also induce V_{TH} shifts, variations in semiconductor mobility and increased leakage current. All of these changes effectively degrade the performance of the system, or induce unexpected system behavior. National Semiconductor® published a manual to assist individuals working in the field of radiation effects and component hardening, reliability, testing and evaluation and presented a summary of the TID radiation tolerance of analog, ASIC, interface, logic, memory and microcontroller systems. It should be noted that this information is disclosed as not being scientifically verified but

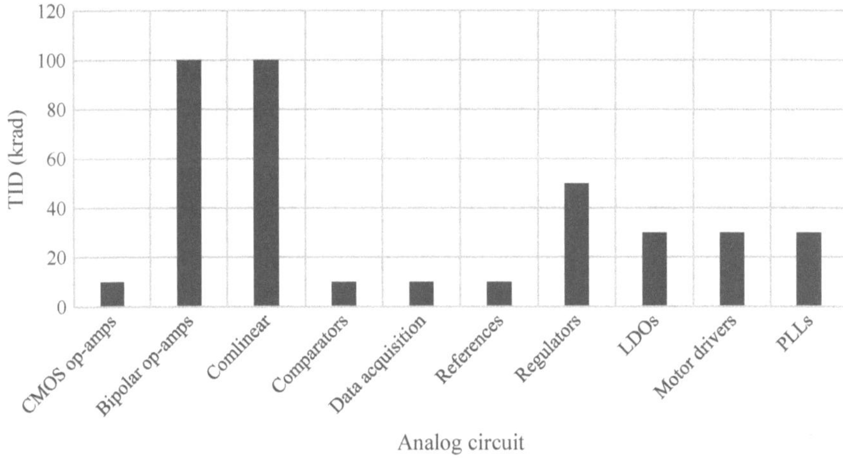

Fig. 7.5 TID radiation tolerance of common analog circuits as published by National Semiconductor®

internally tested within the company; nevertheless, the results give a good indication of the tolerance of analog circuits to radiation. These analog circuits comprise discrete transistors, either CMOS or HBT, and therefore the results of the total circuits also include the additive tolerance based on operating conditions. The analog circuit radiation tolerance results are adapted and presented in Fig. 7.5.

As seen in Fig. 7.5, there is a noticeable difference in TID radiation tolerance between CMOS op-amps and bipolar op-amps. This result is significant for this book, since op-amps are an integral part of EW and SEW equipment. Recall from Chap. 6 that optoelectronic receivers are typically realized with transimpedance amplifiers; this is essentially an op-amp configuration of either CMOS or bipolar devices. The advantages of using SiGe for speed, reliability and potential integration with optoelectronic devices and photodetectors could potentially gain another advantage, radiation tolerance, as is argued in this chapter. The following sections will present a more detailed look into the physical properties of CMOS and bipolar transistors and the mechanics responsible for radiation defects. The results in Fig. 7.5 are, however, disclosed by National Semiconductor® to be under testing and may change, but the difference seen in these experiments already points to a significant difference between these two technologies. Comlinear® op-amps perform equally well in terms of radiation tolerance, followed by voltage/current regulators, low-dropout voltage regulators, motor drivers and PLLs. Comparators, data acquisition circuits and voltage/current references performed worst in these experiments; the small room for error in such devices could be responsible for this performance. Also presented in the results by National Semiconductor® but not shown in Fig. 7.5 are the TID tolerance for bipolar and CMOS data transmission interfaces—bipolar interfaces again reach

100 krad levels, whereas CMOS reaches approximately 10 krad—again showing a significant difference in favor of BiCMOS.

7.4.2 Displacement Damage

Displacement damage, a non-ionization cumulative effect of radiation shown in Fig. 7.4, is generated by highly energized protons, electrons, neutrons and heavy ions and displaces atoms in a material as these particles move through the material. The vacancies and interstitials created by these collisions introduce allowable energy states in the bandgap of the semiconductor material. Displacement damage is therefore particularly prevalent in optical detectors, optoelectronic receivers and bipolar linear ICs. For minority carrier devices such as diodes and bipolar transistors, bulk traps are created in the material, which reduce the carrier lifetime and therefore increase the carrier recombination in a forward-biased junction such as the base-emitter junction. This effectively increases the base current, can reduce the collector current and leads to a decrease in the gain of the transistor (Srour et al. 2003). Displacement damage is often quantified by the non-ionizing energy loss rate (NIEL). The final concentration of defects depends on the NIEL value and not the type and initial energy of the incident particles. The NIEL parameter is defined as the portion of the particle energy T_0 per unit travelling length expressed in mass per area—thus the displacement damage. The expression for NIEL is given by

$$\mathrm{NIEL}(T_0) = \frac{N_A}{A} \int_{T_{\min}}^{T_{\max}} dTQ(T)T \left(\frac{d\sigma}{dT}\right)_{T_0} \tag{7.5}$$

where N_A is Avogadro's number (6.022×10^{23}), A is the atomic mass of the particle, $Q(T)$ is the partition factor that gives the fraction of T to be lost to NIEL and ${}^{d\sigma}/_{dT}$ is the differential partial cross-section for creating a given recoil with energy T. $T_{\min}$ is equal to $2T_d$ where T_d is the displacement energy. For silicon, T_d = 21 eV and for GaAs, T_d = 10 eV. $T_{\max}$ is given by

$$T_{\max} = \frac{4T_0A_1A}{(A+A_1)^2} \tag{7.6}$$

where A_1 is the atomic mass of the spectrum of recoils emitted from the bombardment (Srour et al. 2003). Displacement damage is therefore another cumulative effect, which is dependent on the intensity of the exposure and the surface area of the target. Not all radiation effects are due to constant bombardment and timely degradation; a single particle can potentially penetrate a semiconductor material and immediately cause an upset in the circuit operation. Such an event is called a SEE and can be a transient or a permanent effect, as described in the following section.

7.4.3 Single-Event Upsets

SEEs, as shown in Fig. 7.4, are caused by a single energetic particle, which can take on various forms: protons, electrons, neutrons and heavy-ion particles. In space, heavy ions in GCR and high-energy protons are the dominant causes of SEEs. For semiconductors, three typical sources that lead to soft errors are alpha particles, high-energy neutrons with energies exceeding 1 meV and secondary radiation from the interaction between cosmic ray neutrons and boron. Resulting from SEE are SEUs, which are soft errors and non-destructive. These errors are typically divided into two categories: as transient pulses in logic or support circuitry or as bit-flips in memory cells or registers. Kundu et al. (2005) present a four-phase approach to protect a system from soft errors. This approach encapsulates a logical outline for prevention, testing, assessment and recovery and for implementing methods in a design to effect:

- Prevention: protecting chips from soft errors—using radiation hardening and radiation-tolerant components/technologies.
- Testing: detecting soft errors—having the ability to know when a soft error has occurred in a system, as opposed to a 'normal' error.
- Assessment: estimating the impact of soft errors—determining which systems will be influenced by a soft error if/when this occurs.
- Recovery: recovering from soft errors—using the techniques required to reset a system and redo the calculation/operation, assuming the soft error is not present anymore. Techniques include redundancy, error-correction codes and parity.

SEE may lead to permanent and destructive errors, also referred to as hard errors, where examples of destructive events are single-event latchups from unexpected high operating currents, as well as burnouts of MOSFETs, gate ruptures, frozen bits and noise in charge-coupled devices. Essentially, a SEE will induce an unfamiliar state in a semiconductor component. The dependent circuitry will therefore act upon this unfamiliar state and may drive the subsystem to output a state that cannot be interpreted, or will be interpreted incorrectly. This interpretation can cause the entire system to malfunction and depending on the severity of this malfunction, a system reset might be required, or worse, an unrecoverable state might be initiated.

Petersen (1998) examines one FOM for heavy-ion and proton SEU rates and the typical calculations for this FOM are presented in the following discussion. The FOM for heavy ions determined by Petersen (1998) is presented as

$$\text{FOM} = \frac{\sigma_{hL}}{(L_{0.25})^2} \tag{7.7}$$

where σ_{HL} is the saturation value of the heavy ion cross-section per bit, $\sigma(L)$, given in units of cm^2 per bit, $L_{0.25}$ is the value for which $\sigma(L_{0.25}) = 0.25\sigma_{hL}$ and for most orbits (LEO, medium-earth orbit, geosynchronous, geostationary or high-earth orbit) the upset rate is given by

$$R = C \times \text{FOM} \tag{7.8}$$

where C is the rate coefficient in bit upsets per day in an orbit parameter. Since satellites (and therefore the electronics in the satellite) are usually shielded by an aluminum enclosure, the differential energy spectrum of inner particles $f(E)$ can be expressed by

$$f(E) = f'(E')\frac{S(E')}{S(E)}e^{-ct} \tag{7.9}$$

where $f'(E')$ is the differential energy spectrum of the satellite surface in $\text{cm}^{-2}\text{d}^{-1}\ \text{meV}^{-1}$ and E is the particle energy for the inner satellite in MeV. The value for c is determined by

$$c = \frac{N_A\left(A^{\frac{1}{3}} + 8.6\right)^2}{27} \times 5 \times 10^{-26} \tag{7.10}$$

where A is the mass of the particle and E' is determined by

$$E' = \frac{1}{R}[R(E) + t] \tag{7.11}$$

where $R(E)$ is the transmission distance of the particle with energy E and t is the thickness of the aluminum casing. The heavy ions SEU rate R_h can be expressed through the LET spectrum and SEU cross-section area for various LET values such that

$$R_h = \int_0^\infty f_e(L)\sigma_e(L)\,dL \tag{7.12}$$

where L is the LET value of the heavy ion in $\text{MeV.cm}^2/\text{g}$ and $\sigma_e(L)$ is the effective cross-section averaged on all solid angles and $f_e(L)$ is the heavy ions' differential ion-flux spectrum in a given orbit or for a given shielding. This value is expressed by

$$f_e(L) = \frac{f(L)}{2\pi}\int_{\theta_C}^{\frac{\pi}{2}} \cos\theta\,d\psi \tag{7.13}$$

where θ_C is the critical angle at which SEU occurs and given by

$$\theta_C = \cos^{-1}\left(\frac{L}{L_0}\right) \tag{7.14}$$

where L_0 is the LET threshold of SEU. The relationship in (7.13) can trigonometrically be simplified to

$$f_e(L) = \frac{f(L)}{2}(\cos(\theta_C))^2 \tag{7.15}$$

resulting in

$$f_e(L) = \begin{matrix} \frac{f(L)}{s}\left(\frac{L}{L_0}\right)^2, & L \leq L_0 \\ \frac{f(L)}{2}, & L > L_0 \end{matrix} \tag{7.16}$$

where $f(L)$ is the differential energy spectrum of the heavy ions. For the case of protons, the SEU rate R_p is calculated by

$$R_p = \int_0^{\infty} f(E_p)\sigma(E_p)dE_p \tag{7.17}$$

where E_p are the energies of the protons. In Petersen (1998) the *FOM* for proton radiation is given by

$$\text{FOM} = 4.5 \times 10^4 \sigma_{pL} \tag{7.18}$$

and the upset rate is similar to (7.8). Using these results, the FOM method can also be used to predict the SEU rate by using the upset rate, R, as

$$R = R_h + R_p = (C_h + C_p)F = (C_h + C_p)\frac{\sigma_{hL}}{(L_{0.25})^2} \tag{7.19}$$

where C_h and C_p are the orbit SEU coefficients for the heavy ions and the protons. The FOM (see (7.7)) can be achieved through experimental results. The parameter $L_{0.25}$ as first presented in (7.7) can be calculated by

$$L_{0.25} = L_0 + 0.28^{\frac{1}{s}} W \tag{7.20}$$

where S and W are Weibull parameters for the size and weight of the device. As seen from the above derivations, the SEU of electronics in space are statistically distributed events based on the particle energy and the material through which it moves. This is true for any material in space or which is exposed to radiation of other forms, such as nuclear. A similar conclusion can be drawn for TID, where the energy of the incident particles and the time of exposure play a role in the degeneration of the semiconductor material. As EW move into space, these considerations are becoming more important to ensure effective and reliable transmissions into and from space. Essentially, the cost of launching a vehicle into space also plays a role in the understanding of radiation effects in space for commercial applications as well. The following sections further explore the radiation to which electronic equipment is exposed in space, but it focuses more on the discrete transistors and the mechanisms present, which cause degradation or single events in the components. Bipolar transistors and CMOS transistors are constructed differently in semiconductor materials and it is these differences that distinguish the radiation tolerance attributes of both. If a CMOS and a bipolar transistor are exposed to exactly the same radiation, at the same orbit around the earth, with the same amount of radiation hardening, the cumulative effect and SEEs will be different; the following sections investigate the main differences in the process flow of CMOS and BiCMOS processes to determine the contributing factors for the difference in behavior/effects during and after radiation exposure.

7.5 CMOS and BiCMOS Process Flow

Bipolar components integrate into the core CMOS process for compatibility with analog, digital, RF and mixed-signal SOC ICs. Along with the high-performance SiGe active components in a BiCMOS process, changes and improvements on the passive components are also incorporated, including high-quality factor inductors, varactors and precision resistors. A SiGe *npn*-transistor in a standard CMOS process typically requires approximately four to five additional photolithographic masks in the manufacturing process. In addition, such a process requires high-energy phosphorous implantation for the sub-collector and selective epitaxy for the SiGe base. These additional steps can be considered relatively low-cost and low-impact on a CMOS infrastructure, but it also introduces the inherent advantage of BiCMOS over CMOS in terms of radiation hardening. Figure 7.6 summarizes the process flow variation for CMOS and BiCMOS processes. Figure 7.6 also shows the required process steps for passive component manufacturing.

The starting point for the process flow example in Fig. 7.6 is a *p*-type or *p*+ substrate layer with a lattice-matched buffer layer on top to provide high-quality inductors while preserving modularity as well as density, isolation and latch-up protection requirements for CMOS. Patterns are etched into the silicon where typically oxide is deposited into the trenches to prevent electrical leakage current between adjacent devices, in a process called shallow trench isolation. This is followed by the implantation step to define the *p*-well (or *n*-well) areas for the CMOS transistor. Boron is typically used to create the *p*-well (or phosphorous for the *n*-well) in two steps involving a low-energy implant for isolation and high-energy implants for latch-up prevention. The dopant is diffused into the bulk to define the thickness of the well. The MOS capacitor is also defined and formed during this process step. The gate oxide is thus defined and the polysilicon-gate is

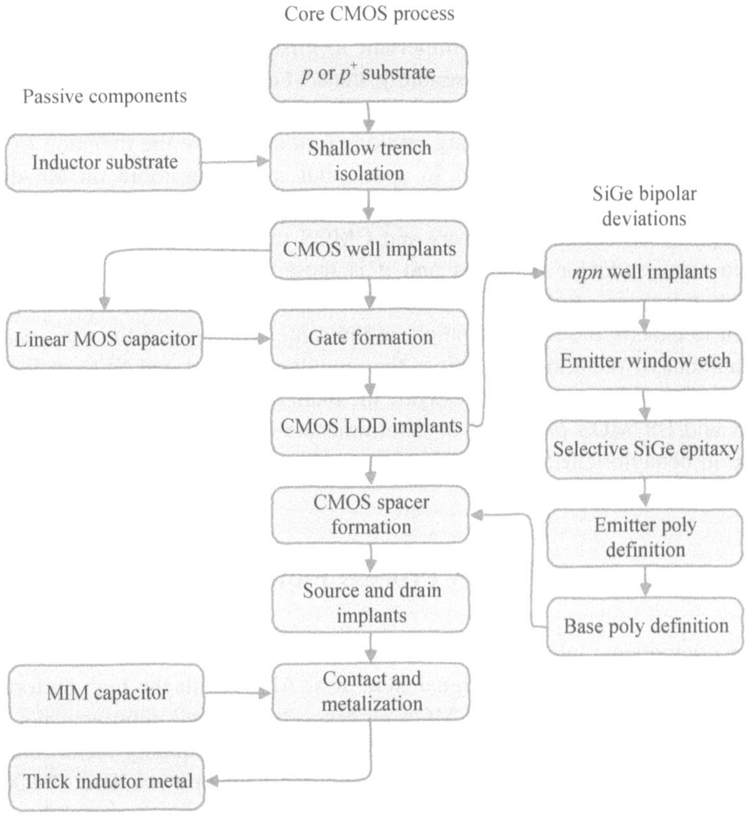

Fig. 7.6 Variations and additions to the core CMOS process flow to manufacture BiCMOS SiGe transistors, including the process steps for passive components (inductors and capacitors)

deposited and its selectively etched. The lightly-doped drain regions for the NMOS and PMOS transistors are formed protruding under the formed gate. At this point the process flow for SiGe structures deviates from the core CMOS process.

In this example, an additional five masks are required to form the SiGe transistor from the core CMOS module. These steps are shown in Fig. 7.6 and include the *npn*-well implants; this is a high-energy and high-dose implant. This is followed by the emitter junction window formation and selective SiGe epitaxy to define the emitter-poly further. The final step is the lithography and etching of both the emitter and base polysilicon, after which the process flow returns to the core CMOS flow.

The process for the CMOS formation is completed by reactive-ion etching CMOS spacers, the source and drain implantations (typically arsenic to form the source/drain of an *n*-channel transistor and contacts to the *n*-substrate) and then the deposition of, for example, Ti/Au for contacts and top-layer metalization; also used to define the metal-insulator-metal capacitors and thick-metal inductors. These process-flow variations between CMOS and BiCMOS are important when considering the radiation tolerance of both technologies.

Radiation can damage either the oxide or the silicon in CMOS transistors, depending on the angle of the incident radiation. If the radiation is incident on the oxide, charge-generation in the interface between the gate and the oxide or between the silicon and the oxide leads to variations in the threshold voltage V_T, the transconductance g_m of the transistor or in the leakage current of the device (Maiti and Armstrong 2001). Incident radiation causes two types of charges in CMOS transistors: oxide-trapped charge and interface-trapped charge. Radiation-induced interface states lower the transconductance of the device and move the *I-V* curve from its default operating conditions as a result of the variations in leakage current and threshold voltage. Device degradation is typically observed after incident radiation has bombarded the device, since the electrons from the generated electron-hole pairs tunnel into the bulk of the transistor and the hole charge remains trapped.

Bipolar transistors rely on junctions for operation where CMOS transistors rely on surface states, making bipolar devices inherently more tolerant to radiation. Another factor that increases the tolerance of bipolar devices is the fact that these transistors are doped up to three orders of magnitude higher when compared to CMOS devices. Bipolar transistors are also, like CMOS transistors, susceptible to increases in leakage current but also to degradation of the current gain h_{fe}. Atomic displacement increases the number of recombination opportunities, which reduces the minority-carrier lifetime and thus increases the current leakage at its base. Displacement is the primary mechanism that causes gain degradation in bipolar devices. Gain degradation also occurs in response to ionization of the passivation layer mainly in the emitter-base junction region where charge trapping and generation of new interface traps occur (Maiti and Armstrong 2001). Across the base-collector junction, ionization of the surface oxide causes an increase in the junction leakage current, usually due to charge build-up in the oxide layer over the junction. This effectively creates a highly conductive surface channel.

The following sections discuss a more in-depth approach to the radiation effects and mechanisms in CMOS and BiCMOS processes, specifically targeted at MOSFET and HBT transistors.

7.6 Radiation Effects on CMOS Transistors

In radiation environments, either in space or near nuclear reactors, MOS devices suffer from significant degradations and failures. Ionizing radiation effects in MOS transistors include buildup of positive charge in the oxide layer and interface state production, which lead to shifts in the threshold voltage and channel mobility degradation (Fossum et al. 1975). The threshold voltage V_t in a MOS transistor is defined by

$$V_t = V_{t0} + \gamma\left[\sqrt{2\phi_f + V_{SB}} - \sqrt{2\phi_f}\right] \quad (7.21)$$

where V_{t0} is the threshold voltage when the source-bulk voltage $V_{SB} = 0$ and ϕ_f is the Fermi level defined by

$$\phi_f = \frac{kT}{q}\ln\left[\frac{N_A}{n_i}\right] \quad (7.22)$$

where N_A is the acceptor doping concentration and n_i is the intrinsic carrier concentration of the semiconductor material. The effect of the higher threshold voltage becomes apparent in the transconductance gain of the MOS transistor, gm, given by

$$g_m = \mu C_{ox}\frac{W}{L}(V_{GS} - V_t) \quad (7.23)$$

where μ is the mobility, C_{ox} is the parasitic oxide capacitance, W and L are the width and length dimensions of the transistor and V_{GS} is the gate-source voltage. From (7.23) it can be seen that if the threshold voltage V_t increases, the gain of the MOS transistor decreases. In CMOS inverter logic, for example, the shift in the threshold voltage will also shift the inverter logic threshold, which can lead to inaccurate transitions and overall system failures. In CMOS current mirrors, the shift in the threshold voltage causes unbalanced transistors, leading to changes in the bias current or increased leakage current of the device and consequently to higher quiescent power consumption. In early experiments such as those of Fossum et al. (1975) it was already shown that the dependence of the voltage threshold shift has a large proportionality to the gate-oxide thickness, t_{ox}. This proportionality is defined by Fossum et al. (1975) as

$$\Delta V_t \propto t_{ox}^3 \quad (7.24)$$

which shows that the change in V_t is proportional to the cubic-power of the gate-oxide thickness. This ratio has been used extensively in ionizing radiation-hardening of CMOS transistors from the buildup of fixed charge in the SiO_2 and the surface states in the silicon-silicon-dioxide interface (Fossum et al. 1975). Gate-oxides have scaled down in thickness with technology; very little charge can be trapped in the thin gate-oxide and the trapped charges can easily be neutralized by thermal emission and electron tunneling (Dodd et al. 2010). Figure 7.7 shows the leakage currents present in CMOS transistors and the gate-oxide thickness effect on the tunneling current.

In Fig. 7.7 the gate-tunneling current I_{TUNNEL} through the gate-oxide layer is shown. Figure 7.7 also shows the charged particle effect as the incident radiation moves through the CMOS transistor, *n*-type substrate with *p*-type drain and source regions. Diffusion currents, tunneling and drift currents are generated in the substrate and the *p*-type drain region as a result of the passing ionized particle. These

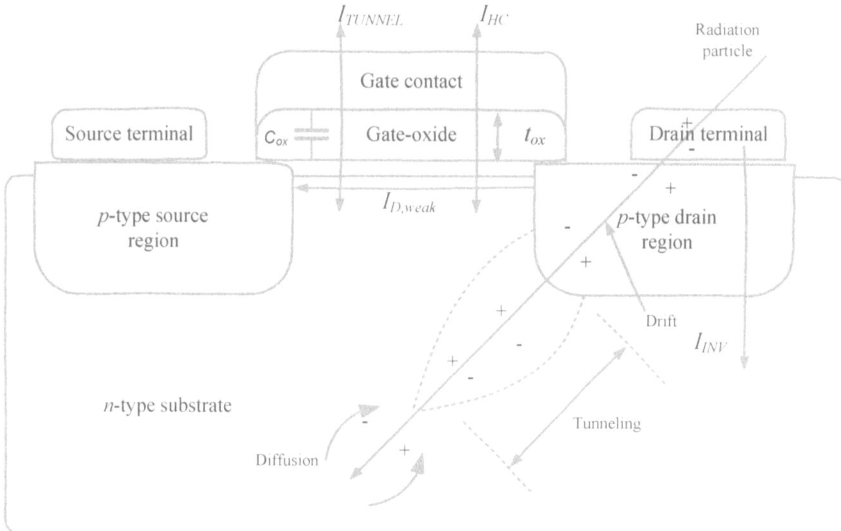

Fig. 7.7 Leakage current in CMOS transistors (*n*-type substrate shown) showing the charged ionization particle path through the transistor, as well as gate tunneling, hot carrier injection, reverse biased current and the weak sub-threshold drain current

effects, in CMOS logic, can lead to SEEs, the mechanisms described by Texas Instruments (1994):

- Drift: Generated in the depletion region and part of the prompt portion of the induced current.
- Tunneling charge: Generated in a *funnel* region, located below the depletion region and resulting from instantaneous distortion of electrical fields deep in the silicon. The effect of this current is larger than the drift component and is quickly drawn back into the affected region (drain or source) contact terminal.
- Diffusion: This is the delayed portion of the event and is located below the tunneling region.

The hot-carrier leakage current I_{HC} is seen if a carrier gains enough kinetic energy and is able to overcome the gate-potential barrier, therefore it is dependent on the biasing conditions of the transistor. If the gate voltage is smaller than the threshold voltage and $V_S = V_B = 0$, the transistor forms a weak inversion layer, has a constant voltage across its channel and the longitudinal electric field across the channel is zero. A leakage current $I_{D,weak}$ is generated and slowly drifts across the channel because of diffusion. The reverse-biased current I_{INV} is the current that flows through the reverse-biased diode junction between the drain and the *p*-region of the transistor and is dependent on the junction area between the source and drain terminals and the bulk of the device. This current is exponentially proportional to the temperature of the device.

In more recent times, Dodd et al. (2010) estimated that the specialized radiation-hardening techniques in CMOS analog circuits are at least three generations behind the standard commercial technologies in terms of node size. This is a significant statement and its relevance is apparent in the types of electronic circuits that can be used for either commercial or military electronic equipment used in space. It also points at a significant difference in cost when manufacturing these radiation-tolerant electronics.

Dodd et al. (2010) also provide a comprehensive list of the challenges and technology trends regarding radiation effects in CMOS transistors. This list covers both TID and SEEs and is adapted and summarized here. For TID, the primary technology challenges in CMOS are listed as:

- Field isolation leakage: This effect relates to the gate-oxide layer and the thinning of this layer with scaling technologies. Sidewall leakage leads to positive charge trapping in the STI and can invert the edges of the channel where the gate polysilicon passes over the STI. These effects are typically witnessed as anomalies in the I-V characteristics of the transistor and lead to higher leakage currents.
- Buried oxide leakage: This effect is observed in silicon-on-insulator technologies (as opposed to bulk silicon processes) when positive charges are trapped in the buried oxide layer near the silicon and oxide interface. A buried oxide layer increases dielectric isolation, reduces parasitic capacitances and enhances radiation performance, but also leads to more complex effects, which are more difficult to model and prevent. Essentially though, a channel is formed between the source and the drain owing to the added buried oxide layer between these regions.
- Technology variations: Most integrated systems used in commercial and military applications use a variety of technologies to realize an integrated solution. These variations introduce additional weak links in the system and might contain components that are not specifically radiation-hardened. Interfacing with these components could mean that the system may fail because of a subsystem failure not necessarily regarded as radiation-prone, or using older and more cost-effective technology, which is inherently less radiation-tolerant.
- Particle microdose effects: These are cumulative TID effects caused by singular energetic particles of radiation. These effects fall between pure TID and pure SEEs. Such events cause single upsets, leading to gradual degradation of the semiconductor, which is possibly due to currents higher than the nominal current flowing for extended periods of time and inflicting damage on the material.

For the case against SEE challenges, Dodd et al. (2010) list the following factors related to CMOS device radiation:

- Proton direct ionization: Lower-energy proton radiation is typically associated with devices that are designed to collect charge from the environment, such as photodetectors and other optical devices. For CMOS memory, this type of ionization was not considered a high-risk threat, but along with semiconductor miniaturization the amount of charge required to store information in memory has also decreased significantly. This smaller charge requirement has led to SEU from proton direct ionization in CMOS ICs, such as memory becoming a larger concern.

- Digital single-event transients: This effect is generated when voltage or current perturbations occur in logic gates due to irradiation. These effects have the potential to be propagated through subsequent advanced CMOS logic circuitry in, for example, very-large scale integrations and cause SEU when it reaches a latch or other memory element.
- Angular effects: The consideration of the angular effects of incident radiation is often neglected when considering the SEU probability from ionizing radiation. An incident particle at an angle θ from the normal to the surface plane leads to higher collection of charge by $1/\cos\theta$ in the sensitive regions near the drain and source and therefore increases the effective particle LET by this same factor. Experimental data and simulated results should therefore include the tilt angle of the normal surface to generate accurate radiation data.
- Single-event functional interrupts: Such an event leads to temporary loss of device functionality, which is specifically prominent in large-scale memory and control registers, as well as micro-processing components, and requires resetting or a power cycle to recover. For space applications where the hardware is not recoverable, software redundancy is critical to enable recovery from these events.

Bipolar transistors (including HBTs) also suffer radiation damage, albeit to a lesser extent when compared to CMOS transistors. The passivation layer in bipolar devices is thicker than in CMOS devices, which does benefit the device in terms of radiation tolerance, but this is not the primary mechanism responsible for the improved performance in high radiation and lower susceptibility to radiation damage. The main radiation effects experienced in bipolar transistors are gain degradation and current leakage. These mechanisms are reviewed in the following section.

7.7 Radiation Effects on BiCMOS Transistors

Cressler (2013) again stresses that radiation-hardening techniques such as thick metal enclosures are undesirable, as they add to the cost, area, weight and power dissipation of electronic circuits. These trade-offs that are monitored especially in space applications and military equipment should ideally be light and mobile. The susceptibility of SiGe components to space radiation events such as TID, displacement damage and SEEs is less when compared to standard CMOS components. This has been established by Cressler (2007, 2013). The regions in an HBT that are prone to radiation strikes are beneath the emitter and base regions, at the edge of the STI region and *pn* junction associated with the sub-collector to substrate

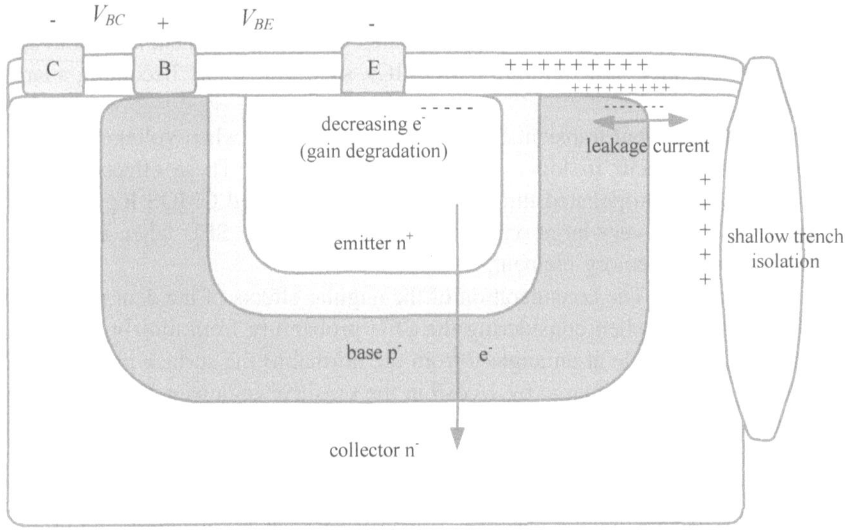

Fig. 7.8 Simplified cross-section of bipolar transistor showing its weakness to radiation damage

junction in bulk Si technologies. The simplified cross-sectional layer structure of a bipolar transistor showing its weaknesses to radiation damage is shown in Fig. 7.8.

SiGe HBTs are multi-Mrad TID-tolerant in their default manufacturing process (Cressler 2013). There is variability between manufactured lots, however, since radiation tolerance is not a controlled process. Typically, lower dose rates of the base region lead to more variability and the tolerance is determined by the manufacturing process (related to the impurities in the semiconductor). This variability leads to higher cost and longer time required for testing batches for space application readiness. TID damage in SiGe degrades the base current and produces an excess base leakage current, which degrades the current gain, similar to silicon BJTs but to a lesser extent. The radiation response of the common-emitter DC current gain, h_{fe}, of a bipolar transistor is described by the Messenger-Spratt equation (Messenger and Spratt 1958; Srour et el. 2003) as

$$\frac{1}{h_{fe}(\Phi)} - \frac{1}{h_{fe0}} = \frac{K(E)\Phi}{\omega_T} \quad \text{(a)}$$

where h_{fe0} is the initial current gain, $h_{fe}(\phi)$ is the final (after radiation) current gain, $K(E)$ is the particle- and energy-dependent displacement damage factor typically determined experimentally, ϕ is the incident particle fluence and $\omega_T = 2\pi f_T$ is the frequency where the common-emitter amplifier has unity gain. The damage factor generally depends on the type of dopant and junction depth of the bipolar transistor.

An important statement in Cressler (2013) is the fact that it is not the germanium in SiGe HBTs that leads to the higher tolerance of these transistors to radiation damage. It is rather the physical construction of the device; a structure emanating from the requirement to embed a strained SiGe alloy in the base region of the HBT. The results achieved by Lu et al. (2003) are freely available online and provide a comprehensive layered structure of SiGe HBTs and the radiation-susceptible regions within these transistors.

7.8 Radiation Effects on Optoelectronic Components

The mechanisms of space radiation effects on optoelectronic devices are briefly presented in Fig. 7.9, adapted from the summarized effects first presented by Srour and McGarrity (1988).

In Fig. 7.9, the radiation effects and mechanisms present in optoelectronic devices are given. These devices include photodetectors, laser diodes, LEDs and optical fibers. Laser diodes (see *Types of Lasers* in Chap. 2) have been reviewed and discussed in this book and the mechanisms that generate laser light have been presented by means of energy-level diagrams. The mechanisms present in laser diodes in terms of radiation effects relate to these discussions through the current density principles in laser cavities. Incident radiation of lasers effectively increases the current density threshold and can decrease the efficiency of the device. Photodetector design considerations are presented in Chap. 4, radiative principles for transmission from an optical source towards photodetectors in Chap. 5 and photodetector bandwidth considerations in Chap. 6. According to these discussions,

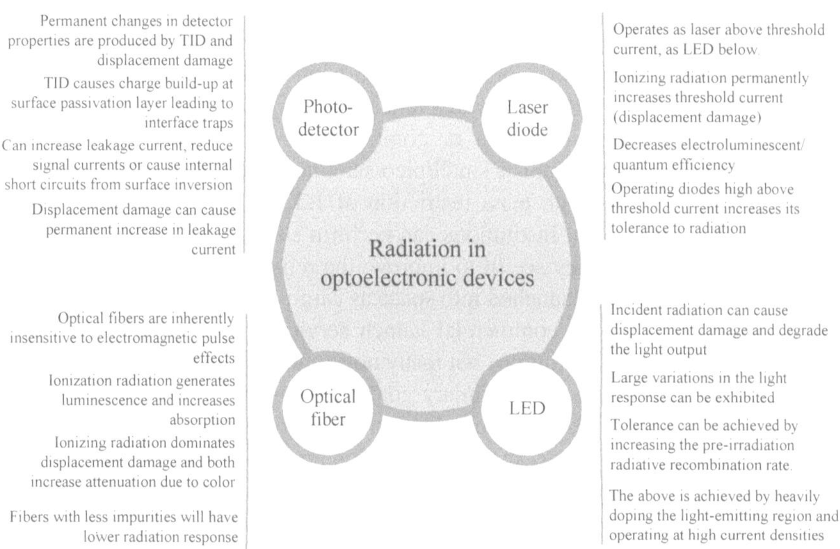

Fig. 7.9 Summary of radiation effects in optoelectronic devices. Adapted from Srour and McGarrity (1988)

the radiation effects presented in Fig. 7.9 agree, the main effect from contributing radiation being increased leakage current from charge build-up. Photodetectors are effectively diodes; therefore, the effects are similar to the effects described in this chapter for active devices. LEDs, effectively similar to lasers operating below their current threshold, also experience lower efficiencies of light output and variations in their output spectrum from ionizing radiation. Optical fibers, which are inherently immune to EMI, can be degraded by ionizing radiation, which dominates displacement damage and effectively increases attenuation of light in the fiber towards the output.

7.9 Space Radiation Effects Program

The Space Radiation Effects Program (SPACE-RAD) is a space measurement and modeling program to advance understanding of the harsh radiation environment of near-earth space and its degradation effects on space systems (Gussenhoven and Mulle 1993). SPACE-RAD offers tools for estimating ionizing radiation environments on any orbit and for computing shielding effects, single-event error rates and total dose in spacecraft systems. A list of features that can be used and calculated by this tool is given in Table 7.1.

Gussenhoven and Mulle (1993) give a list of experiments from various institutions on different equipment. This list is adapted for this book, where only the experiment number and agencies performing the experiments are omitted; these details can be found in the paper published by Gussenhoven and Mulle (1993). These experiments are listed in Table 7.2.

Since the publication by Gussenhoven and Mulle (1993) up to the present, thousands of space radiation experiments have been done by various institutions around the world. A large driver towards relatively inexpensive satellite capabilities for almost any research institution or commercial and military instance is the CubeSat programs. A CubeSat is a small/pico-satellite with dimensions in multiples of $10 \times 10 \times 11.35$ cm with mass restriction of 1.33 kg and COTS components are used in its construction. Institutions can perform additional radiation hardening on these components as they see fit to improve the reliability of the subsystems on board. These satellites are launched into space as cargo on larger launching vehicles and by companies offering commercial launch services. These satellites are typically used by academic institutions, but many non-academic institutions have joined the program in recent years. The primary goals are typically earth observation and amateur radio, but the types of sensors that can be integrated into these satellites are endless—including sensors to measure radiation in LEO. From these relatively low-cost space/satellite capabilities that many institutions now own, experimental data on radiation are extensively available from various sources on the internet.

Table 7.1 List of features of the SPACE-RAD radiation analyzing tool—features listed are applicable to effects on electronic equipment

Feature	Description
Shell shielding distributions	Evaluate the shielding distributions anywhere inside a spherical shell, cylindrical shell or box
Environment grids	Compute latitude and longitude grids of fundamental radiation environment data
Heavy-ion SEU test data fitter	Automatically fit heavy-ion test data to find the sensitive volume dimensions, funnel length, LET threshold and Weibull parameters
Dose-depth functions	Compute dose versus aluminum shielding depth curves for trapped protons, trapped electrons and solar protons
Solar cell damage equivalents	Compute the 1-MeV electron equivalents for solar cells using trapped proton, solar proton, and trapped electron environments
Heavy-ion SEU	Standard heavy-ion SEU models
Extreme solar proton models	Includes several models of the most intense solar proton events. Models include both peak flux and total fluence for most events
Solid shielding distributions	Evaluate the shielding distributions of an infinite slab, or anywhere inside a solid sphere, cylinder or box
Proton, electron and neutron spectrum imports	Import trapped proton, electron and neutron or solar proton, electron and neutron spectra from other sources
Atmospheric neutrons (Avionics)	Atmospheric neutron spectrum model from Avionics. Allows integration along any trajectory in mean or worst-case environments
Neutron-induced SEU	Standard neutron-induced SEU models. Input options include LET threshold and saturation cross section and Weibull parameters
Proton-induced SEU	Standard proton-induced SEU models. Input options include one- and two-parameter Bendel (A&B) model and Weibull parameters
Charged particle trajectories	Follow the trajectories of electrons, protons and heavy ions of any energy
Exact geomagnetic shielding	Incorporates all known physics into an accurate calculation of the geomagnetic shielding at a point or along an orbit

The full list is available at www.spacerad.com

Table 7.2 List of experiments on satellite systems and electronic equipment conducted up to 1993 as published by Gussenhoven and Mulle (1993)

Experiment	Measurement
Microelectronics	Microelectronic equipment characteristics
Internal discharge monitor	Deep dielectric charging discharges
GaAs solar panel cell	Solar cell performance
High-energy electron spectrometer	Electrons 1–10 meV
Spectrometer for electrons and protons	Electrons 40 keV to 5 meV; protons 800 keV to 40 meV
Medium-energy electron spectrometer	Electrons 100 keV to 2 meV
Medium-energy electrons and protons	Electrons 20–250 keV
Spectrometer	Protons 40–30 keV
Low-energy plasma analyzer	Electrons 100 eV to 30 keV
Isotope in solar flares	Ions 20–500 meV
Relativistic proton detector	Protons > 440 meV, electrons > 80 meV
Proton switches	Protons 20–105 meV
Proton telescope	Protons 1–100 meV
Magnetosphere ion composition spectrometer	Ions 30–300 keV
Low-energy magnetosphere ion composition spectrometer	Ions 40 eV to 40 keV
Heavy-ion telescope	Ions 100 keV to 15 meV
Low-energy ion mass spectrometer	Ions 0.1–32 keV
Medium-energy ion mass spectrometer	Ions 20 keV to 8 meV
Dosimeter	Total dose/nuclear stars
MOS dosimeter	Total dose
Fluxgate magnetometer	DC to 10 Hz magnetic field
Search coil magnetometer	5.6 Hz to 10 kHz magnetic field
Thermal plasma probe	DC to 1 kHz electric field
Passive plasma sounder	5.6 Hz to 400 kHz electric field

7.10 Conclusion

The advantages of SiGe BiCMOS technology over standard CMOS processes are being exploited in commercial, military and space applications. High-density circuits with fast response and low noise characteristics are crucial in many military and space applications and BiCMOS HBTs provide an ideal starting point for such designs and also for research. In space, the X-band frequency spectrum has been specifically allocated to military activity requiring high-speed and low-interference communications. The allocated spectrum between 8 and 12 GHz therefore requires fast-switching transistors to avoid frequency down-conversion and takes full advantage of this broad spectrum. BiCMOS has become an important choice for many semiconductor foundries to focus on and provides high-yield and dependable devices. There are relatively few process variations when converting a pure CMOS process to BiCMOS and implementing HBTs along with CMOS transistors. The benefits are of course the ability to design with both transistors on the same die and using the advantages of each for particular subsystems. For space application, however, HBTs have shown higher inherent radiation tolerance compared to MOS transistors. This chapter lists several examples found in the body of knowledge on results obtained for BiCMOS processes in terms of radiation tolerance. These results are introductory to the sections describing the actual effects from radiation, cumulative and single events, and the mechanisms in CMOS and BiCMOS transistors that lead to failures or unexpected circuit operation. It is important to differentiate between the two if space applications and electronic warfare through the use of satellites and supporting equipment in space are to be employed. Radiation-hardening of electronic devices is costly and places additional limitations on the system in terms of size, weight and technology. Although currently radiation-hardening techniques are widely used as an extra barrier against incident radiation, which is highly advisable to ensure operability in space (or at least to decrease the probability of failure), there are also inherent advantages when using HBTs. This book does not advise against using radiation-hardening, but aims to highlight the benefits of using HBTs in space and for SEW. Optoelectronic devices also suffer from radiation effects and require radiation-tolerance enhancement when used for SEW and these effects are briefly listed in this chapter. Several space radiation effects programs at universities and institutions focusing on space applications are available on the internet and it is advisable to visit these pages to determine the type and intensity of radiation at the planned orbit.

References

Adell, P. C., Yager, J., Pannell, Z., Shelton, J., Mojarradi, M. M., Blalock, B., et al. (2014). Radiation hardening of an SiGe BiCMOS Wilkinson ADC for distributed motor controller application. *IEEE Transactions on Nuclear Science, 61*(3), 1236–1242.

Cressler, J. D. (2007). SiGe BiCMOS technology: An IC design platform for extreme environment electronics applications. In *45th Annual International Reliability Physics Symposium* (pp. 141–149).

Cressler, J. D. (2013). Radiation effects in SiGe technology. *IEEE Transactions on Nuclear Science, 60*(3), 1992–2014.

Curtis, S. B., & Letaw, J. R. (1989). Galactic cosmic rays and cell-hit frequencies outside the magnetosphere. *Advances in Space Research, 9*(10), 293–298.

DeBlois, B. M., Garwin, R. L., Kemp, R. S., & Marwell, J. C. (2005). Star-crossed [space-based weapons]. *IEEE Spectrum, 42*(3), 40–49.

Dodd, P. E., Shaneyfelt, M. R., Schwank, J. R., & Felix, J. A. (2010). Current and future challenges in radiation effects on CMOS electronics. *IEEE Transactions on Nuclear Science, 57*(4), 1747–1763.

England, T., Chatterjee, C., Lourenco, N., Finn, S., Najafizadeh, L., Phillips, S., et al. (2014). Cold-Capable, radiation-hardened SiGe BiCMOS wireline transceivers. *IEEE Aerospace and Electronic Systems Magazine, 29*(3), 32–41.

Fossum, J. G., Derbenwick, G. F., & Gegory, B. L. (1975). Design optimization of radiation-hardened CMOS integrated circuits. *IEEE Transactions on Nuclear Science, NS-22* (6), 2208–2213.

Gaucher, B., Floyd, B., Reynolds, S., Pfeiffer, U., Grzyb, J., Joseph, A., et al. (2007). Silicon germanium based millimeter-wave ICs for Gbps wireless communications in radar systems. *Semiconductor Science and Technology, 22*(1), S236–S243.

Gussenhoven, M. S., & Mullen, E. G. (1993). Space radiation effects program. An overview. *IEEE Transactions on Nuclear Science, 40*(2), 221–227.

ICRP. (2007). The 2007 recommendations of the international commission on radiological protection. *ICRP Publication, 103*, 37, 2–4.

IEEE. (2003). IEEE standard for letter designations for radar-frequency bands. *IEEE-SA Standards Board*. Approved January 21, 2003.

Jung, S., Lourenco, N. E., Song, I., Oakley, M. A., England, T. D., Arora, R., et al. (2014). An investigation of single-event transients in C-SiGe HBT on SOI current mirror circuits. *IEEE Transactions on Nuclear Science, 61*(6), 3193–3200.

Kundu, S., Rajesh, G., Vijaykrishan, N., Raina, R., & Pia, S. (2005). Is the concern for soft-error overblown? In *2005 IEEE International Conference on Test* (pp. 1–2), 2005

Kuo, W. L., Krithivasan, R., Li, X., Lu, Y., Cressler, J. D., Gustat, H., et al. (2006). A low-power, X-band SiGe HBT low-noise amplifier for near-space radar applications. *IEEE Microwave and Wireless Components Letters, 16*(9), 520–522.

Label, K. A. (2004, April). NEPP Webex presentation—Radiation effects 101: Simple concepts and new challenges. *NASA*.

Lu, Y., Cressler, J. D., Krithivasan, R., Li, Y., Reed, R. A., Marshall, P. W., et al. (2003). Proton tolerance of third-generation, 0.12 μm 185 GHz SiGe HBTs. *IEEE Transactions on Nuclear Science, 50*(6), 1811–1815.

Maiti, C. K., & Armstrong, G. A. (2001). *Applications of silicon-germanium heterostructure devices*. CRC Press. ISBN 1420034693, 2001.

McLean, F. B., & Oldham, T. R. (1987). Basic mechanisms of radiation effects in electronic materials and devices. Harry Diamond Laboratories Technical Report. HDL-TR, 2129.

Messenger, G. C. (1969). Radiation hardening of electronics systems. Invited paper. *IEEE Transactions on Nuclear Science, 16*(6), 160–168.

Messenger, G. C., Spratt, J. P. (1958, June). The Effects of Neutron Irradiation on Silicon and Germanium. *Proceedings of the IRE, 46*(6), 1038–1044.

Ohyama, H., Vanhellemont, J., Takami, Y., Hayama, K., Sunaga, H., Poortmans, J., et al. (1994). Germanium content dependence of radiation damage in strained $Si_{1-x}Ge_x$ epitaxial devices. *IEEE Transactions on Nuclear Science, 41*(6), 2437–2442.

Oldham, T. R. (1984). Analysis of damage in MOS devices in several radiation environments. *IEEE Transactions on Nuclear Science, 31*(6), 1236–1241.

Petersen, E. L. (1998). The SEU figure of merit and proton upset rate calculations. *IEEE Transactions on Nuclear Science, 45*(6), 2550–2562.

Preston, B., Johnson, D. J., Edwards, S. J. A., Miller, M., & Shipbaugh, C. (2002). Space weapons: Earth wars. *Published by RAND Corporation*. MR-I209-AF, 2002.

Rice, J. S., Ullan, M., Brooijmans, G., Cressler, J. D., Damiani, D., Diez, S., et al. (2008). Performance of the SiGe HBT 8HP and 8WL technologies after high dose/fluence radiation exposure. In *2008 IEEE Nuclear Science Symposium Conference*. N30-134, 2206–2210, 2008.

Shah, T., Danziger, S., Moores, K., & Joshi, Y. (1998). Cyanate ester die attach material for radiation hardened electronic packages. *Adhesive Joining and Coating Technology in Electronics Manufacturing*, 49–54.

Space Foundation. (2016). U.S. defense Space-based and—Related systems fiscal year 2016 budget comparison. *Update 2*. Retrieved May 3, 2016 from http://www.spacefoundation.org/

Srour, J. R. (1983). Basic mechanisms of radiation effects on electronic materials, devices, and integrated circuits. In *1983 Annual International Nuclear and Space Radiation Effects Conference, Short Course*.

Srour, J. R., Marshall, C. J., & Marshall, P. W. (2003). Review of displacement damage effects in silicon devices. *IEEE Transactions on Nuclear Physics, 50*(3), 653–670.

Srour, J. R., & McGarrity, J. M. (1988). Radiation effects on microelectronics in space. *Proceedings of the IEEE, 76*(11), 1443–1469.

Teply, F. E., Venkitachalam, D., Sorge, R., Scholz, R. F., Heyer, H., Ullán, M., et al. (2011). Radiation hardness evaluation of a 0.25 μm SiGe BiCMOS technology with LDMOS module. In *12th European Conference on Radiation and Its Effects on Components and Systems* (pp. 881–888), 2011.

Texas Instruments. (1994). Application note AN-926. Radiation considerations using CMOS logic. *Literature Number SNOA254A*.

Thrivikraman, T. K., Cheng, P., Phillips, S. D., Comeau, J. P., Morton, M. A., Cressler, J. D., et al. (2008). On the radiation tolerance of SiGe HBT and CMOS-based phase shifters for space-based, phased-array antenna system. *IEEE Transactions on Nuclear Science, 55*(6), 3246–3252.

Thrivikraman, T. K., Kuo, W. L., Comeau, J. P., Sutton, A. K., Cressler, J. D., Marshall, P. W., et al. (2007). A 2 mW, Sub-2 dB noise figure, SiGe low-noise amplifier for X-band high-altitude or space-based radar applications. In *2007 IEEE Radio Frequency Integrated Circuits Symposium* (pp. 629–632), 2007.

Yuan, J., Cressler, J. D., Krithivasan, R., Thrivikraman, T., Khater, M. H., Ahlgren, D. C., et al. (2009). On the performance limits of cryogenically operated SiGe HBTs and its relation to scaling for terahertz speeds. *IEEE Transactions on Electron Devices, 56*(5), 1007–1019.

Chapter 8
Microwave Photonics: Complementing Light-Wave Technology with High-Speed Electronics

8.1 Introduction

A discipline that is fast emerging as an important component in modern high-speed data-transfer and communications technology is microwave photonics (MWP). MWP studies and applies the interaction between microwave and optical waves (light- and radio-waves) for high-speed applications, typically aimed at the millimeter-wave (mm-wave) frequency spectrum. Conversely, the transmission of analog microwave signals over optical fiber is known as radio-over-fiber (RoF) and utilizes the advantages of optical fiber in terms of high bandwidth, low signal losses, reduced size, weight and cost (per unit bandwidth) and immunity to EMI (Wiberg 2008). MWP also enables the realization of key functionalities in microwave systems such as

- high-order filtering,
- arbitrary waveform generation,
- frequency up- and down-conversion, and
- instantaneous measurement of functions that either require complex electronic circuitry or might be impossible to achieve in the RF domain (IDST 2016).

The RoF technique modulates the RF sub-carrier (used to modulate the baseband signal, typically in the GHz range) onto an optical carrier (able to accommodate several GHz carriers on a typically THz optical carrier) for distribution over a fiber network. By combining various RF sub-carriers on a high-bandwidth optical carrier, multiple RF signals carrying information at different frequencies can simultaneously be transmitted across an optical fiber to a central location; the primary difference here being that the central location can convert all these signals from a single fiber, where traditional methods require a channel or a medium for each RF carrier. If, for instance, a WiFi signal, cellular transmission and LTE data are transmitted at the same time, all these signals can *fit* onto a single optical carrier and be transmitted via the fiber simultaneously over large distances.

W. Lambrechts and S. Sinha, *SiGe-based Re-engineering of Electronic Warfare Subsystems*, Signals and Communication Technology,
DOI 10.1007/978-3-319-47403-8_8

The relatively recent commercialization of fiber-optic technology to provide global access to broadband internet and television to residential areas and businesses (especially prevalent in developed countries rolling out fiber-to-the-home (FTTH) and becoming more commonplace in certain developing countries) shows that there is significant driving force behind this technology. From a military perspective, in EW for example, fiber-optic communications systems are effective components for current and future applications such as

- EW decoys, jammers and receivers,
- signal intercept and recognition,
- location services,
- combat identification,
- signal acquisition, and
- point to multi-point direction finding.

A light-wave is essentially an electromagnetic wave, so it has an internal oscillating frequency. The wavelength of the light typically used in commercial and military applications has extremely high frequencies. A 1 μm wavelength light-wave translates to approximately a 300 THz oscillation frequency, whereas a 12 μm light-wave translates to about 25 THz. The advantages of speed and bandwidth that come with the inherent oscillation frequency of light also have trade-offs when considering the electronic circuits that are required to process this information, ideally at the same speed. Current microelectronic components are unable to process data at these high-THz frequencies and are limited by the switching speeds of the active components such as transistors. This means that the data should possibly be up- and down-converted on the transmitting and receiving side for the sake of the electronic equipment. This is unfortunate when working in the infrared spectrum where lasers and radiation-detecting photodetectors already operate in military applications such as radar, communication, laser rangefinders and laser detonators and in various commercial applications.

MWP aims to make use of the already integrated optical and electrical equipment, which has proven critical in many applications and implements techniques to integrate both domains on small, lightweight and low-power systems. Integrated microwave photonics (IMWP) apply integrated photonics and electronics to microwave photonics systems on a single substrate and has been successfully presented in materials such as InP, silicon-photonics, $LiNbO_3$ and Si_3N_4. Typical system components used in MWP and IMWP include lasers, modulators, photodetectors, optical amplifiers, optical switches and optical filters, along with passive components such as couplers, splitters and delay lines.

Another advantage of MWP apart from its high bandwidth capabilities is that it is tunable over a large frequency range. A change in light-wave wavelength of 1 nm when shifting the laser threshold current (see Chap. 2) translates into a 300 GHz change in frequency. This level of tunable oscillation can be useful if implemented correctly. Controlled variable oscillations in electronic jamming equipment, function generators, phase-locked loops and frequency synthesizers have long been critical components in various electronic systems, traditionally realized by the VCO. These oscillators, depending on their topology, are typically defined by their tuning range, tuning gain and phase noise; low phase noise is required in a VCO. The ability to modulate a baseband signal onto a carrier generated by a light-wave has distinct bandwidth and tuning advantages and this chapter explores the techniques used to realize purely optoelectronic oscillators (OEOs) for MWP systems. In electronic jamming equipment in EW, a large tuning range of the local oscillator is beneficial, as it allows jamming over a large spectrum. Since MWP is a combination of photonic and microwave technologies, the viability and performance limitations of enabling microelectronic technologies are explored in this chapter along with the unsurpassed advantages of light-waves.

8.2 Distinguishing Between the Microwave and Optical Domain

MWP is a relatively new discipline, active for approximately the past 20 years (Jäger 2009), using photonics to generate, process, control and distribute microwave signals. Applications of MWP include radar, missile guidance, antenna arrays, astronomy, spectroscopy, high-speed communication, wireless sensor networks, increased performance of traditional EW systems and instrumentation (Yao 2012) such as medical imaging, since certain THz radiation is not harmful to biological samples. Figure 8.1 summarizes MWP by displaying its fundamental components, the typical tasks required from an MWP system and also the applications where MWP has had a large influence in recent years.

Figure 8.1 shows, at the top, a basic Venn diagram of the core components that define MWP. As the name suggests, these core components are optical light-waves, typically transported by optical fibers, combined with more traditional radio-wave wireless links to process and transmit data. Combining optical technology with high-speed electronic components, such as devices manufactured in SiGe for example, present many advantages in terms of high-speed data processing, low noise and low susceptibility to interference, as well as long-distance transmissions. Light-waves travel at the speed of light, which makes them ideal to send information between two points with very short delays; having the ability to process information at the transmitter and receiver with microelectronic circuits without the need to down-convert the data-rate makes MWP a contender in various applications. Figure 8.1 furthermore shows the typical tasks required from microwave photonics systems where these tasks must be shared between the optical and the microwave domains. These tasks include the generation of high-frequency signals, processing and manipulating the data to achieve a usable format, distributing the data from a transmitting device to a receiver and converting the received data into a format that the receiver can display or store. At the bottom of Fig. 8.1 are the typical applications that can benefit, and already benefit from, MWP systems. These

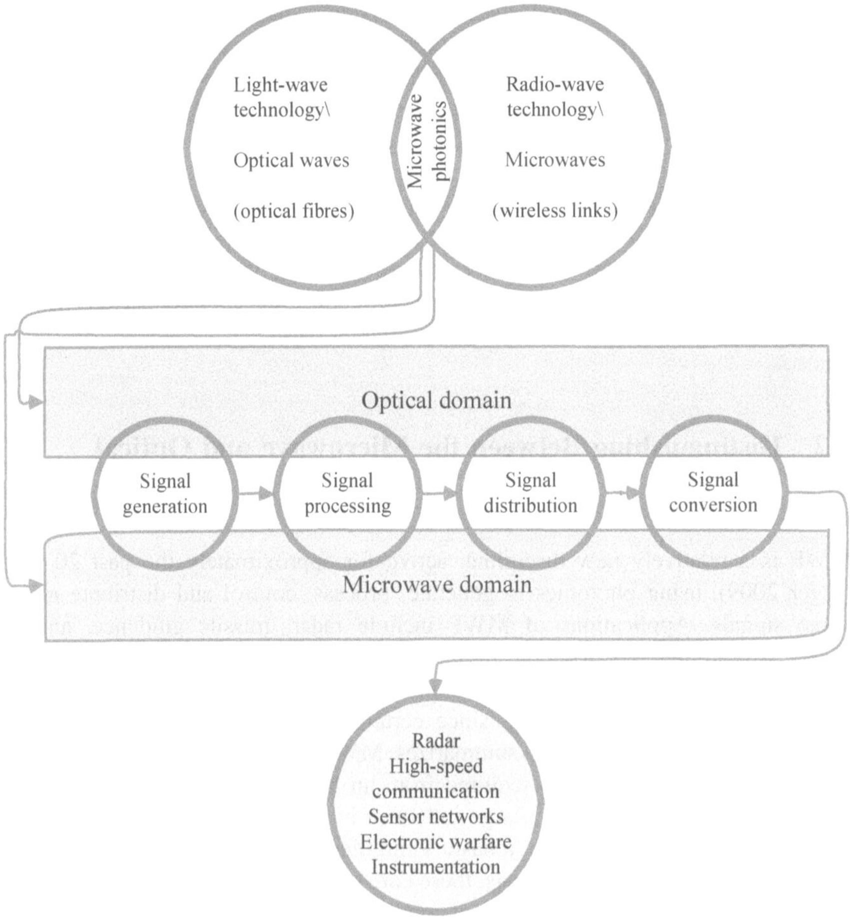

Fig. 8.1 Simplified Venn-diagram (*top*) indicating microwave photonics domain; required techniques to realize microwave photonic communication (*center*) and typical applications of microwave photonics (*bottom*)—adapted from Yao (2012)

disciplines or applications include radar systems, high-speed communication, sensor networks, EW and instrumentation.

Microwave photonics technology, as applied to EW systems, involves the modulation of broadband RF signals onto an optical carrier for transmission through fiber-optic links or guided-wave optical devices for signal processing. The optical signals are usually converted back to electrical signals for further processing either in the analog or digital domains (Aulakh 2013). The ability to distribute broadband signals is becoming more relevant to EW receivers because of the emerging trends in contemporary radar technology requiring larger amounts of information being transmitted between the source and the target. The major trends are the emergence of mm-wave threats transmitting, in particular between 32 and

36 GHz, and radars employing low probability of intercept (LPI) techniques. The ability to transmit the full received bandwidth to on-board receivers, rather than processing at the antenna location, is a key enabling technology in the detection and characterization of LPI emitters. While the term LPI is used to describe a wide range of radar techniques, many of these systems employ very low power and use spread spectrum techniques involving complex modulation schemes. Accurate analysis and classification of these threats require high-speed digitization and digital signal processing. While down-conversion and digitization at the antenna location are possible, practical considerations such as power consumption, clock distribution and electromagnetic interference make implementation difficult. It is in these applications that photonic distribution can become the optimum solution (Aulakh 2013).

Generating light-waves at frequencies compatible with microelectronic components involves beating two light sources together and using the resultant signal, a frequency-difference between the frequencies of the original source. This technique is discussed in the following section.

8.3 Two Light Sources; One Microwave Frequency

The biggest advantage of microwave photonics lies in its ability to generate high-frequency (in the THz spectrum) and frequency-tunable signals by beating two optical waves at a photodetector (Rolland et al. 2011). SiGe comes to mind when discussing optical signal detection and high-frequency analog signal processing on integrated solutions (on the same wafer).

If two coherent light-waves of marginally different frequencies are concurrently incident on a photodetector, they yield a photonic signal modulated at a frequency Δf, a technique that is called optical heterodyning and reviewed here. This resultant frequency is called the beat note or beat frequency and can be observed, depending on whether the process meets some criteria. Before discussing the technique and the mechanisms to achieve optical beating, the criteria that must be realized to observe the beat frequency are summarized:

- The spatial distributions of the two light fields must overlap and must not be orthogonal.
- The polarization states must not be orthogonal.
- The optical frequency difference must be within the bandwidth of the photodetector.
- The wavelengths of both light signals must be within the range where the photodetector is light-sensitive.

Once these criteria have been met, optical beating can be realized at a photodetector. Optical beating allows microwave frequencies to be generated in the THz and high-GHz band and is limited only by the bandwidth of the photodetector.

For example, in order to generate a microwave signal at 1 THz, consider the following two optical waves: If the first wave of wavelength λ_1 of 1.55 µm, therefore having a frequency f_1 of 193.4 THz, is combined with a second light-wave with λ_2 of 1.542 µm, therefore f_2 of 194.4 THz, the beat frequency is equal to

$$\Delta f = f_1 - f_2 \tag{8.1}$$

resulting in Δf = 193.4–194.4 THz = 1 THz. If f_1 is detuned slightly, for example by 0.8 nm, the resulting beat frequency changes by a significant 100 GHz (Iezekiel et al. 2015). This example shows the significance of optical beating to generate highly tunable and stable microwave signals using optical waves; it also shows the large dependence on the stability of the laser diode in terms of its frequency output, which can change with variables such as temperature. A disadvantage of this technique is that the generation of the microwave signal at a frequency equal to the difference between the two optical signals inherently produces high phase noise, since the two signals are uncorrelated. Phase noise is specified in dBc/Hz at a specified offset frequency. For instance, an arbitrary phase noise figure of −150 dBc/Hz means that for 1 Hz bandwidth, at a specified offset frequency (typically 10, 100 kHz or 1 MHz) from the carrier frequency f_0, the single sideband noise power is 10^{-15} of the total signal power. Modeling phase noise in an oscillating circuit is typically presented using Leeson's model (Leeson 1966). The expression to determine the phase noise of an oscillating signal is given by

$$\mathcal{L}\{2\pi f_m\} = 10\log\left[\left[\left(\frac{f_0}{2Q_L f_m}\right)^2 + 1\right] \times \frac{FkT}{P} \times \left(\frac{f_c}{f_m} + 1\right)\right] \tag{8.2}$$

where f_m is the offset frequency (measured at 1 MHz) from the carrier f_0, $L\{2\pi f_m\}$ is the noise level at f_m in dBc/Hz, Q_L is the loaded quality factor of the oscillator tank circuit, k is Boltzmann's constant at 1.38×10^{-23} J/K, F is the noise factor, T is the operating temperature, P is the carrier power in mW and f_c is the corner frequency for flicker noise. Various techniques to generate low phase noise microwave signals exist and are categorized in three main classes, namely optical injection locking, optical phase-lock-loops and generating microwave signals using external modulation. These techniques fall outside the scope of this book and are presented in more detail in the work of Yao (2012).

Essentially, an MWP system comprises the same core components as a traditional communication system. These components, however, differ in their functionality, frequency requirements and the domain in which they operate (optical or microwave). These core components include the source, channel, amplifier and receiver stages. MWP systems have changed in their architecture, topology, functionality and efficiency since first being realized, but the fundamental principle of these systems has remained unchanged from first-generation systems. An example of a first-generation microwave photonics receiver is shown in Fig. 8.2. The system

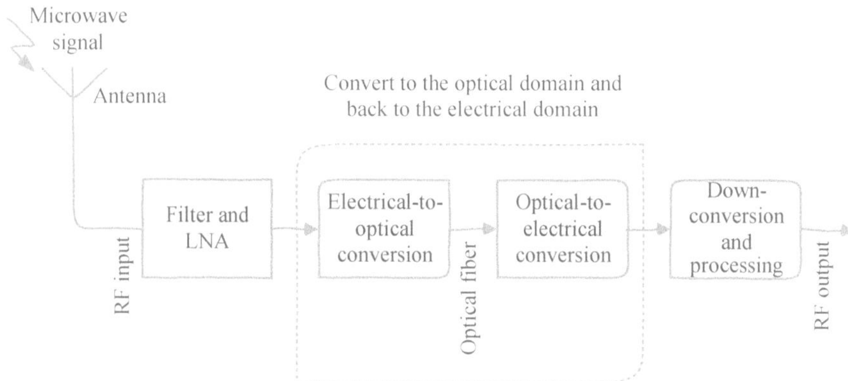

Fig. 8.2 Simplified diagram of a typical, first-generation, microwave photonics system

shows how a microwave signal is received and used in an MWP system to relay the received information at high speed across an optical fiber channel.

Figure 8.2 shows a simplified MWP system where an incoming microwave wireless signal is received at an antenna, filtered to remove any unwanted spurious signals and amplified by the low-noise amplifier in preparation for its conversion to the optical domain. This signal is fed to an electrical-to-optical converter (E/O) and the resulting optical signal is distributed to the end-point through an optical fiber. At the receiving site the optical signal is converted back to an electrical signal with the optical-to-electrical converter (O/E) and if necessary (depending on the operating speed limitations of the electronic components) down-converted and processed based on the final requirements of the system and the required format of the data. An important consideration to take note of is that during the stage where the electronic signal is converted to its optical equivalent, the possibility of combining several electronic signals of limited bandwidth into a single optical signal with significantly superior bandwidth enables large *chunks* of information to be sent across the fiber medium. A schematic representation of this technique, called sub-carrier multiplexing is given in Fig. 8.3. Figure 8.3 only shows the transmitting side of a communication system for simplicity; at the receiver side, the inverse of the representation can be used to present the reverse process.

E/O and O/E are essentially lasers as reviewed in Chap. 2 and perform optical detection with photodetectors as reviewed in Chap. 6. Figure 8.4 summarizes these processes in a simplified optical data communication system.

As shown in Fig. 8.4, the electrical input signal, either an analog or digital representation of the data, is amplified by the driving circuitry of the laser. This generated light pulse is transmitted, ideally through a low-loss optical medium such as an optical fiber, to the receiver. A photodetector receives the incident incoming light pulses and a transimpedance amplifier converts these signals to usable voltage signals, which recreates the input signal back to its electrical equivalent at the output. The frequency of the signal received by the photodetector depends on the

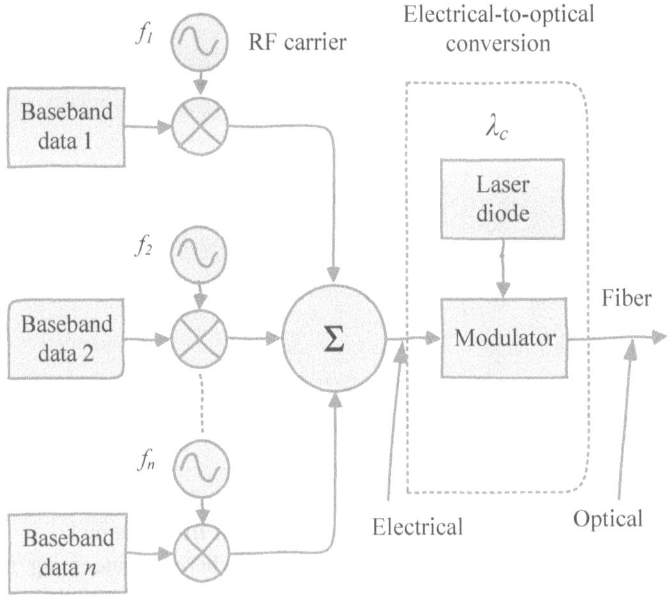

Fig. 8.3 Functional representation of multiple RF carrier-modulated baseband data signals multiplexed onto a single, extremely high-frequency optical carrier transported across a fiber. At the receiver side, a similar procedure is followed but inverted to de-multiplex the individual signals

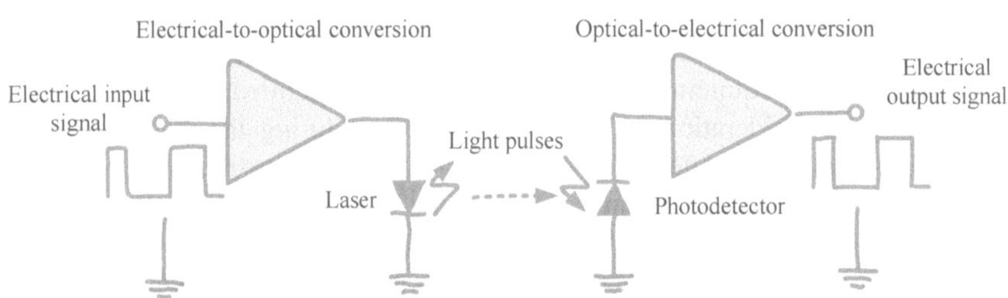

Fig. 8.4 Optical data communication system showing the essential components required for electrical-to-optical conversion (*left*) and optical-to-electrical conversion (*right*)

carrier frequency at which it is modulated (mixed), typically achieved by the local oscillator of the electronic modulator. In MWP, an OEO can modulate the baseband signal at much higher frequencies, comparable to the frequency of the light-wave itself. These frequencies are typically higher than can be processed by current electronic equipment, spanning hundreds of THz for typical IR light-waves, except if the two frequencies are beaten together to generate a lower frequency, as discussed in the following section.

8.3.1 Achieving THz Microwave Signals

To beat two optical frequencies together, two separate light-waves with frequencies ω_1 and ω_2 (where $\omega = 2\pi f$ and f is specified in Hz) are required. These light-waves can be realized by lasers for example, making them ideal for small, lightweight and relatively low-power devices. Figure 8.5 shows the basic arrangement required to beat two optical frequencies.

As shown in Fig. 8.5, two optical sources, laser diodes for example, with different optical frequencies are required to achieve optical beating. The optical signals from these two sources are combined in an optical coupler and fed to a photodetector, which is used to convert the optical signal to its electrical equivalent at the resultant frequency. This technique forms the basis of optical signal generation, where the optical frequency can be varied substantially, depending on the requirements of the application. Before discussing the optical beating technique in further detail, the optical coupler component critical to achieving the output light at frequency $\omega_2 - \omega_1$ is reviewed.

8.3.2 Optical Couplers

Fiber- and integrated optical couplers are crucial components in photonics and specifically in microwave photonics applications. Optical couplers are typically passive devices used to distribute light entering the device to various configurations of outputs, depending on the wavelength and polarization. These components are generally four-port devices and their operation depends on distributed coupling between two individual waveguides in close proximity, resulting in gradual power transfer between modes supported by both these waveguides. This power transfer and cross-coupling at the coupler output ports can essentially be described as the beating between Eigen modes of the composite two-waveguide structure along the length of the composite coupler cavity. Fiber-optic coupler definitions that should be understood in order to effectively define optical couplers include (adapted from AFOP 2002):

- Center wavelength: As with many electronic and optical components, the performance of the system is dependent on the wavelength of operation. The frequency of the incoming signal is typically a limitation of any system and the electronic or optical components must be able to compute incoming data at the speed at which the data arrive at its input. In the case of optical couplers, the devices are generally specified over a specific wavelength window and in some cases can be specified over multiple wavelength windows. The nominal wavelength of the optical coupler is the center wavelength of the specified window(s).
- Bandpass [Hz]: The bandpass of the optical coupler is the range of wavelengths at which the coupler is guaranteed to operate nominally.

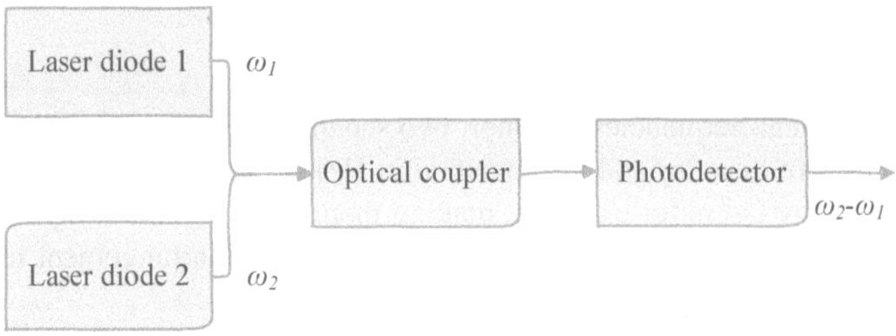

Fig. 8.5 Simplified diagram for beating two optical frequencies to produce a microwave/electrical output signal

- Coupling ratio (splitting ratio) [% of output power at the center wavelength]: The ratio of the optical power from one output port of the coupler to the sum of the total power from all input ports.
- Excess loss [dB at the center wavelength]: The ratio of the optical power launched at the input port of the optical coupler to the total optical power measured from all output ports is known as the excess loss.
- Insertion loss [dB at the center wavelength]: The ratio of the optical power launched at the input port of the optical coupler to the optical power from any single output port is called the insertion loss.
- Uniformity [dB]: The uniformity is the measure of how evenly the output power is distributed at the output ports of the optical coupler.
- Directivity (near-end isolation or near-end crosstalk) [dB]: The ratio of the optical power launched into an input port to the optical power returning to any input port is called the directivity.
- Return loss [dB]: The ratio of optical power launched into an input port to the optical power returning to the same input port is known as the return loss.
- Reflectance: The reflectance is similar to return loss but is specified as the negative quantity of return loss.
- Wavelength isolation (far-end crosstalk) [dB]: The ratio of the optical power at the two output ports of a de-multiplexer.

Optical fiber coupling is typically a passive device and most fiber coupling systems use fused bi-conical taper technology. This discussion is adapted from Alegria (2001). For this process, two parallel fibers are positioned close to each other, annealed and drawn by a micro-machine to form a bi-conical narrowing tube. Figure 8.6 represents a typical four-port fused fiber-coupler system.

Figure 8.5 shows the four-port optical coupler configuration, a 2 × 2 coupler system (two inputs and two outputs). In Fig. 8.6, L_C is the coupling region and comprises two tapered regions, L_{T1} and L_{T2}; the coupler waist is denoted by L_W. If a beam of light is launched into Port 1, the normalized field amplitudes $A_1(0)$ and $A_2(0)$ of the even (A_e) and odd (A_o) Eigen modes at the input, therefore at $z = 0$, can be approximated by

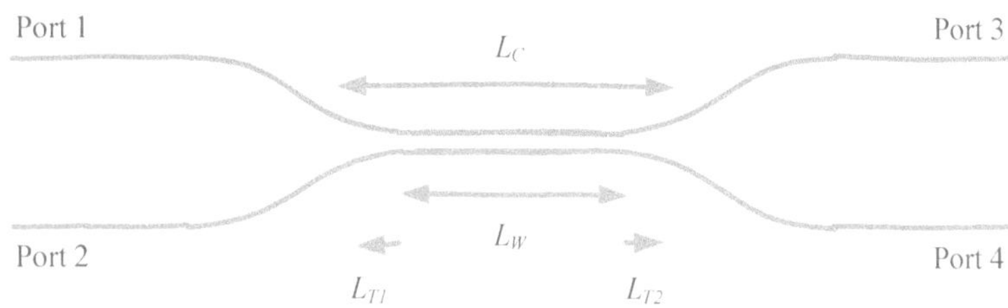

Fig. 8.6 Simplified four-port optical coupler schematic

$$A_e(0) = \frac{A_1(0) + A_2(0)}{\sqrt{2}} \tag{8.3}$$

and

$$A_o(0) = \frac{A_1(0) - A_2(0)}{\sqrt{2}}. \tag{8.4}$$

For single-port excitation, $A_1(0) = 1$ and $A_2(0) = 0$ or $A_1(0) = 0$ and $A_2(0) = 1$ and in this case the even and odd Eigen modes are equal, given by

$$A_e(0) = A_o(0) = \frac{1}{\sqrt{2}} \tag{8.5}$$

and if light is launched into only one port of a 2 × 2 port device, it excites the lowest-order (even and odd) Eigen modes along the region where coupling takes place equally. Two Eigen modes propagate adiabatically along the entire coupling region with propagation constants $\beta_e(z)$ and $\beta_o(z)$ respectively (Alegria 2001). The beating between these two modes is responsible for the coupling laterally with the optical coupler. The propagation total electric field at z is described by

$$E_t(z) = E_e(z) + E_o(z) \tag{8.6a}$$

such that

$$E_t(z) = A_e(z)e^{-i\int_0^z \beta_e(\xi)d\xi} + A_o(z)e^{-i\int_0^z \beta_e(\xi)d\xi} \tag{8.6b}$$

where during adiabatic propagation of the light waves, the even and odd Eigen modes retain their respective amplitudes and vary only their relative phase. This results in spatial beating along the coupler waist and power redistribution between the individual waveguides comprising the optical coupler. The peak field amplitudes for each individual waveguide along the coupling region can be approximated by

$$E_1(z) = \frac{E_e(z) + E_o(z)}{\sqrt{2}} = \cos\left(\frac{1}{2}\phi(z)\right)e^{-i\frac{1}{2}\int_0^z[\beta_e(\xi)+\beta_o(\xi)]d\xi} \tag{8.7}$$

and

$$E_2(z) = \frac{E_e(z) - E_o(z)}{\sqrt{2}} = -i\sin\left(\frac{1}{2}\phi(z)\right)e^{-i\frac{1}{2}\int_0^z[\beta_e(\xi)+\beta_o(\xi)]d\xi} \tag{8.8}$$

where $\phi(z)$ is the relative accrued phase variance of the even and odd Eigen modes and β_e and β_o are the odd and even propagation constants. The equivalent normalized peak power of the individual waveguides is given by $P_{1(2)} = |E_{1(2)}|^2$ where

$$P_1(z) = \cos\left[\frac{1}{2}\phi(z)\right]^2 \tag{8.9}$$

and

$$P_2(z) = \sin\left[\frac{1}{2}\phi(z)\right]^2 \tag{8.10}$$

and the strength of the interaction between the Eigen modes can be described by $k(z)$ where

$$k(z) = \frac{\beta_e(z) - \beta_o(z)}{2} \tag{8.11}$$

and the minimum interaction length of the coupler to induce constructive interference is determined by

$$L_\beta = \frac{2\pi}{\beta_e - \beta_o} \tag{8.12}$$

to ensure that the two Eigen modes are again in phase after travelling the distance of L_β. The mathematical derivations to achieve optical beating and the resulting light-wave frequency are discussed in the following section.

8.3.3 Optical Beating

A microwave signal can be generated by beating the longitudinal modes of a mode-locked laser where the phases of the longitudinal modes are identical (Deng and Yao 2006). The mathematical expression of the output of a mode-locked laser is given by

$$E(t) = \sum_{p=0}^{N} A_p \cos[(\omega_0 + p.\omega_c)t + \theta_0] \tag{8.13}$$

where *E*(*t*) is the electrical field of the laser output in W/cm, *N* is the number of longitudinal modes in the laser, ω_c is the frequency spacing between adjacent modes and A_p and $\omega_0 + p.\omega_c$ are the amplitude and frequency of these modes. The longitudinal modes in a laser cavity are the solitary frequencies that are self-regenerating and permissible to oscillate in the cavity where all additional frequencies are attenuated by destructive interference. In, for example, a plane-mirror laser cavity, the permissible modes must be a multiple of half the wavelength of the light, given as

$$L = \frac{q\lambda}{2} \tag{8.14}$$

where *L* is the length of the cavity, *q* is an integer value known as the mode order and λ is the wavelength of the light-wave. The frequency separation between two adjacent modes Δv is given by

$$\Delta v = \frac{c}{2L} \tag{8.15}$$

where *c* is the speed of light in m/s. Therefore, if two laser beams are emitted continuously at respective frequencies separated by *c*/2*L*, the light output due to interference of the two waves will be modulated by a sinusoidal term, with a frequency *c*/2*L*. This modulation is typically very fast, relating to a few nanoseconds for a one-meter-long cavity and can only be detected by a sufficiently sensitive photodiode or frequency-comparable oscillator. This technique forms the basis of optical beating resulting from interference between light beams with different frequencies, called optical heterodyning or photo-mixing in photodetectors.

8.3.4 Optical Heterodyning

The central physical characteristic in photonic THz OEOs is to down-convert the optical frequency of IR light to the lower-THz domain by using a frequency-sensitive photodetector. The technique of optical heterodyning is based on the wave equations for a wave in a plane travelling in the *x*-direction in space, given its first principles by Iezekiel et al. (2015)

$$\frac{\partial^2 y}{\partial x^2} = \frac{1}{c^2}\frac{\partial^2 E}{\partial t^2} \tag{8.16}$$

where c is the speed of light (the optical light-wave) and E is the electric field. The same form is valid for the magnetic field B, since both the electric and magnetic fields are perpendicular to the direction of travel. The wave equation for electromagnetic waves arises from Maxwell's equations and the form of a plane wave solution for the electric field is given by

$$\mathbf{E}(\mathbf{x}, t) = E_0 \cos(\mathbf{k}.\mathbf{x} - \omega t) \tag{8.17}$$

where E_0 is the amplitude of the magnetic field, $\mathbf{k}$ is the wave-vector and ω is the angular frequency. The magnetic field can be represented by

$$\mathbf{B}(\mathbf{x}, t) = B_0 \cos(\mathbf{k}.\mathbf{x} - \omega t + \phi) \tag{8.18}$$

where ϕ is the phase difference between the electric and magnetic field. Since the electric field and magnetic field follow sinusoidal behavior and are always in phase, when solving the partial derivatives of (8.17) and (8.18), the relationship

$$\frac{E}{B} = c \tag{8.19}$$

holds true on all accounts. To determine the optical beating frequency, assume the two optical light-waves $E_1(t)$ and $E_2(t)$ are defined by (Yao 2012)

$$E_1(t) = E_{01} \cos(\omega_1 t + \phi_1) \tag{8.20}$$

and

$$E_2(t) = E_{02} \cos(\omega_2 t + \phi_2) \tag{8.21}$$

where E_{01} and E_{02} are the amplitude terms, ϕ_1 and ϕ_2 are the phase terms of the two optical waves with time (t) dependence and ω_1 and ω_2 are the angular frequencies for the two waves, respectively. Pointing vectors define the intensities of the constituent waves (I_1 and I_2), where

$$I_1 = \frac{1}{2}\sqrt{\frac{\varepsilon_r \varepsilon_0}{\mu_0}} |E_1|^2 \tag{8.22}$$

and

$$I_2 = \frac{1}{2}\sqrt{\frac{\varepsilon_r \varepsilon_0}{\mu_0}} |E_2|^2 \tag{8.23}$$

are true if the two incident optical waves have equal polarization and are perfect plane waves. The resulting electric field E_0 of the optical interference signal is

$$|E_O|^2 = |E_1 + E_2|^2 = |E_1|^2 + |E_2|^2 + E_1E_2^* + E_1^*E_2 \quad (8.24)$$

which becomes

$$|E_O|^2 = |E_1|^2 + |E_2|^2 + 2|E_1||E_2|\cos[(\omega_2 - \omega_1)t + (\phi_2 - \phi_1)]. \quad (8.25)$$

The intensity of the electric field incident on a photodetector is found by

$$I_O = I_1 + I_2 + 2\sqrt{I_1 I_2}\cos[(\omega_2 - \omega_1)t + (\phi_2 - \phi_1)] \quad (8.26)$$

where the phase of the electric field is determined by

$$\phi(t) = \tan^{-1}\left\{\frac{Im[E(t)]}{Re[E(t)]}\right\}. \quad (8.27)$$

If this optical interference signal is launched into a photodetector, a photocurrent I_{PD} is generated, which can be expressed as

$$I_{PD} = \eta_0 \frac{q}{hf_1} P_1 + \eta_0 \frac{q}{hf_2} P_2 + 2\eta_{f_c} \frac{q}{h}\sqrt{\frac{P_1 P_2}{f_1 f_2}}\cos[(\omega_2 - \omega_1)t + (\phi_2 - \phi_1)] \quad (8.28)$$

where q is the electron charge and P_1 and P_2 denote the optical power levels of the two constituent optical input waves. The photodetector DC and high-frequency quantum efficiencies are represented by η_O and η_{fc} respectively, and both are dependent on frequency. By further assuming, for simplicity, that the optical input waves have equal power, the photodetector current can be derived by

$$I_{PD} = 2\eta_0 \frac{q}{hf_1} P_{eq} + 2\eta_{f_c} \frac{q}{h} P_{eq}\cos(2\pi f_c + \Delta\phi) \quad (8.29)$$

where $f_c = f_2 - f_1$, $\Delta\phi = \phi_1 - \phi_2$ and (8.29) is the fundamental equation describing optical heterodyning in a photodetector. This photodetector current is amplified and converted to a voltage in the TIA, as discussed in Chap. 4. The resultant signal is applied to various MWP systems, utilizing the optical signal carrier frequency in fiber-wireless applications.

8.4 Fiber-Wireless Networks

In regions where radio-wave transmissions are impractical or not secure or where temporary high-speed communication is required, typically in warfare applications, dense wave division multiplexing (DWDM) is employed to create open-air transmissions similar to microwave data transmission technology, but by using light rather than radio waves to transmit data. An mm-wave fiber-wireless system, for example, has the advantage of high bandwidth occupation in an unlicensed spectrum available for scientific research (radio astronomy and remote sensing), telecommunications (WiFi and future 5G implementations), weapons systems (radar and active denial systems), security screening and medicine (un-intrusive radiation for diseases). A hybrid architecture combining mm-wave signal transfer and the high-bandwidth capabilities of optical fiber leads to optically fed mm-wave wireless technologies, also known as fiber-wireless networks. The transport of mm-wave wireless signals over an optical link can be done by three main mechanisms, namely RF-over fiber, IF-over-fiber and baseband-over-fiber. These three mechanisms are reviewed in the following sections.

8.4.1 RF-over-fiber

RF-over-fiber essentially allows the high-frequency, high-bandwidth mm-wave carrier frequencies to be directly transported over a fiber connection. Therefore, it does not require any frequency translation at a base station and the mm-wave signal is externally modulated on the optical carrier wavelength, generating an optical double-sideband modulated wave. High-speed photodetectors are required to recover the wireless (mm-wave) signals at the receiver and demodulate the information signal back to its original form. Since no frequency translation is required, this method significantly reduces circuit complexity, but it also imposes design constraints and drawbacks. The most significant complexity arises from generating mm-wave local oscillators with microelectronic circuits. These circuits must be able to modulate the mm-wave signals with precision, stability and low noise and also demodulate the signals back to the mm-wave spectrum at the receiver. This places stringent design limitations and inevitably higher cost and specialized design techniques on the modulators and detection equipment. Operating in the 60 and 77 GHz frequency spectrum requires many current technology nodes to operate well within their maximum operating conditions and is typically avoided, especially in military applications where stability and accountability are required. To lessen the rigorous requirements on the electronic equipment in fiber-wireless networks, the mm-wave signals can be down-converted to longer wavelength, intermediate (IF) frequencies before being transmitted through the optical medium.

8.4.2 IF-over-fiber

Implementing IF-over-fiber rather than RF-over-fiber relieves the electronic component from its high-speed requirement, but this method also increases the complexity of the system and reduces the opportunity for expansion at a later stage. In order to down-convert the mm-wave signal to an intermediate value, defined by the system requirements and/or communication standards, requires a stable LO to be introduced into the system. The LO interacts directly with high-speed mixers to modulate the RF signal to a lower carrier frequency. A major advantage of this method, apart from being able to use matured and cost-effective technologies due to the lower frequency requirements, is that lower levels of chromatic dispersion are observed in the optical medium, leading to smaller pulse spreading at the receiver. Chromatic dispersion is due to different wavelengths (since light sources emit light at different wavelengths by definition) travelling at different velocities through an optical medium and is much more prevalent at higher frequencies. The delay perceived by varied wavelengths causes pulse spreading, a function of the length of the fiber and the wavelength. At a constant length, shorter wavelengths undergo higher effective pulse spreading. The wavelength dependence is seen from the power distribution $P(f)$ equation given by

$$P(f) = \left[\cos\left(\frac{\pi LD}{c}\lambda_c^2 f^2\right)\right]^2 \tag{8.30}$$

where L is the length of the optical fiber, D is the group-delay dispersion parameter, λ_c is the wavelength of the optical carrier, f is the microwave frequency and c is the velocity of light in a vacuum (Yao 2012). To reduce the effects of chromatic dispersion further and decrease the stringent requirements on the technology used, baseband-over-fiber can be implemented to transport wireless information across an optical fiber.

8.4.3 Baseband-over-fiber

The alternative to RF-over-fiber and IF-over-fiber is to send the baseband data (unmodulated) directly across the optical medium towards the base station, where the signal is further processed, i.e. up-convert the baseband data to be wirelessly transmitted on an mm-wave carrier. Essentially the base-station electronic equipment is significantly reduced in terms of cost and complexity; however, the receiving station requires more complex equipment to process the incoming data. Chromatic dispersion is at a minimum when using this technique but the available bandwidth in the fiber-optical medium is underused, although many baseband signals can be multiplexed onto the fiber network. Optical communications generally have an advantage over microwave communications in terms of high-quality cumulative bit rates and long-distance transmission. The modulation and multiplexing schemes to harness these advantages efficiently require somewhat different considerations compared to traditional RF schemes.

8.4.4 Modulation

Optical modulation concerns the techniques to superimpose information, whether from baseband, IF or RF, onto an optical carrier that is transmitted through an optical medium. Two main techniques modulate high-frequency electrical/microwave signals onto an optical carrier, namely direct and indirect methods (Breed 2007). Direct methods generally require modulation applied directly to the source of the light-wave, whereas indirect methods only manipulate the light beam.

Direct modulation requires that the laser diode be modulated at very high frequencies, therefore be switched on and off by external circuitry at the rate of the modulation frequency. This method introduces high mechanical and thermal stress in the laser diode and can potentially wear out the laser diode. The modulation rate is also limited by the frequency capabilities of the switching circuitry; although it is possible to achieve high-GHz rates with modern technology, this method is still a limitation when compared to the available optical bandwidth. The laser diode transient response time is another limitation of this technique; rapid successive on-off commands to the laser diode depend on the manufacturing technique, intrinsic capacitance and drift times.

Indirect modulation uses a continuous light beam and modulates the light by allowing it to pass through an external medium or forbidding it. This technique introduces lower SNR compared to the actively switching laser diodes in direct modulation. The technique is, however, more complex in terms of required components to achieve external modulation of the continuous light-wave. Improvements to this technique are allowing the light source and modulation medium to be integrated on a single wafer, also allowing active control circuitry in close proximity to the source and modulator, further decreasing the added noise.

8.4.5 Multiplexing

A multiplexer is a device that combines multiple analog or digital input signals into a single line of transmission (Winzer 2009). De-multiplexers at the receiving end of the transmitted stream perform the inverse task and decompose the stream into its individual components. Multiplexing effectively enhances the efficiency of bandwidth usage in a given system by the number of data streams. Optical multiplexer and de-multiplexer systems are essentially passive optical filters arranged to process specific wavelengths in and out of the transport system, typically an optical fiber. The process of filtering the wavelengths can be achieved by prisms, thin film filters, dichroic filters or interference filters. These materials selectively reflect a single wavelength of light and allow other wavelengths to pass freely; they are therefore transparent to these wavelengths. Each filter is specifically tuned for a specific wavelength (Winzer 2009).

Various multiplexing techniques for light-waves can be used. These include optical time division multiplexing, wavelength division multiplexing (WDM), coarse wavelength division multiplexing, DWDM and code division multiplexing. The advantages of optical multiplexing are high bandwidth and data throughput opportunities, low attenuation, low propagation delay and low BER.

WDM is in principle similar to frequency division multiplexing, but is applied only to optical signals. This multiplexing technique (also true for all other types of optical multiplexing) increases the capacity of a network without the requirement of adding physical optical fibers in the system. With optical multiplexing, several generations of technology development can be accommodated, which increases its popularity in telecommunications and other permanent network fixtures.

The following section reviews the realized and potential applications of MWP, specifically in EW, and reviews these possibilities taking into account the optical and electrical requirements.

8.5 MWP EW Applications

The traditional means of transmitting and processing data and information in EW systems has been challenged by the emerging MWP fiber-wireless transmission technology and integrated photonics. The performance improvements and additional benefits of optical communication cannot be ignored, as is seen in many commercial applications such as FTTH internet connectivity, which is receiving large investments and a capital drive to realize it globally. Similarly, in military applications, these advantages can be harnessed and implemented in traditional systems to improve bandwidth capabilities and decrease the chance of signal interference or signal intercept, yielding improved signal security. Applications that can benefit from MWP in EW are mission-critical fiber-based RF communications, positioning systems, telemetry, radar, jamming countermeasures and electronic support, to name a few. In military applications there are several vital factors that are not necessarily equally imperative in commercial applications. These include

- data and transmission security,
- integrated system flexibility,
- scalability, and
- survivability.

"*RF-over-fiber technology offers unprecedented performance and mechanical advantages for radar, SATCOM, JTRS/SDR platforms, navigation, and SIGINT/COMINT/ELINT technologies. Delay lines are a key component to many EW, radar and telemetry systems for calibration and timing*" (Fulton 2003). Fulton

(2003) lists various EW applications that benefit from MWP, specifically RF-over-fiber. These benefits, along with additional examples, are adapted and presented briefly in the following list.

8.5.1 Remote Transmitters

In a remote transmitter scenario, it is assumed that very little to no signal pre-processing or signal pre-amplification is available before the microwave signal is converted to the optical domain (Lindsay 1992). In such a situation, most of the signal processing and amplification are done at the receiver side of the communications system. A typical example of remote transmitting is within the EW mast of a military submarine, where the mast itself leaves little area available for signal processing and complex circuitry, systems that potentially use up an entire room in a standard-sized building. Not only is physical space a limitation, but also the risk of water leakage in the event that the mast is damaged. It is therefore preferred to house the crucial and mission-critical components in a more secure and less exposed housing and to limit the signal processing that takes place in the mast itself. This technique moreover makes the mast/periscope architecture with little to no active electronic components in either the antenna or physical link

- secure,
- covert, since no electromagnetic leakage can be intercepted from the protruding periscope, and
- reliable as a result of its resilience to EMI and its lower complexity.

The receiver, where the signal-processing and amplification are done, can be located in a safe and spacious area that is hidden from the enemy, where EW support is activated whenever the submarine is at periscope depth and mission tasks such as

- threat warning to avoid counter-detection and collision,
- determining the number and location of targets, and
- conducting intelligence, surveillance and reconnaissance

can be undertaken with minimum electronic complexity at the transmitting antenna. Since the mast only performs basic functions (no amplification) and the sensitivity of a fiber-optic link would be relatively low in such a case when transmitting signals with low amplitude, remotely deployable and expendable EW masts for submarines can also be used where at least modulators and certain pre-amplifiers are connected to a wideband antenna or antenna arrays. This deployable mast can rise to the surface of the water to sample the electromagnetic environment without requiring the submarine to rise to periscope level.

Another situation where fiber links can effectively be used in military applications is in a remote receiver scenario.

8.5.2 Remote Receivers

Similar to the remote transmitter situation, but reversed in complexity in terms of the transmitter and receiver electronic circuits, the remote receiver principle assumes that very little to no signal processing is available at the receiver, but that the transmitter is able to perform most of the signal pre-conditioning and amplification. In such cases, link insertion loss becomes an important parameter because of the lack of signal post-processing at the receiver. An example of such a situation in EW is in an airborne towed decoy. The towed decoy is typically an anti-missile countermeasures decoy with a launcher and launch controller situated on the primary aircraft. The towed decoy lures incoming RF-guided threats away from the intended target by re-radiating decoy signals from its antenna. The primary receiving antennas are situated in the aircraft and a fiber-optic link connects the decoy and the signal-processing components in the aircraft. Signal amplification is also performed in the aircraft and relayed to the towed decoy, where is it re-radiated. In such a system, the signal pre-amplification and high input power requirements lead to the requirement of low insertion loss between the aircraft and the towed decoy. Directly modulated laser diodes are used to modulate the RF input and the use of fiber-optic links to the towed decoy becomes more attractive if impedance matching of the laser is also employed to further reduce insertion loss. The advantages of a fiber link to a towed decoy are therefore (Lindsay 1992)

- reduced channelizing of the original RF signal,
- phase coherence of the re-radiated signal being maintained,
- low electromagnetic leakage from the tow cable, and
- minimization of circuit complexity by not having to use digital RF memory owing to the low-loss characteristics of optical delay lines.

Another important application of fiber links in military environments is its use in antenna remoting, where fixed antenna arrays are placed at a distance from the central processing equipment and environmentally sensitive terminal stations (Fulton 2003).

8.5.3 Antenna Remoting

Antenna remoting involves the transport of analog RF signals between a central location where the primary signal processing occurs and remote antenna sites where signal quality is low or completely invisible (Fulton 2003). With antenna remoting it is therefore possible to deploy temporary, lightweight antenna sites placed in accessible locations where the primary RF signal integrity is compromised. Typically, the primary signal processing and RF transmission base station requires large, permanent and high-powered transmitters, which cannot easily be relocated. Antenna remoting through fiber-optical channels provides a convenient technique to repeat the signal without limiting the bandwidth performance and ensuring secure and interference-robust implementations. In military radar systems, antenna remoting allows the operators to be situated a significant distance from the primary base station and RF antennas. Several remote antennas can be implemented across large areas to increase the overall range from which the RF signals can be controlled or received and analyzed, typically ideal for increasing battlefield communications coverage. Antenna remoting can also be used for fixed and tactical satellite ground stations at IF for telemetry tracking, where fiber-optic links make it possible to locate receivers at a convenient and safe distance from the primary antennas while maintaining secure links between the systems (Fulton 2003). Efficient, low-loss interfaces between the guided optical medium and the free-space microwave link are also necessary for maximizing overall system performance (Novak 2009). In order to support wideband fiber-optic antenna remoting links, broadband antenna structures that can be directly integrated with optical transmitters and receivers are typical requirements where the main challenges of overall antenna bandwidth and efficiency of the RF photonic link are automatically addressed by using light-waves and optical links (Novak 2009).

A somewhat more traditional approach with added benefits of using optical technology is within network distribution and navigation in aircraft and naval vessels.

8.5.4 Aircraft and Naval Vessel Information Distribution

Optical technology provides various advantages in aircraft and naval vessels that are inherently exposed to high levels of potential EMI, have size and weight constraints for many of their subsystems, requirements for network flexibility and necessary adaptability. Data security and interference-tolerant systems are paramount in these applications and size and weight reduction directly translates to longer flight times and reduced payload.

Fulton (2003) lists additional advantages of fiber links and fiber-wireless communications in military applications, which include easy implementation of delay lines using light-waves, replacing electrically matched and temperature-controlled coaxial cables for phased array and direction-finding antenna arrays with optical links, bandwidth and signal re-routing for secure cellular telephony and integrated networks by fiber-wireless connections, as well as applications in jamming and countermeasures. Manka (2008) also expands on the use of MWP links applied to EW applications where the development of these systems relies heavily on advances and requirements in the commercial telecommunications industry (in contrast to where historically, the telecommunications industries and commercial institutions depended on technology maturity from the military industry).

8.5.5 Radar and EW Receivers

Future generations of radar aim to be primarily based on software-defined radio to adapt to the increasingly variable electromagnetic environments and crowding of the electromagnetic spectrum. These systems are required to operate at higher carrier frequencies and therefore make use of smaller antennas (typically patch antennas) and wide bandwidth operation to increase the resolution of the radar. Microwave components typically operate less-than-ideally (non-linearly) when frequency and bandwidth are increased and simultaneously become more prone to noise, which detrimentally affects low-amplitude signals. Fully digital radars are limited to a few GHz speeds by the signal-processing equipment and analog-to-digital converters must be able to suppress noisy environments at the expense of bandwidth. MWP essentially aims to provide an integrated solution to these problems experienced in radar with flexible generation of stable RF signals (with techniques such as optical beating discussed in this chapter) and allows direct digitization without the requirement to down-convert the signals. Photonics-based coherent radar is being researched to harness the advantages of photonics in historically microwave-only systems (IDST 2016).

MWP also plays a significant role in next-generation wideband EW receivers and simultaneous transmit and receive systems. Silicon-carbide photonic filters and low-noise lasers with high-saturation photodetectors can be used to achieve compact wideband channelized ELINT receiver architectures performing both channelization and photonic frequency down-conversion for X-band to baseband, offering high image rejection using advanced photonic filtering (IDST 2016). The typical photonic-based integrated circuits required to achieve wideband receivers and simultaneous transmit and receive systems comprise lasers, modulators, filters, amplifiers and photodetectors, yielding complete MWP systems, which are becoming more commonplace with the maturity of photonics on integrated substrates and increases in maximum transistor frequency. Challenges that are still delaying MWP systems replacing traditional EW and military applications include the power generation requirements of these optical systems on the integrated semiconductor dies. Silicon and InP, for example, are not capable of generating the amounts of power required by fully integrated MWP, which typically requires 10–15 times more power compared to traditional analog circuits. Alternative substrates such as glass composites and lithium niobate ($LiNbO_3$) are being researched to fulfill these tasks, but the unique integration and assembly of multiple materials such as glass, $LiNbO_3$, InP, GaAs, silica, Si and rare-earth materials make these systems complex, challenging and expensive. $LiNbO_3$ is extensively used in the telecommunications industry and is the material of choice for surface acoustic wave devices and optical waveguides. The following section briefly discusses $LiNbO_3$ as an enabling material for optoelectronic devices within the commercial and military environment.

8.5.6 $LiNbO_3$ and the Mach-Zehnder Principle

High-speed electro-optic modulators based on the electro-optic material $LiNbO_3$ can be found at the core of many microwave photonic systems (Mitchell 2007). These modulators have exceptionally low noise when employed in analog photonic transmission systems. $LiNbO_3$ allows optical transmission of signals with wavelengths ranging from 400 to 5000 nm (5 μm) with an OH^- absorption at 2.87 μm, has a very high melting point (1240 °C) and is a ferroelectric material. Its usefulness is due to the electro-optic, linear and piezoelectric characteristics of the intrinsic material. It can be grown consistently because of its relatively large crystal structure. The most typical applications of $LiNbO_3$ are optical modulation and *Q*-switching of IR wavelengths, because the crystal is non-hygroscopic,[1] has a low half-wave voltage and is often used for *Q*-switches in military applications. $LiNbO_3$ is also an optimal solution for the external modulation of light-waves using Mach-Zehnder interferometers/modulators at high data rates and offers ideal tradeoffs in terms of low insertion loss, low driving voltage, broad bandwidth and low chirp. Other electro-optic materials that are commonly used are potassium dihydrogen phosphate and GaAs. The Mach-Zehnder modulator principle is presented in Fig. 8.7.

The two classifications of electro-optic modulators are longitudinal and transverse. This classification depends on how the voltage is applied relative to the direction of light propagating in the device. A Mach-Zehnder transverse modulator as shown in Fig. 8.7 is used for adjusting the amplitude of an optical signal and depending on the output signal, the output can be read out as either an on or off (1 or 0) state. Light with intensity $I(t) = |E(t)|^2$ at the input waveguide is divided up into two individual waveguide interferometer paths. If an electric field is applied over one of the paths, a shift in its phase occurs. The varied phase shifts (of each path) at the output are defined by

$$\phi_1 = 2\pi \frac{L}{\lambda} \tag{8.31}$$

where L is the length of the electrode, λ is the optical wavelength and

$$\phi_2 = 0. \tag{8.32}$$

When the two paths are recombined along its route, the phase difference is transformed to an amplitude modulation and the output is detected by a photodetector sensitive at the optical wavelength. The principle of the Mach-Zehnder configuration is typically used for interferometers along with Michelson and

[1]'Hygroscopic' describes the ability of a material to draw and sustain water molecules from its immediate environment.

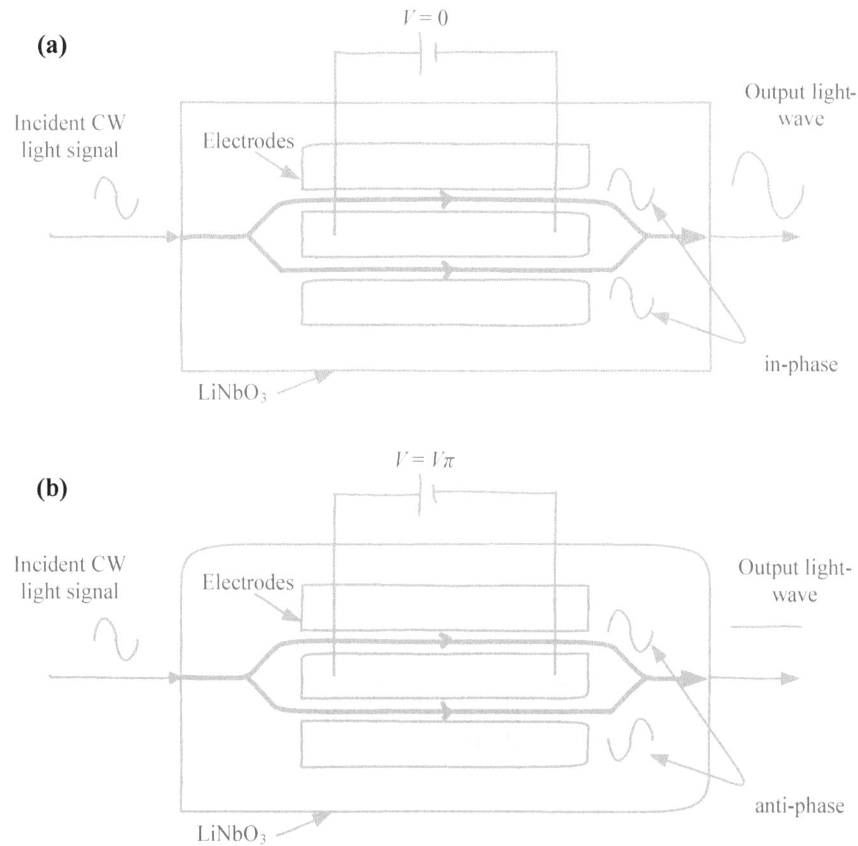

Fig. 8.7 The Mach-Zehnder modulator principle **a** shows two in-phase optical signals, which generate an exact but amplified representation of the input at the output with 0 V across the splitting electrode and **b** two out-of-phase optical signals, which generate no optical output when the half-wave voltage $V\pi$ is applied to the splitting electrode

Fabry-Perot interferometers. The theoretical value for the half-wave voltage that is applied to the electro-optic modulator to infer a 180° phase shift is given as

$$V\pi = \frac{\lambda s}{L n_e^3 r_{33} \Gamma} \tag{8.33}$$

where s is the physical gap distance between electrodes, n_e is the optical refractive index, r is the electro-optical coefficient (discussed in the following paragraph) and Γ is the electric-optic overlap coefficient.

At an atomic level, an electric field causes small an-isotropic deformation of the crystal lattice. As a result, the inverse dielectric constant (impermeability) tensor changes in the structure and crystals that are not symmetric exhibit a linear electro-optic effect. This linear electro-optic effect can be denoted by a third-rank tensor r_{ijk}. This third-rank tensor can be denoted by a 6 × 3 matrix due to the permutation symmetry of the tensor. Therefore,

$$r_{ijk} \equiv r_{ij} \tag{8.34}$$

with $i = 1, \ldots, 6$ and $j = 1, 2, 3$. Typically, the r_{ij} coefficients have very little dispersion in the optical transparent region of the crystal. Bass et al. (1995) provide an in-depth review of the linear (and quadratic) electro-optic effect and calculating the tensor coefficients, which falls outside the scope of this book. In addition, the modulation of light parameters with electro-optic modulators using polarizers and passive birefringent[2] elements can be phase, polarization, amplitude, frequency or position modulation. These techniques are also discussed in detail by Bass et al. (1995). Essentially, the critical parameters in an electro-optical modulator are

- low optical insertion loss,
- optimal characteristic impedance of the RF electrodes,
- a low driving voltage $V\pi$,
- low microwave losses in the coplanar RF electrodes,
- effective microwave index, and
- low chirp.

These parameters are dependent on the electro-optical material and thus materials such as $LiNbO_3$ have been researched and applied in various disciplines and applications requiring electro-optical modulation. The abilities and performance figures of the electronic components in an electro-optic system are equally important. The optical devices and light-wave technology can produce efficient signals for modulation, but the processing of the signals at electronic level requires fast, low-noise and high-gain transistors. The following section discusses the significant contributions that a technology such as SiGe BiCMOS makes to MWP systems.

8.6 SiGe HBTs and SiGe HPTs in MWP

Transmitters and receivers based on optoelectronic and fully electronic components operating at high-Gbit/s data rates are typically realized using multi-chip module packaging technology and MMICs (Krozer et al. 2009). Silicon photonics, as one of the various photonic integrated technology platforms, has attracted attention to achieve microwave photonic system integration because of its compatibility with

[2]Materials that have a refractive index dependent on the polarization and propagation direction of incident light.

the current CMOS technology and its potential of seamless integration with active electronic components (Zhang and Yao 2016).

High-speed voltage swing operation with high current capabilities to drive the optical components are critical requirements for MMIC optoelectronic systems. For the analog front-end of these systems, several technology alternatives can be used to realize the active components in OEICs, including SiGe HBTs, InP HBTs, GaAs mHEMT, InP HEMT and GaN HEMT technologies. High-efficiency power amplification for higher order modulation schemes and high-breakdown voltage are typical requirements in OEICs. InP HBTs are well-suited for these operations but are not as compact as SiGe HBTs and from a digital baseband perspective, SiGe HBTs are preferred for their bandwidth capabilities, despite their lower power generation abilities. The laser driver circuitry of optoelectronic equipment such as modulators requires high-bandwidth and high-voltage switching to low-impedance loads, which results in high dynamic currents (Krozer et al. 2009). Photodetectors, the primary active component in optoelectronic receivers, are typically packaged using coplanar waveguides to connect the photodetector/IC to the output and to the TIA.

SiGe heterojunction bipolar phototransistors (HPTs), also referred to as photo-bipolar transistors, can be used in the front-end of the optical receiver in OEICs. The SiGe HPT is a photosensitive transistor based on the vertical structure of SiGe HBTs with raised base-doping concentration and the enlargement of the lateral size for optical coupling to a fiber. Incident photons generate electron-hole pairs in the floating base, creating a photo-voltage $V_{BE,ph}$ approximately given by

$$V_{BE,ph} \approx \frac{2q}{\varepsilon N_A}\left(\eta \frac{P_{opt}\tau_p}{Ahv}\right)^2 \tag{8.35}$$

where q is the electron charge, ε is the permittivity of the material, A is the cross-sectional area of the floating base, N_D is the emitter doping concentration, τ_p is the base-recombination rate and P_{opt} is the optical power incident on the phototransistor. The collector current generated in the transistor is given by

$$i_C = \beta i_B = \beta I_S e^{\frac{V_{BE,ph}}{V_T}} \tag{8.36}$$

where β is the phototransistor current-gain parameter and V_T is the thermal voltage. The phototransistor base transport factor α_T, used to determine the current gain, is given by

$$\alpha_T = 1 - \frac{W^2}{2L_n^2} \tag{8.37}$$

where W is the base width and L_n is the electron diffusion length in the base junction. The current gain β of the phototransistor can be approximated by

$$\beta \approx \frac{1}{1 - \gamma \alpha_T} \tag{8.38}$$

where γ is the emitter injection efficiency of the heterojunction bipolar phototransistor. Phototransistors offer advantages for low-power optical detection and have been manufactured as photo-heterojunction JFETs and in CMOS and BiCMOS processes. Phototransistors are useful for various reasons, such as their ability to offer

- higher current outputs compared to photodiodes,
- output voltages, compared to simply current output in photodiodes,
- decreased circuit complexity, since no TIA is required at the output, as is the case with photodiodes, for example,
- instantaneous output from optical input, and
- relatively low cost and small footprint, allowing multiple HPTs to be placed on a single OEIC.

Unfortunately, though, these devices cannot operate at high voltages, do not have particularly high frequency response (approximately 250 kHz for homo-junction phototransistors and 1 GHz for heterojunction phototransistors) owing to the large junction capacitance associated with the base-collector junction and are extremely vulnerable to electrical surges or electromagnetic energy.

A typical OEIC representation, adapted from Ko et al. (2012) is presented in Fig. 8.8. Figure 8.8 aims to provide insight into the requirements of the on-chip components when using OEICs and the technology requirements, which ultimately leads to choosing the optimum technology to perform the required tasks.

As depicted in Fig. 8.8, the optoelectronic system receives baseband data at a central office, which is modulated by the OEIC, indicated in the dashed rectangle. The baseband data are fed to the OEIC through optic fibers, where these are received and converted to usable voltages by the photodetector and TIA combination. The voltage is fed to a variable gain amplifier after which it is modulated by the carrier frequency generated by the LO. The LO can be placed off-chip or on-chip, depending on the application and frequency requirements. Once modulated, the signal is amplified by the power amplifier (PA) and sent to the transmitting antenna.

On the receiver side, an external antenna receives the RF signal and is then fed to an on-chip LNA. The signal is demodulated by the same carrier frequency and digitized. An external laser is used to optically modulate the received signal, which is fed back to the central office through an optical fiber link.

The entire on-chip set of sub-systems can potentially be realized by SiGe BiCMOS technology providing high-bandwidth and low-noise performance, also allowing the optical components such as the photodetector and interconnects to be realized in the Si substrate, depending on the wavelength of the optical modulator. Typically, without Ge-doping, the optical wavelength is in the 850 nm range. If

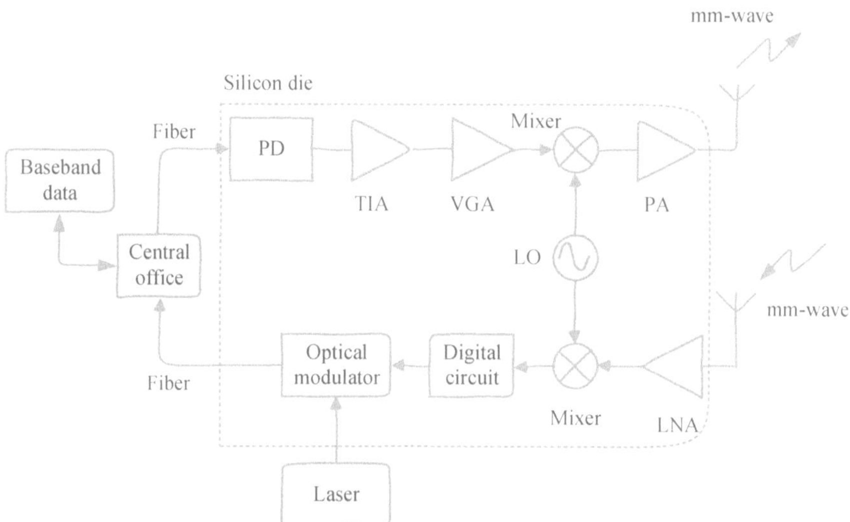

Fig. 8.8 Simplified diagram of an optoelectronic system highlighting the requirements and components integrated on-chip, adapted from Ko et al. (2012). The electronic components should (ideally for future-generation communications and EW applications) be able to operate in the mm-wave domain

Ge-doping is performed, as is possible with SiGe BiCMOS, the optical wavelength can be extended to 1550 nm.

Optoelectronic clock converters can also be realized in integrated photonics SiGe BiCMOS technologies, converting optical pulses into low-jitter electrical switches with sharp transitions. Such a system is presented in Krueger et al. (2015) and a simplified schematic of this system is given in Fig. 8.9.

In Fig. 8.9, the core component of the clock converter is the current-mode logic RS-latch circuit. The differential pair transistor configuration of the optically clocked feedback loop, highlighted and encircled in Fig. 8.9, uses a photodetector to modulate its tail current source by an external optical source transported by a waveguide. This effectively operates as a D-flip-flop and the present state in the RS-latch is fed back to the data input and the subsequent photocurrent pulse will toggle the state. The supporting circuitry, the driver circuit, buffer and inverter are required to condition the output signal, in this case an optically clocked square wave with 50 % duty cycle (Krueger et al. 2015). Krueger et al. (2015) implement 0.25 μm SiGe BiCMOS technology with f_T and f_{max} of 180 GHz and 220 GHz, respectively, five metal layers, MIM capacitors and polysilicon resistors to realize the full circuit, without any modifications to the standard 0.25 μm process.

Washio (2003) presents an optical transmission and wireless communications system implemented also in 0.25 μm SiGe BiCMOS technology (f_{max} of 180 GHz) to obtain high speed and decreased power through the SiGe HBTs. A simplified schematic of the communications system is adapted and presented in Fig. 8.10.

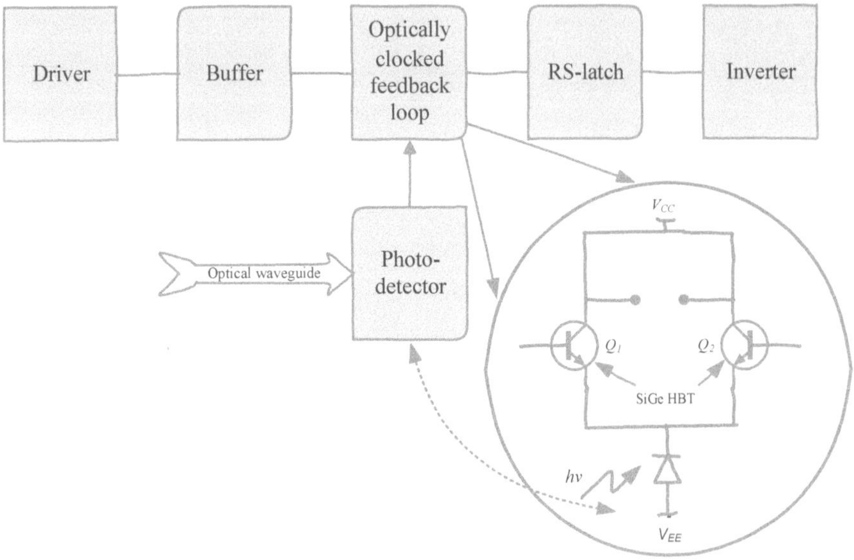

Fig. 8.9 Simplified schematic of an optoelectronic clock converter as presented in Krueger et al. (2015)

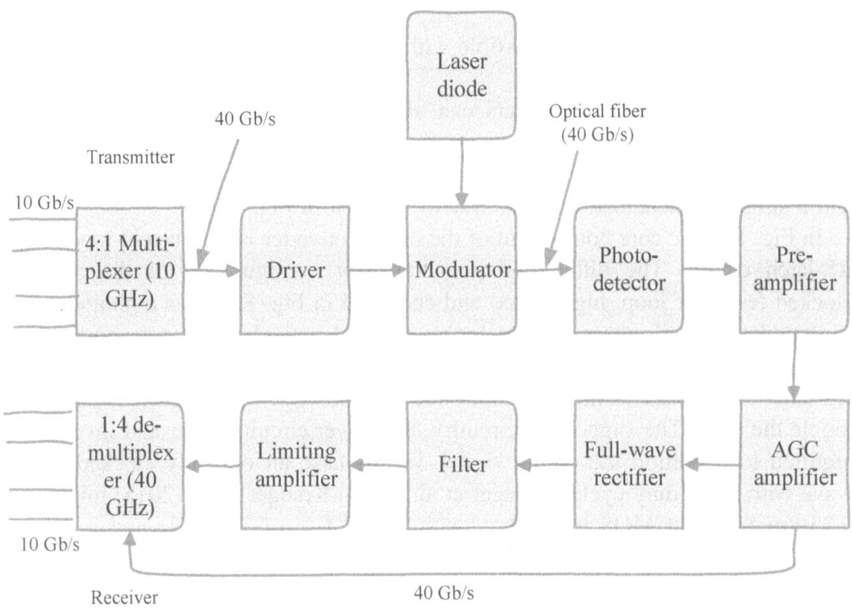

Fig. 8.10 Simplified schematic of an optoelectronic wireless communication systems as presented in Washio (2003)

Washio (2003) presents a similar approach to realizing a fully integrated OEIC with the only off-chip component being the laser diode. The method essentially uses a multiplexer to combine four 10 Gbit/s input streams and condition the signals, on-chip, at 40 GHz, without having to down-convert any signals. It is also important to note that this article was already presented in 2003; since then the 0.25 μm BiCMOS processes have matured and became more cost-effective, with current state-of-the-art technologies already reaching 12 nm nodes, therefore offering much better performance in many areas.

From these examples, the interest in and abilities of BiCMOS technologies for OEICs and MWP are apparent. In EW, especially in the case of MWP, updates and upgrades to current technologies can be achieved by combining optical light-wave technology and high-speed microelectronics. This section aims to highlight these possibilities through practical and researched work. The following section concludes this chapter.

8.7 Conclusion

Microwave photonics combines the broad bandwidth and low losses of modern photonics and the increasingly fast and small microelectronic technologies to realize integrated microwave and photonic systems on single substrates. From these technologies, SiGe provides an already matured approach to high-speed BiCMOS systems and Si-based photonics for commercial and military applications. The interaction of photonic and microwave signals provides research and development opportunities in radar, communications, sensor networks, warfare systems and instrumentation (Yao 2012). This chapter reviews the MWP discipline from both the optical and microwave perspective and provides discussions on current and future technologies and applications that can benefit from MWP. MWP is relatively new, considering the time that microelectronic and analog components have been available to process microwaves and photonics has been used in the processing of data and information. For this reason, various opportunities exist to adapt current microwave/photonic-only systems to obtain more integrated MWP systems. Common tasks such as signal generation, processing, control and distribution can be shared between the optical and microwave domain and this chapter reviews these tasks in both domains in order to provide insight into the shared functions.

Photonics operates at very high frequencies; its wavelength dependent on the wavelength of light, typically in the nanometer-range. This translates to high-THz frequencies, which if applied correctly, allow signal modulation in the THz range without the requirement of analog components to generate oscillations; a component limitation of even the most advanced technology nodes available. To harness these frequencies, light-waves should be beaten together to generate lower, but more practical frequencies (still in the THz range) for current electronic equipment to process. This chapter reviews the process of optical beating used to realize this requirement. O/E and E/O conversion are required to switch between the optical and microwave domain and allow components in each domain to perform the tasks optimum to the domain itself and to the components. These conversions are essentially performed by either photodetectors or lasers and have been reviewed in this book. Once these signals are converted, they form part of an OEIC offering advantages of speed, immunity to EMI, low noise and reduced footprint with smaller size, weight and space occupied in typically limited areas when used in military vessels.

Fiber-wireless networks use MWP to realize high-speed data transmission. The mm-wave frequency spectrum is ideal for future-generation communications, radar and sensor networks and is currently already used for various applications, such as airport security and collision avoidance radar. MWP enables the transmission of these signals, modulated on optical carriers, between transmitter and receiver, with abundant bandwidth and immunity to EMI.

The chapter is concluded with a review and examples of the practical uses and potential implementations of MWP in EW, as well as how SiGe BiCMOS technology is used to facilitate the advance and maturity of this technology.

References

AFOP. (2002). *Fiber optical coupler definitions*. Alliance Fiber Optic Products, Inc. REV.A 8/02. Retrieved May 21, 2016 from http://www.afop.com

Alegria, C. F. G. (2001). *All-fibre devices for WDM optical communications*. University of Southampton: Thesis at Faculty of Applied Science; Dept. of Electronics and Computer Science (Chapter 4).

Aulakh, S. K. (2013, September). Application of microwave photonics in electronic warfare. *International Journal of Computer Science and Telecommunications*, *4*(3), 53–58.

Bass, M., Van Stryland, E. W., Williams, D. R., & Wolfe, W. (1995). *Handbook of optics: Volume II. Devices, measurements, and properties* (2nd ed). In Maldonado, T. A. (Ed.) Chapter 13—Electro-optic modulators. New York: McGraw-Hill, Inc.

Breed, G. (2007, May). A tutorial introduction to optical modulation techniques. *High Frequency Electronics*.

Deng, Z., & Yao, J. (2006, February). Photonic generation of microwave signal using a rational harmonic mode-locked fiber ring laser. *IEEE Transactions on Microwave Theory and Techniques*, *54*(2), 763–767.

Fulton, R. L. (2003). RF over fiber architectures and EW applications. *Fiber-Span: White Paper*.

IDST. (2016). Microwave photonics in next generation communications, radars and electronic warfare systems for spectrum dominance. *International Defence, Security and Technology*. Retrieved May 25, 2016 from http://www.myidst.com

Iezekiel, S., Burla, M., Klamkin, J., Marpaung, D., & Capmany, J. (2015). RF engineering meets optoelectronics: Progress in integrated microwave photonics. *IEEE Microwave Magazine*, *16* (8), 28–45.

Jäger, D. (2009). Microwave photonics—Past, present, and future [Member Benefits]. *IEEE Microwave Magazine*, *10*(4), 154–156.

Ko, M., Youn, J., Lee, M., Choi, K., Rücker, H., & Choi, W. (2012, July). Silicon photonics-wireless interface IC for 60-GHz wireless link. *IEEE Photonics Technology Letters*, *24*(13), 1112–1114.

Krozer, V., Johansen, T. K., & Jiang, C. (2009). Wireless and photonic high-speed communication technologies, circuits and design tools. In 2009 SBMO/IEEE MTT-S International Microwave and Optoelectronics Conference (IMOC 2009) (pp. 770–774), 2009.

Krueger, B., Makon, R. R., Landolt, O., Krune, E., Knoll, D., Lischke, S., et al. (2015). A monolithically integrated opto-electronic clock converter in photonic SiGe-BiCMOS technology. In 2015 IEEE Bipolar/BiCMOS Circuits and Technology Meeting (pp. 129–132), 2015.

Leeson, D. B. (1966). A simple model of feedback oscillator noise spectrum. Proceedings of the IEEE. *54*(2), 329–330.

Lindsay, A. C. (1992). *Wideband guided-wave photonics for electronic warfare applications.* Electronics Research Laboratory: Electronic Warfare Division (Australia). AR-006-950, Unclassified and Approved for Public Release, March 1992.

Manka, M. E. (2008). Microwave photonics for electronic warfare applications. In 2008 Asia-Pacific Microwave Photonics Conference (pp. 275–278), 2008.

Mitchell, A. (2007). $LiNbO_3$ devices for microwave photonics. In The 20th Annual Meeting of the IEEE Lasers and Electro-Optics Society (pp. 184–185), 2007.

Novak, D. (2009). Enabling microwave photonic technologies for antenna remoting. *IEEE LEOS Newsletter: Research Highlights*, February 2009.

Rolland, A., Loas, G., Frein, L., Vallet, M., Brunel, M., & Alouini, M. (2011). Optoelectronic down conversion: A novel approach for optical beat note stabilisation up to the THz domain. In Proceedings of the 2011 IEEE MWP (pp. 33–36), 2011.

Washio, K. (2003, March). SiGe HBT and BiCMOS technologies for optical transmission and wireless communication systems. *IEEE Transactions on Electron Devices*, *50*(3), 656–668.

Wiberg, A. (2008, March). *Generation, modulation, and detection of signals in microwave photonic systems.* Sweden: Chalmers University of Technology.

Winzer, P. J. (2009). Modulation and multiplexing in optical communication systems. *Bell Labs: IEEE LEOS Newsletter: Research Highlights*, February 2009.

Yao, J. (2012). A tutorial on microwave photonics. *Microwave Photonics Research Laboratory: Research Highlights.* University of Ottawa, Canada, April 1–4, 2012.

Zhang, W., & Yao, J. (2016). Silicon-based integrated microwave photonics. *IEEE Journal of Quantum Electronics*, *52*(1), Article #: 0600412.

Chapter 9
The Future of Electronic Warfare: Potential Contributions by SiGe

9.1 Introduction

The future of EW has been described by Thompson (2015) as a cat-and-mouse game. EW drives technology and technology drives EW; however, in most military-based activities, a significant limitation is brought about by secrecy of development of new weapons, technology and research. Keeping technologies, protocols, radar strategies and research secret makes contributing to EW from an outside perspective difficult. Institutions such as the United States DoD often subsidize research on subsystems without making the final implementation and targeted applications known, which is understandable. The difficulty comes in when attempting to determine which technologies are currently used, have been declared obsolete or are future contenders in military applications. This chapter identifies traditional applications that are likely to be upgraded in the near future with advances in technology, which can enable performance increases of orders of magnitude or change the way applications are applied in the field. Figure 9.1 represents the essential components where research and development, new innovations and technology advancements contribute to the future of EW.

From Fig. 9.1, it is seen that the future of EW depends on optimizing the subsystems and the technologies used to realize these subsystems, as well as dependence on integrating these subsystems into applications; this can also be applied to traditional military technologies and the commercial sector. Subsystem optimization entails addressing the stability of these systems and ensuring that each technology, whether SiGe, GaN, InP, InGaAs, GaAs or AlAs, is correctly used, integrated and operated according to its strengths. Parasitic components, especially at high frequencies, are always present in microelectronic circuits and analog interfaces, especially where amplifiers such as op-amps are concerned. Lower parasitics in advancing technologies mean less parasitic capacitance, inductance and impedances that influence the operation of the circuit without the consent of the designer, often leading to oscillations, shifts in frequency operation and circuit

W. Lambrechts and S. Sinha, *SiGe-based Re-engineering of Electronic Warfare Subsystems*, Signals and Communication Technology,
DOI 10.1007/978-3-319-47403-8_9

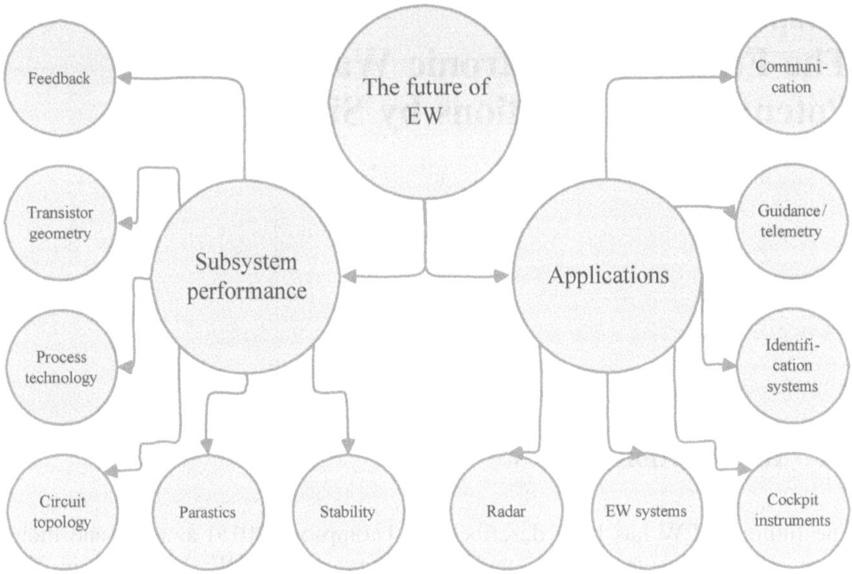

Fig. 9.1 The future of EW—technology performance and applications

instability. Circuit topology, therefore circuit layout, is an art on its own in high-frequency circuits and must be considered during the layout phases of new developments. The importance of circuit topology is often overlooked and it becomes difficult and often impossible to overcome non-optimized topologies, which can also lead to oscillations in circuits and non-ideal circuit operation. Spending more time on circuit topologies and layout techniques and conducting more critical reviews of them are cost-effective, compared to redesigning and re-manufacturing of ASICs and PCBs. Subsystem performance is also dependent on the choice of process technology and transistor geometry, especially when considering the performance criteria of a subsystem, such as power generation, gain, noise levels, stability and temperature dependence. Identifying the crucial performance parameters in an application or at sub-circuit level is essential before determining the optimal enabling technology for any new development. Chapter 10 of this book reviews the differences, advantages, disadvantages and strengths of typical technologies and aims to identify the role of each technology in EW and military applications. Finally, feedback in any subsystem, especially in elements such as amplifiers, cognitive electronics and sensing equipment, requires careful consideration of many of the abovementioned criteria and can in many cases be treated as a separate entity during circuit design and layout. Feedback loops are typically sensitive to disturbances and interference from external sources and layout techniques to enhance the performance of subsystems often require complex feedback-loop topology strategies.

All of these considerations influence the overall application and intended performance of the application to which the subsystems are applied; as shown in Fig. 9.1, the typical applications in EW often involve radar, conventional EW systems, cockpit instruments such as head-up-displays, identification of allied and enemy vehicles and structures, guidance (of autonomous vehicles and of weapons) and telemetry systems, as well as all communication among subsystems, other vehicles, soldiers and typically anywhere where data transfer is required. Many of these subsystems require seamless integration of analog and digital equipment, where both technologies can benefit from the advances in technologies and subsystems, as shown in Fig. 9.1.

Possibly the most notable changes in the field of general electronics and in military applications are the use of digital subsystems to perform functions previously only attributed to analog circuits and using increasingly powerful microelectronic components to achieve cognitive warfare. Digital electronic warfare systems (DEWS) protect new and legacy aircraft by providing radar warning, situational awareness, offensive targeting support and self-protection. System capabilities include all-aspect broadband radar warning, emitter location, multispectral RF/IR countermeasures and response management. These capabilities are accomplished through a digital radar warning receiver, digital RF memory as application in jamming (Lichtman et al. 2016) and an integrated countermeasures dispenser to provide the latest-generation EW capability (BAE 2014). A typical DEWS implementation has advantageous features such as

- modular, scalable and open-source hardware and software,
- interoperability with AESA, and
- enabling simultaneous jamming without interfering with the traditional radar and radar warning receivers.

Such a system integrates radar warning and countermeasures into a single system, providing faster countermeasure response, accurate threat geolocation and increased situational awareness. These measures are obtained though efficient and complex digital processing of analog signals. The integrity of the analog signals from sensors on an aircraft or naval vessel, for example, is crucial to the digital subsystems and the analog circuitry should have wide bandwidth, high-gain and low-noise capabilities. Among the most crucial enabling technologies of adaptable radar and EW are the ADC and DAC modules. These devices are situated between the transceiver antennas and the digital signal processors (DSPs) of these analog incoming signals. In software defined radios (SDR), some or all of the physical layer functions are software-defined. Digital speed is essential in these radios where real-time processing is required. High-speed DSPs and field programmable gate arrays (FPGAs) are becoming extremely powerful and cost-effective, which reduces the cost and technical skills required to design and program ASICs for specific hardware and software combinations. Precision-guided munitions (PGMs) are also benefiting from the increases in computing resources of miniaturized processors, which quickly and effectively solve complex algorithms in-flight, guiding them to their targets.

Cognitive warfare essentially entails combining hardware (analog and digital) with effectively engineered algorithms implemented in typically open-source software to teach weapons how to adapt to their environment and the immediate scenario. Cognitive warfare allows analog sensors and low-level circuits (such as amplifiers), digital equipment able to process information at unparalleled speeds and algorithms that constantly learn from feedback provided by the sensors, to estimate, adapt and defeat or hoax the enemy by generating unexpected but meticulously defined waveforms. Cognitive EW is highly dependent on cognitive radar, so much so that Haykin (2006) made the well-defended statement that cognitive EW cannot exist without cognitive radar. This chapter briefly reviews the case study presented and in addition aims to provide some context to demonstrate how enabling technologies can benefit the realm of cognitive warfare, specifically cognitive EW. For a device, weapon, machine or tool to be cognitive, it only requires feedback (typically from sensors) and memory to store data and learn. The measure of cognitivism is relative and arbitrary, but still quantifiable by determining the absolute error (a human-defined parameter) by the end of the computation. This provides the system with a goal to achieve, based on the cognitive abilities and traced back to the quality of the analog inputs, digital processing, available memory and effectiveness of the algorithms.

Defensively, RF countermeasures, such as IR flares, therefore effectively IF countermeasures, are becoming more crucial for defense against IR-guided missiles with sophisticated tracking algorithms. Potenziani (2006) presented research on the role thin films such as ferrites, ferroelectrics, magneto-static films and metamaterials can play in devices such as filters, phase shifters, limiters and smart antennas to increase the effectiveness and useful lifetime of EW weapons with minor impact on redesigning and re-imagining already qualified and operational weapons. Also in development are futuristic, but prototyped, electric cannons (or railguns) able to fire a shell at 5600 miles per hour (9000 km/h) far more cheaply than a conventional missile. A shell of tungsten inside a metal casing is loaded into the railgun and projected with large amounts of capacitive energy as opposed to an explosive charge. Up to 32 MJ have been experimented with to create an electromagnetic field that propels the shell down the rails of the gun at these speeds.

Cyber EW is an increasing threat and simultaneously an advantage to nations and militaries that have the skills and infrastructure to deploy cyber EW attacks. A cyber EW action uses EM to control the EM spectrum/domain where electronic equipment operates. Similar to traditional EW, cyber EW can be categorized into three primary categories, namely cyber EA, cyber EP and cyber ES. Cyber EA uses EM energy to attack the electronic circuitry of the enemy with the intent to destroy information or the physical infrastructure. Cyber EP entails the actions taken to protect one's own information in the EM domain as well as to protect the infrastructure and integrated networks that store, transmit and analyze data and information. Cyber ES primarily locates sources of EM energy that are unknown or unwanted for immediate threat recognition and early warnings to the cyber EA and/or cyber EP systems.

Cyber EW presents many challenges in both offensive and defensive strategies. Defensively, early warning of cyber EW attacks is crucial, entailing a process of identifying non-legitimate or abnormal activity in the EM domain. The sheer amount of EM activity inherently leads to false alarms or incorrectly identified threats and complex layers of hardware and software criteria are required to handle the complexity and sophistication of a cyber EW attack.

As more technologies in EW merge and systems become more integrated, condensed and multi-functional, it becomes more difficult to distinguish between the enabling technologies that can efficiently and cost-effectively serve all these functions. In each system, whether low noise, high gain, good optical properties, high-power or high-temperature operation or radiation-hardened components, each technology presents its own advantages and disadvantages and it is important to be able to distinguish between these properties. This chapter looks primarily at the future of EW and where technology advancements contribute to EA, EP or ES operations. Highly integrated microelectronic circuits implemented in CMOS or BiCMOS technologies or on optoelectronic substrates often require high electron mobility within the substrate, where GaAs, InAs and InSb still prove most efficient. In terms of high-temperature operation, technologies such as GaN and SiC are often the preferred choice and optoelectronic devices require modified compounds of Si, such as SiGe, to improve performance.

Essentially, there is no single technology that will define the future of EW, but rather a combination of technologies in applications that harness the advantages of each. By highlighting the most prominent areas where the future of EW is heading, this chapter aims to review and identify the specific areas where SiGe can be of benefit. Reducing the number of subsystems that need to be supported by any mission platform and the overall size, weight and power is beneficial, but in view of the further need to support cognitive and real-time configurability, the challenges can seem daunting. However, a new generation of high-performance, wideband components is potentially providing a solution to this challenge, supporting the high performance levels required for each system, but with a broad enough operating range to meet the multifunction challenge. The ultimate goal for many of these future systems is a completely software-determined architecture (Analog Devices 2015); an analog subsystem will, however, always be present. The magnitude of its contribution may vary, but its presence cannot be overlooked. As digital technology advances and more transistors are able to fit on small form-factor ICs, the analog supporting circuitry must advance at a similar, or at least comparable, pace. Advances in CMOS and SiGe processes, among others, are enabling a significant increase in digital functionality to be incorporated into next-generation devices. Advanced signal-processing capabilities are able to provide calibration or digital compensation functionality, in addition to flexibility, enabling the overall system performance levels to be closer to their narrow-band counterparts, while maintaining the ability to be reconfigurable and use wider bandwidths for mode of operation as needed (Analog Devices 2015).

The commercial and military worlds are moving towards smarter implementations of technology, as opposed to the historical 'chasing the GHz' strategy. By using current-generation technology more efficiently and using software, feedback and additional sensors to define the operation of the subsystems, many applications are moving towards adapting to the environment and using past and present events as baseline for future events. A measure of cognitive processing is being sought rather than pushing the limits of the available technologies, or at least the current implementation of systems. Algorithms, for example, enable current technologies to achieve tasks only imagined by powerful supercomputers and large, expensive and high-maintenance equipment. The power of Google® for example, is based on applying an algorithm (PageRank) by using existing technology in an innovative way. The achievements of the technology far surpass its processing capabilities through efficient implementation. In a similar way, cognitive EW, although in early development phases, can learn from past events and become more efficient through learning from the environment. Cognitive computing, artificial intelligence and neural networks are becoming more attractive in EW applications such as radar. The following section explores cognitive electronics and where it can be implemented in EW.

9.2 Cognitive EW

Cognition is a noun that describes the activities of thinking, understanding, learning and remembering to acquire knowledge. If directly compared to any electronic equipment capable of processing and storing information, parallels can be drawn between machine learning and cognitive learning. The activities are related in respect of:

- thinking: a set of interconnected active devices (transistors) capable of processing information it is supplied with. The information is received from an external source and the device is able to do computations on this information;
- understanding: a machine could be able to understand information only if it is given a set of instructions and possible outcomes. The outcomes are based on the set of inputs and if a machine understands, it is able to determine all the possible outcomes and decide which outcome is optimal;
- learning: if a mistake was made, or the outcome was not ideal for the given situation, a cognitive machine should be able to learn from the mistakes, pointed out again by a set of input versus output possibilities, and adapt itself to improve or completely redefine the output. The device would therefore be dependent on a feedback system, which is used to vary the input parameters based on the achieved output;
- remembering: if an optimal output is achieved, as well as non-optimal outputs, a cognitive system must be able to store the output and the set of inputs resulting in the output for future reference. If the same or a similar problem is presented to the machine, it must be able to recall the input and output combination and process it accordingly.

From the parallels drawn in the above discussion, the term cognitive electronics (in reality any cognitive system) should indicate particular abilities. A cognitive system, as listed in Haykin (2006), comprises three primary ingredients, namely

- intelligent signal processing, which builds on learning through interactions of the system with the surrounding environment,
- feedback from the receiver to the transmitter, which facilitates the intelligence, and
- preservation of the information content of information returns, which is realized by the Bayesian approach to target detection through tracking.

Most importantly, cognition in electronic systems is dependent on sensors and the data gathered from these sensors. This is similar to human cognition, using sight, smell, sound, taste and touch to make cognitive decisions in real-time. Adaptive behavior results from cognitive intelligence; in the case of electronic equipment and other military and EW applications, the electronic processing of information can be adapted based on the cognitive learning of the system. Possibly the most important requirements of a cognitive system are

- accurate and relevant inputs from sensors or programmed by human operators,
- high-speed data processing, and
- large storage.

Without these three components, an electronic system cannot be cognitive, or can only be cognitive to a limited degree. These requirements directly translate to technology capabilities and limitations.

Radar is a commonly used technique in traditional warfare and in EW and uses radio waves to detect objects and determine the range, angle and/or velocity of the object. Since radar is so commonly used and encompasses many systems in warfare (and commercial applications), it serves as an ideal example to review and discuss cognitive EW.

EW using conventional radar systems typically transmit or receive fixed waveforms, making them relatively easy to spot within the EM spectrum, learn the commonly used waveforms and the information modulated in these signals and develop tactics against these waveforms, such as intercepting or jamming them. Radar systems typically use a feed-forward system, with initial low-level data processing of sensor data, passed on to higher-level processing, which extracts additional information from memory and processes it according to the system objective. The primary goal of cognitive radar, for example, is to generate new and unique waveforms that have not been seen by the intercepting radios and that are difficult to identify and theoretically impossible to defeat. Digitally programmable radars are at the core of these cognitive radar waveforms. Intelligently adaptive radars can be achieved through knowledge-aided processing with environmental dynamic databases and adaptive transmit technologies and architectures, which is a relatively recent approach to achieving powerful cognitive radar. Baylis et al.

(2014) present a possibility of a transmitter for cognitive radar. Importantly, Baylis et al. (2014) highlight the fact that research and literature, as well as possible applications of cognitive radar, predominantly focus on software requirements for cognitive learning. The enabling RF technology required to achieve high-speed processing and large storage with limited power and low susceptibility to interference is typically overlooked and could potentially be the limiting factor in cognitive systems. On mobile, lightweight and space-limited systems, high-powered computers, large memory banks and high-bandwidth data transmission must be realized by electronic and/or optical technologies capable of achieving all these requirements in limited space with limited power.

Programmable radar and adaptive EW sense the RF environment and should adapt in nanosecond timescales, which places stringent demands on the digital signal processing and digital conversion technologies (Keller 2013). Modern radar systems briefly inspect the immediate radar environment, detect sources of electronic noise such as RF jamming or co-location antenna interference and adapt the output of the radar in real-time. Radar using these techniques are referred to as digitally programmable or adaptive radar and uses DSPs to sense and compensate for jamming and interference rapidly, adapt its transmit-receive modes and continue to perform its primary mission. The ability of the radar to sense the environment and adapt relies on its speed; it should be able to perform these tasks faster than the threat-waveform can adapt. These techniques, if applied to radar, follow similar principles compared to traditional RF and microwave techniques used to sense and adapt waveforms to protect military communications, conversations and data transmission. The primary difference is that radar signals are transmitted across a distance, R, and re-radiated from their intended target back to the receiver, if co-located as is typically the case with airplanes and naval vessels, travelling the distance R back to the receiver again. The signal is therefore exposed to $2R$ distance and R^4 atmospheric attenuation and is susceptible to interference, jamming or spoofing during this time.

Frequency-agile radar can broadcast across a wide range of RF frequencies and also denies other users certain bands of the RF spectrum. This wideband availability and constant changing of the RF signal in frequency makes it difficult for the enemy radar to follow and keep up with the varying signal it is aiming to interfere with. A smarter, adaptive and cognitive radar is able to first sense the RF environment before transmitting on a spectrum; this is achieved through ELINT subsystems, which adapt their radar frequency digitally, as well as the amplitude and phase of the waveform. Baylis et al. (2014) present a future radar transmitter for a cognitive system. This concept is adapted and presented in Fig. 9.2.

Achieving the cognitive radar concept in Fig. 9.2 is the main focus of Baylis et al. (2014), who review the areas requiring detailed reviews, critical design, performance trade-offs and enabling technologies. These areas include the load-matching network optimization techniques, waveform optimization and matching, real-time search optimization, dynamic radar spectral masking and on-chip functionality of SDR technology.

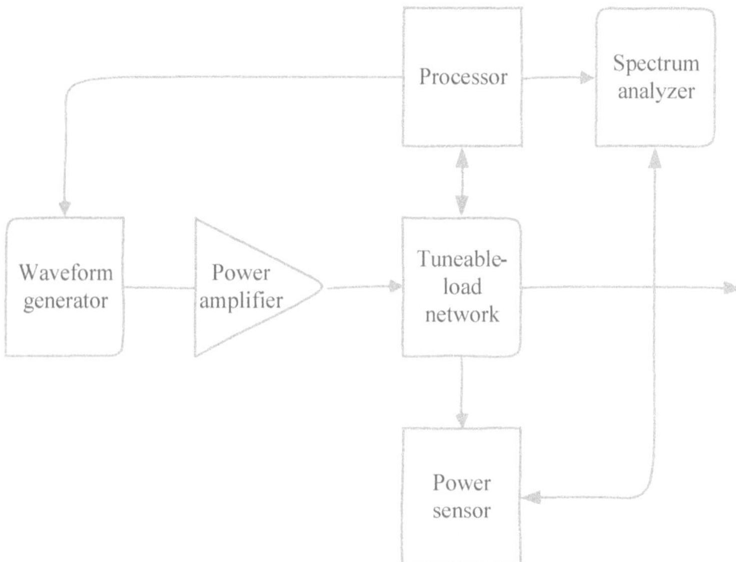

Fig. 9.2 Concept of a future radar transmitter to achieve cognitive learning, adapted from Baylis et al. (2014)

Cognitive processing in radars requires powerful computing capabilities and phased-array antennas for rapid waveform scanning in the EM environment. The base of achieving cognitive radar is using Bayes' theorem. This theorem is mathematically represented by

$$P(A|B) = \frac{P(B|A)P(A)}{P(B)} \tag{9.1}$$

where A and B are events and $P(B)$ cannot be equal to zero. From (9.1), $P(A)$ and $P(B)$ are the probabilities of observing A and B when these events are in no way related to each other. $P(A|B)$ is the conditional probability of observing event A given that B is true and $P(B|A)$ is the conditional probability of observing event B given that A is true. The true power of a Bayesian network, such as the problem given in (9.1), is its accuracy when new information and data become available within the network. It is the basis of a learning system, improving with accuracy through feedback of the output towards the input and recalculating the network based on the new information. Applying Bayes' theorem to radar, Haykin (2006) aptly explained this relationship as: "*For a given search area, radar returns are collected over a certain period of time. 2) For each range-azimuth resolution cell in the search space, the probability that the cell contains a target is computed. 3) With the evolution of target probability distribution resulting from the recursive computation of step 2 over time, target tracks are detected, and corresponding hard decisions on possible targets are subsequently made.*" Essentially, applying cognitive learning to EW radar, information of real-time and past missions is required to build, model and predict digitally generated radar waveforms in real-time, with high-speed electronics and large memory banks to hold all information in an easily accessible location for the processor.

Cognitive radar has been researched and defined and its first principles have been logically and practically explained in various research works. The most prominent and well-known, most frequently referenced work was done by Haykin (2006). Additional publications reviewing cognitive radar include those of Guerci (2010), Li et al. (2014), Griffiths and Baker (2013), Baylis et al. (2014) and Bell et al. (2015) and relevant case studies presented by Scott et al. (2011). This chapter does not aim to reproduce the information from these publications, but rather aims to highlight the fact that cognitive radar is a technology gaining much traction in EW.

The Lockheed Martin F-35 Lighting II developed in the Joint Strike Fighter program in the United States is believed (although, in keeping with the nature of military projects, not confirmed or advertised) to be one of the first projects incorporating cognitive EW elements (Tucker 2016). Although information about the complexity of the cognitive systems is also not publicly available, it would make sense to employ some level of cognitive radar in the F-35. The F-35 Lighting II employs AESA radar, also referred to as APAR, for improved performance over passive electronically scanned array (PESA) radar by electronically focusing the power of the radar signal in the required direction through solid-state technology and electronic scanning, as opposed to PESA, which mechanically focuses the radar and is prone to mechanical malfunction. AESA radar not only emits radar signals, but also can be used for non-traditional incoherent scatter radar and EA. Furthermore, when designed with a modular approach in mind, AESA radars can be gradually upgraded by replacing the solid-state transceiver modules based on GaAs semiconductor technology with more advanced elements, thus significantly improving performance. The AESA radar is briefly discussed in the following section, with the focus on the enabling technologies and the potential of improving performance based on the choice of technology.

9.3 Active Electronically Scanned Array

In AESA, the dynamic range of the scanned array is improved by incorporating transmitters and receivers with phase shifters behind every antenna element in the array. AESA offers the ability for multiple simultaneous beams to track multiple targets simultaneously. Losses between the amplifiers and antenna elements are reduced with this topology, phase noise between elements are uncorrelated and individual failures of elements do not cause dramatic failures in the entire system. A fundamental challenge is to reduce the cost of individual transceiver modules and to physically fit the components between the lattice spacing of the phased array, typically placed at $\lambda/2$ spacing. As the operation frequency increases, the spacing between arrays decreases and the space limitation becomes more stringent. These limitations and requirements are technology-dependent, therefore advances in technology are required to realize enhancements in AESA performance. A simplified transceiver module in an AESA implementation is given in Fig. 9.3, adapted from Sturdivant and Parrish (2015) and Song (2014).

In Fig. 9.3, the fundamental differences between the traditional PESA radar and future-generation AESA radar are highlighted. PESA systems use a central PA implemented with travelling wave tube (TWT) amplifiers. As shown in Fig. 9.3, the PA of the signal to be transmitted and the LNA of the received signal are shared between all phased array elements. Each antenna element is equipped with a phase shifter between the transmit/receive antenna and the power divider network. PESA systems in general have lower reliability because of the TWT topology and introduce noise and attenuation to the signal between the amplifiers and the antenna. AESA addresses these issues by introducing PAs and LNAs for each received or transmitted signal between the antenna and the power divider network. By switching each element individually, the AESA topology allows for multiple signals to be received or transmitted simultaneously and increases the dynamic range of the overall system. The reliability of the AESA topology is also increased when compared to the traditional PESA arrays by introducing transmit/receive modules for each element, removing the TWT network, which is effectively a single point of failure for the entire array. The technology used in each transmit/receive element should ideally be highly reliable, capable of fast switching speeds and affordable, since many of these elements are required in a phased array architecture (Huang et al. 2015). Receiver modules should have a low noise figure per element, since the total noise figure of the system is an additive function of each element. The total noise figure of the receiver determines the sensitivity of the array for small-amplitude incoming signals or radiation. The transmitting module's requirements are strictly defined by its power capabilities, both the attainable output power and the input power requirements. For mobile systems, such as the systems fitted to fighter planes, low-power (high-efficiency) electronic systems are crucial in view of the inherent power constraints.

Phased array antenna architectures rely on beamforming or spatial filtering to achieve directional transmission on reception of a radar signal. Phased arrays work on the principle of transmitting multiple signals where the relative phase of these signals either cancel each other out or combine together, to form an effective radiation pattern. Changing these patterns in real-time is where a phased array system is able to dynamically change the waveforms to confuse enemy radar and allow for cognitive radar to be implemented. Beamforming is therefore the technique used to steer the resultant beam towards its intended target.

Beamforming can be achieved either in the analog domain, or digitally, depending on the application. Analog beamforming (ABF) occurs when the received echoed signals from each portion of the phased array antenna are combined at the RF carrier frequency level. Up to four centralized receiver channels are fed together and the signal is down-converted to baseband or IF and fed to the ADC. A simplified representation of the ABF architecture is given in Fig. 9.4a).

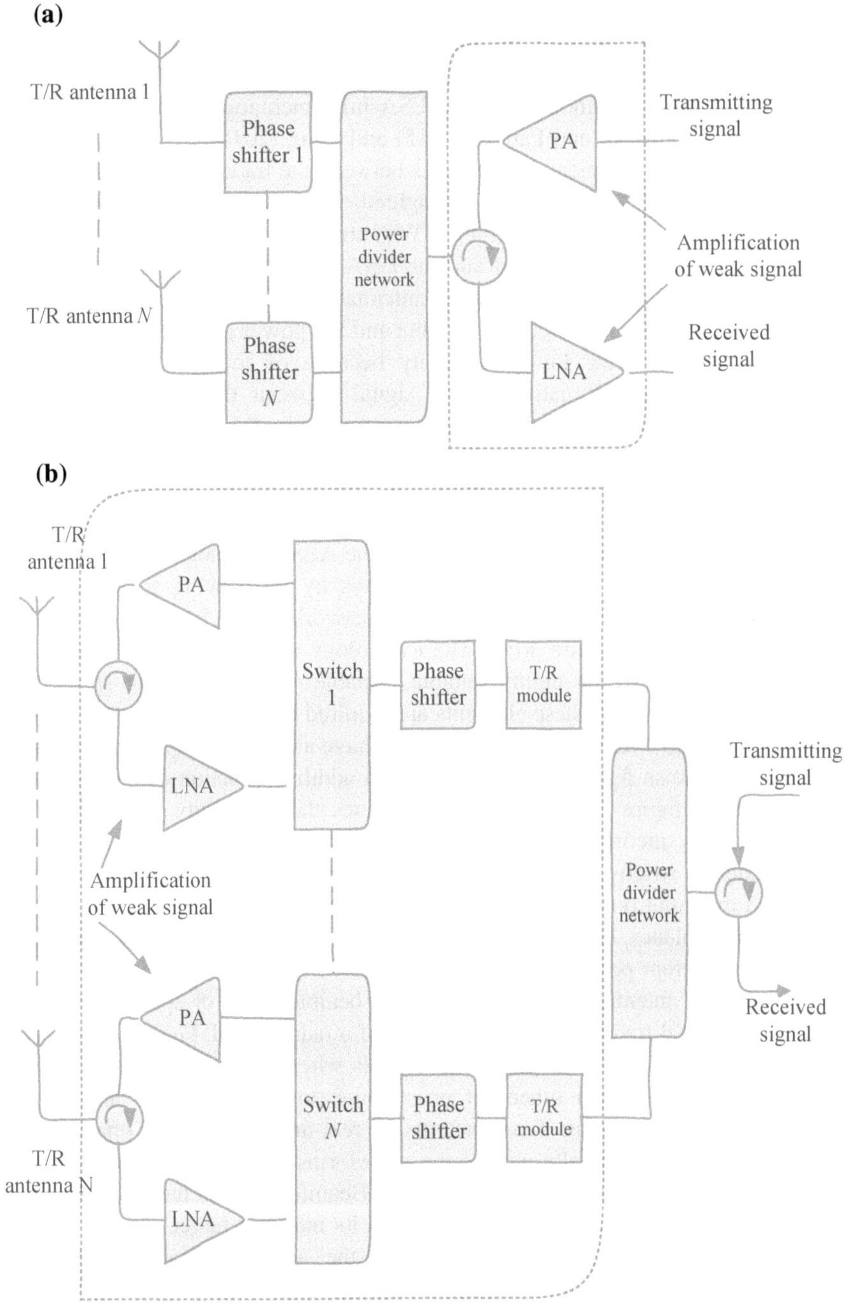

Fig. 9.3 Simplified concept of **a** PESA antenna array compared to the **b** AESA antenna array

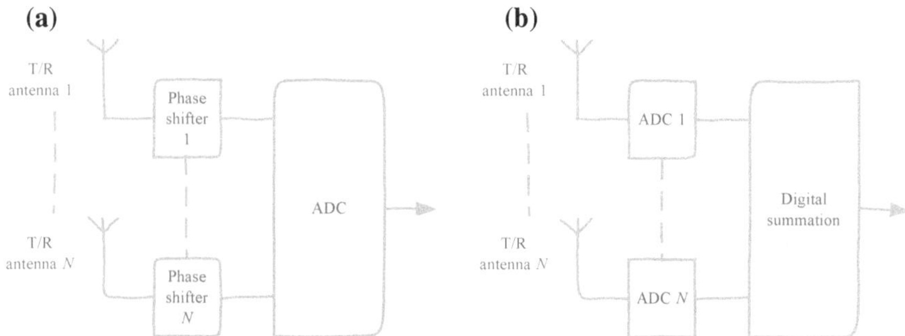

Fig. 9.4 Simplified architecture of **a** an analog beamforming technique where four receiver channels are down-converted and digitized and **b** a digital beamforming technique where each receiver channel is down-converted and digitized

Conversely, digital beamforming (DBF) can be realized at element-level or at sub-array level (Wolff 2016). In a DBF system, each receiving channel is equipped with an ADC at each of the radiating elements of the antenna array, depicted in Fig. 9.4b. In DBF, the down-converting to IF is realized at each element and signal processing at each element can include noise cancellation, for example, which is more effective when compared to processing a set of analog signals based on the additive noise in each channel. Multiple independent beams steered in all directions can be formed in the digital beamforming processor. DBF has improved dynamic range, control of the beams and control over the amplitude and phase of each beam, when compared to ABF systems (Wolff 2016).

The transmit/receive modules are the key components in an AESA DBF system and consist of the PA and driver amplifier (DA) for the transmitter and the LNA for the receiver. For military and space applications, these components operate primarily in the X-band frequency spectrum and therefore require discrete active devices capable of either generating carrier frequency for down-conversion or processing incoming signals without down-converting the signals, therefore at X-band switching speeds. This puts stringent requirements on the technology in terms of noise performance and power capabilities. GaAs and GaN traditionally provide good trade-offs between power requirements and operating speed; however, advances in SiGe technology have led to potential re-engineering and re-evaluation of GaAs and GaN as the enabling technology. As the complexity of these microelectronic circuits increases, space and cost also become important trade-off considerations when realizing larger arrays capable of high-speed signal processing in smaller spaces (due to the $\lambda/2$ spacing requirement). Complete multifunctioning ASICs are realizable on BiCMOS and/or CMOS technologies, with optical integration (see Chap. 8) additionally realizable at optimum wavelengths in Si-based substrates such as with the SiGe options. Table 9.1 indicates the crucial components in a common transmit/receive module, the commonly used technologies to realize these subsystems and the advantages these technologies offer over alternatives.

Table 9.1 Crucial components in transmit/receive modules for AESA applications, the preferred technologies and the advantages thereof

Subsystem	Preferred technology	Comments
PA	GaN	GaN offers highest power capabilities and high breakdown voltages
DA	GaN/GaAs/SiGe	Can use GaAs at lower cost; lower power requirements than PA. SiGe can also be used for cost-effectiveness if output power requirements are low
LNA	GaAs/GaN/SiGe	GaAs is low-cost with a good noise figure but lower breakdown voltage (susceptible to damage). GaN can withstand larger power leakage but has lower noise figure compared to GaAs. SiGe can also be considered
Receiver protection	GaAs	Typically uses PIN diode limiters in GaAs. SiGe can also be considered
Phase shifter	GaAs/SiGe	GaAs has high third-order intercept (IP3). SiGe costs less and offers additional functionality due to smaller size and temperature compensation

As shown in Table 9.1, each technology (GaN, GaAs and SiGe) offers unique advantages for each subsystem in AESA transmit/receive module. As in most applications, the requirements of the subsystem are dependent on the subsystem itself and choosing a technology that is optimum for each subsystem is a daunting task. Power output and power efficiency are possibly the most crucial requirements for AESA, considering all subsystems, and GaN remains optimal in offering premium size, weight, power efficiency and high-power performance. SiGe, as an established technology, can be incorporated into AESA subsystems with requirements for high circuit complexity, reliability and low cost.

The digital domain in coherence with analog signals also provides advantages and opportunities to enhance and further develop the capabilities of SDR. The principles of SDR are reviewed in the following section.

9.4 On-Board Digital Systems (Software Defined Radio)

Radios in which some or all of the physical layer functions are software-defined are called SDR. In this, SDR is not limited to the hardware and protocols it is defined for, but can adaptively change with the requirements of the user or the environment (Du Plessis 2014). In military applications, the SDR operating system is called the open-systems software communications architecture (SCA), which is based on COBRA, COBRA services, COBRA Component Model and POSIX operating systems and expressed in COBRA IDL, UML or XML to coordinate different SDR waveform software modules (Keller 2011). SCA is not a system specification, but a set of rules to ensure standardized compatibility across systems. SCA was designed to satisfy the need to provide real-time information exchange on the battlefield and overcome limitations in communication due to radio-specific design. It was set out by the Joint Tactical Radio System (JTRS) Joint Program Office (JPO) to acquire capabilities with a mission clearly stated that such a technology had to be modular, scalable and backwards-compatible. SDR uses embedded systems (software upgradable to extend the lifetime of these radios) to realize components that are typically designed in hardware, such as the mixers, filters, amplifiers, modulators and demodulators. It can also consist of additional functions such as encryption to safely transmit voice or data across the wireless spectrum. The functionality that SDR provides is summarized as providing

- software applications that provide waveform generation (new waveforms and legacy waveforms), processing, encryption, signal processing and cognitive abilities,
- multiple software units to allow operation of diverse standards in a single radio system and advanced networking features,
- programmability and ability to use multiple physical layer formats and protocols,
- flexibility through software-upgradable or added functions by providing standardized application programming interfaces, and
- decreased maintenance cost, since radios are reconfigurable over the air, high component reuse and primarily implementation with off-the-shelf components.

In military applications, these functions are critical to ensure that the communication between allied forces is secure. The primary difference between SDR and traditional analog radio is that the information is immediately digitized and can be processed before transmission. Since the digitization, modulation, transmission, reception, demodulation and conversion, if necessary, must be done with as short as possible delay (for instance in the case of voice communications), the speed and data handling abilities of the embedded systems must be high. This is another example where a technology, the active components performing the operations, is a crucial component. The principal enabling technology in SDR is therefore the embedded systems, realizable through DSPs, FPGAs or general processors. The accompanied software further extends the potential applications of SDR. Applications of SDR vary greatly, as do the functions performed in hardware and in software (in the digital signal processing subsystem). Figure 9.5 shows one example of an implementation of SDR, where the analog front-end and the software-defined subsystem are indicated, including the functions performed in each domain.

The SDR implementation in Fig. 9.5 shows the different signal-processing steps achieved in the analog front-end, as well as processing techniques done within the DSP or FPGA (or general processor, depending on its capabilities and requirements). The processor is typically also equipped with memory to store waveform patterns and modulation techniques, has the ability to up- or down-convert digital signals, encode or decode the signals and digitally filter the processed signal. These

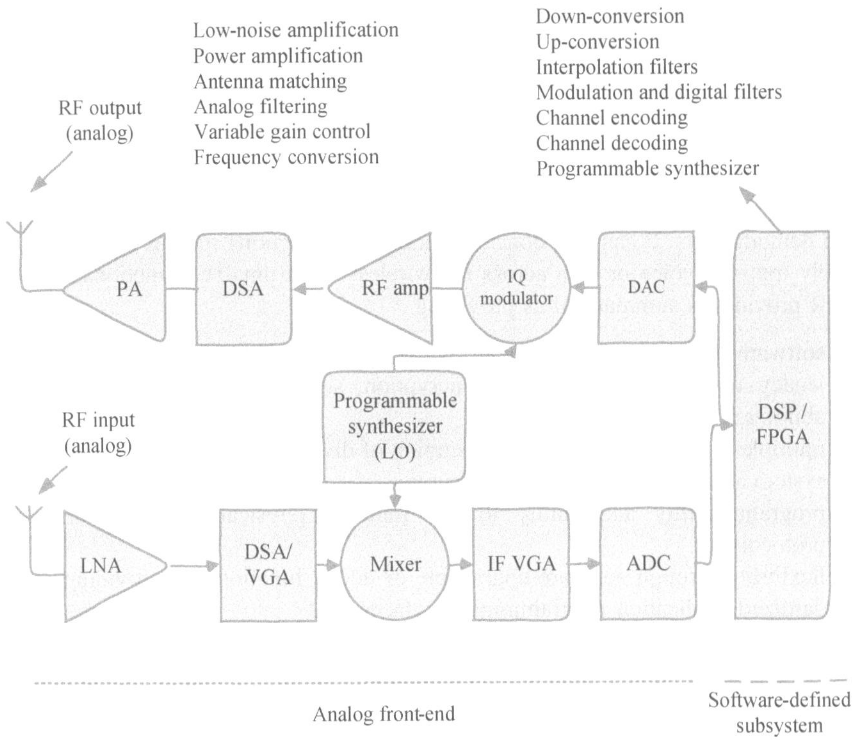

Fig. 9.5 The analog front-end and software-defined subsystem of an SDR implementation with the functions performed in each domain

functions are dependent on the application and requirements of the SDA. For SDA applications, a key requirement is configurability and flexibility (Altera 2007). In the White Paper presented by Altera (2007), the recommendation to use both DSP and FPGA modules in SDR is advantageous for several reasons. Ideally, an ASIC to perform the specific tasks required by the SDR would be preferred, but for cost-effectiveness off-the-shelf DSPs and FPGAs can suffice. The advantages of using both modules, according to Altera (2007) include:

- The DSP is an ideal platform to implement complex algorithms.
- The FPGA provides high processing rates and bit resolution required for repetitive functions, such as finite impulse response and fast Fourier transform.
- With optimal interconnections, the FPGA and DSP can function as co-processors to realize a multi-core system where each module performs the tasks optimal to its design (complex algorithms in the DSP and repetitive tasks in the FPGA).
- For both off-the-shelf implementations, mature, maintained and after-market services are available.

Table 9.2 Performance comparison between SiGe BiCMOS and GaAs pHEMT technologies for L- and S-band LNAs

Performance parameter	SiGe BiCMOS	GaAs pHEMT
Noise figure (dB)	≥ 0.9	≥ 0.4
Gain (dB)	10–17	12–21
OIP3 (dBm)	≥ 31	≥ 41
Breakdown voltage (V)	<15 V	15
Inductor quality factor	5–10	15
f_T/f_{max}	Comparable (similar)	

Adapted from Das (2013)

On the analog front-end, lower-level signal processing is done, including the crucial high-powered PA for transmitted signals and the LNA for low-amplitude incoming signals, as well as eliminating inherent EM interference before feeding the signal to the software-defined subsystem. LNA design, for example, is a subsystem where SiGe presents distinct advantages in performance and the SiGe BiCMOS process and GaAs pHEMTs are among the most popular implementations (Das 2013). Table 9.2 is adapted from Das (2013) and lists the performance parameters of SiGe BiCMOS and GaAs pHEMT technologies for L- and S-band LNAs.

The ADC and DAC components are crucial in SDR implementations and require high-resolution (high dynamic range) and high-speed operation to process the incoming, typically X-band, signals. Conversion should ideally be done close to the transmit/receive antennas to reduce the probability of signal degradation. The selection of ADC and DAC components is influenced by the required IF, SNR, sampling rate, jitter and two-tone intermodulation distortion. To maximize the dynamic range of the SDR, Clarke and Kreitzer (2014) provide a typical checklist to use when starting the design of the SDR, or optimizing an existing SDR. This checklist includes the questions to be asked concerning:

- the requirements of the implemented standard,
- the desired minimum and maximum signal levels,
- the amount of filtering required,
- types of image filters, channel filters and anti-aliasing filters that are available,
- the effect of group delay in the filters,
- the system architecture (zero IF or single, dual or triple conversion),
- the technique for generating quadrature, and
- the primary domain, analog or digital.

For the ADC (and DAC), the requirements have their own set of questions, predominantly the dynamic range, which influences the architecture of the converter, or if the architecture is predetermined, the requirements of the dynamic range that can be achieved. The signal bandwidth and sampling frequency influence the maximum attainable SNR; however, the dynamic range is the fundamental trade-off between analog and digital (within the DSP or FPGA) filtering. Additional margins should also be considered where atmospheric attenuation, EM interference noise and adjacent-channel interference during demodulation are among these unplanned signal degradations.

Similar to cognitive radar, cognitive radio can be achieved by implementing SDR. Radios, along with software defined antennas, can inspect the RF environment and change the communication protocols, waveforms, power and modulation based on the immediate environment, add encryption, or even transmit across various bands, making the transmission difficult to intercept. Additional functions such as situational awareness by sharing positioning coordinates or relaying the vital signs of a soldier to medical personnel (Keller 2011) are possible with SDR.

Software defined antennas, or reconfigurable antennas, are essentially antennas that can change their frequency of operation, radiation pattern or polarization modes through external inputs such as RF switches, varactors, mechanical actuators or tunable materials. Various innovative schemes to dynamically change the physical properties of the antenna have been reported, such as the individually controllable pixels to form a patch antenna-array, as presented in Walton et al. (2008).

Another discipline with many benefits in the EW and military application environment in the use of radar, software and low-noise analog equipment, is the use of precision-guided munitions. This topic is reviewed in the following section, highlighting the basic principles, enabling technologies and subsystems and the advantages of using such weapons.

9.5 Precision-Guided Munitions

A PGM is a missile, bomb or artillery shell armed with a thermal guidance system with electrical and electronic components to guide the munitions towards the final phases before impact. This terminal guidance system senses the EM environment within its field of view. Guidance can be achieved by target acquisition systems such as radio-waves, IR or electro-optical devices, laser, satellites and other advanced techniques such as combining laser guidance with GPS, for example. A PGM can therefore be aimed in the direction of the target and relies on a combination of external guidance and its own internal guidance system at close proximity, for example to reach the target. Implementations of PGM weapons include

- precision-guided joint stand-off weapons,
- conversion of unguided bombs to guided joint direct attack munitions,
- conversion of cluster bombs to precision-guided munitions such as wind corrected munitions dispensers,
- sensor-fused weapon follow-on, and
- joint air-to-surface stand-off missile,

which are autonomously guided and are able to function in adverse weather. Including modern techniques of cognitive radio and/or cognitive radar into PGM adds to the functionality of these weapons. Raytheon® produces a handheld, 17 in. (43 cm) long, 1.57 in. (40 mm) diameter PGM weighing two pounds (900 g) that is fired from a rifle-mounted grenade launcher. The PGM can travel 1.5 miles (2.4 km), with five yards (4.57 m) or less accuracy to reduce collateral damage. Using a laser designator that resembles a pistol, one soldier points at a target while another fires the munition. The device can be mounted on multiple platforms, including Class I and II UAVs, all-terrain vehicles, ground mobility vehicles, small boats and remotely operated weapons stations.

PGM can be considered a subclass of cognitive EW and therefore also requires complex algorithms based on Bayes' theorem to identify, track and label its target. To perform these calculations, powerful processors are required and are mounted on the weapon itself. These are ideally small, have low power consumption and are not susceptible to external interference. The choice of technology again depends on the strategy used to guide the munitions, the complexity of the system and the analog and digital requirements that affect the accuracy and resolution of the weapons. In Maurer et al. (2001) a low-cost gun-launched seeker concept design for naval fire support is presented with reviews of its required functional capabilities, the seeker design, which includes the detector and optics of the munitions, as well as the processing and algorithm requirements. PGM depends equally on the hardware and software engineering and the hardware must be able to perform the software algorithms in the time constraints specified. Target detection algorithms must be able to detect desired targets with a high probability of success, with limited processing time, dependent on the capabilities of the microprocessors. Maurer et al. (2001) depend on parallel operations to provide detection capabilities under a larger range of image conditions and present a baseline target detection architecture. This detection algorithm can be modified and applied to various applications and environments, depending on the conditions of the environment and the capabilities of the hardware. The parallel processing is done when detecting the contrasting intensity levels and enhance/detect edges in the image. The results of these processing streams are fused before the features are extracted and the target is then detected, classified and labelled. Maurer et al. (2001) implement a microprocessor on a DSP where the algorithms do not lend themselves to pipeline operations and are data-dependent, where an FPGA would be preferred if the algorithms could be decomposed into many small independent computations and take advantage of the configurable architecture of FPGAs. Typically, algorithms can be implemented on DSP and FPGA and computational times can be compared to determine the best architecture if possible. The algorithm structure has multiple pipelined and non-pipelined threads and it is difficult to determine which architecture is optimal.

Various techniques and models to solve three-dimensional interception geometrical problems, which can be converted to practical algorithms, can be researched, used and adapted for different applications of PGMs. The vast number of possible implementations, open-source hardware and software and availability of standard geometrical algorithms make PGMs a large contender in the future of EW, cognitive EW and adaptable multi-functioning projectile weapons. Possibly the most crucial consideration in these applications is the velocity of the projectile towards the target, the distance from the target (therefore the time until impact), the complexity of the algorithms and the limitations on the processing speed of the microprocessor.

UAVs are also dependent on multiple subsystems, efficient interconnects between subsystems, different communication techniques, radar systems and mechatronic control, remote control and autonomous systems (with respect to hardware and software) and are distinct advantages when long-range communication with these vehicles are possible. In the military environment, and also more recently in the commercial sector, UAVs and drones are becoming more crucial and are receiving an increase in market share, investment opportunities and technology research projects. This is because of the various advantages they offer and the decreases in the cost of powerful processors, high-quality communication systems and the potential of adaptable applications through software and digital subsystem integration. UAVs are briefly reviewed in terms of their operation, potential applications and the advantages offered to EW in the following section.

9.6 UAVs

UAVs (also referred to as UAS or drones (Barton 2012)) have been used in military activity in one form or another for many years. As electronic technology is advancing in terms of its capabilities, speed, reduction in size and power consumption and cost-effectiveness, the applications of UAVs are also expanding rapidly. All UAVs rely on advanced electronic systems to gather information about surrounding objects and obstacles (Silicon Radar 2016). Such sensor systems have very special requirements from the enabling technology, where SiGe, GaAs, GaN and InP all contribute to satisfy each requirement. Importantly, the requirements that are primarily considered in UAVs include

- miniaturized electronic components,
- low weight and material usage,
- low power consumption,
- high-frequency operation with broad usable bandwidth,
- high integration of mixed and digital circuits, and ultimately
- low cost (Silicon Radar 2016).

Essentially, UAVs can contribute to EW in many aspects requiring the transmission and manipulation of information signals. The advantage of long-distance flight capabilities, their unmanned nature not directly putting a soldier at risk by flying into enemy territory, relatively large payload (since there is no requirement for a human pilot on board) and potential for the integration of various electronic subsystems make UAVs ideal vehicles for EW in hostile military environments. Among the most effective tasks and payloads UAVs can deal with are situational awareness, SIGINT, COMINT and ES. Possibly the largest disadvantage of UAVs

Table 9.3 Five classes of UAVs as classified by the United States DoD

Category	Size	MGTW (lbs/kg)	Normal operating altitude (ft/m)	Airspeed (knots/km.h^{-1})
Group 1	Small	0–20 (0–9)	<1200 (365) AGL	<100 (185)
Group 2	Medium	21–55 (9.5–25)	<3500 (1066)	<250 (463)
Group 3	Large	<1320 (599)	<18,000 (5486) MSL	<250 (463)
Group 4	Larger	>1320 (599)	<18,000 (5486) MSL	Not defined
Group 5	Largest	>1320 (599)	>18,000 (5486)	Not defined

currently, not overshadowing the loss of life in traditional vehicles, is the fact that there are very few political consequences of shooting down an unwelcome UAV spotted in the local airspace. This results in significant losses in resources and opportunities and may increase hostility by enemies if the mission and payload of the UAV are unknown to the enemy. However, shooting down a manned aircraft or vessel has much more serious consequences and can be accompanied with loss of life, which means that UAVs are still much safer and are essential for reconnaissance, defensive, offensive and situational awareness missions requiring autonomous and remote operation.

UAVs are classified by the DoD in terms of either their physical size and thus the potential size of the payload or their travelling range. According to the United States DoD, UAVs are classified into five categories, summarized in Table 9.3.

Table 9.3 groups the UAV classifications by category, size, maximum gross takeoff weight in pounds, normal operating altitude above ground level or at mean sea level in feet and airspeed in knots. In addition, UAVs can also be classified, as outlined by the United States DoD, by their range of travel as being

- very low-cost and close range (range of 5 km and endurance time of 20–45 min),
- close range (range up to 50 km and endurance times of 1–6 h),
- short range (range up to 150 km or longer and endurance times between 8 and 12 h),
- mid-range (high-speed and operating radius of 650 km) or
- endurance UAVs (operating radius of 300 km and endurance time of 36 h or more).

Furthermore, UAV classification typically falls into one (or more) of six main categories (DHL Trend Research 2014), namely

- civil and commercial applications in industries such as mining, civil construction, energy transmission and distribution, water, agriculture, oil and gas, energy services, emergency response, film and photography and telecommunication;
- logistic support in the service delivery and transport sector,
- research and development for weather patterns, environmental effects due to global warming, sustainable development and movement of ocean water,

- armed UAVs used in combat situations,
- exploration of enemy areas to gain reconnaissance information, and
- copying the flight and physical profile of aircraft to confuse the enemy's air defense systems, thus used as a decoy.

Apart from the application categories of UAVs listed above, because of the diverse range of systems it is becoming difficult to classify these systems into a standardized set. The US military, for example, uses a tier system with prerequisites pertaining to range, endurance and application. UAVs can furthermore be classified by their physical dimensions, maximum take-off weight, service ceiling, price and build types (DHL Trend Research 2014). UAV build types include

- fixed-wing (long-range and endurance flights but with low maneuverability and requiring catapult launch, typically used in military reconnaissance and situational awareness missions),
- tilt-wing (complex technology and expensive),
- unmanned helicopter (highly maneuverable and carrying large payloads, but expensive and high maintenance) and
- multi-copter (cost-effective, easy to launch and low weight, but with limited payloads and susceptible to wind).

UAVs have a large combination of computing capabilities, sensors and actuators and software requirements, since these vehicles are remotely controlled, have primarily electronic-based mission requirements and are used to sense the environment for various elements, including the RF spectrum, visual confirmation of points of interest, or if used in meteorological missions, sensing the environment for pollution, weather patterns or natural disasters.

UAVs are not exclusively flying vehicles and also include unmanned sea surface vehicles (USSVs), which exploit the potential of naval remotely operated vehicles to achieve similar EW goals as UAVs, but from a different platform and with different requirements in terms of payload, operating conditions and missions. Also falling under the unmanned category are unmanned ground vehicles, unmanned spacecraft and unmanned underwater vehicles. The EW payload for the HS-USSV includes, for example, a wideband DRFM-based EA system in addition to more traditional payloads such as jamming equipment with high-gain antennas (Ramasamyraja et al. 2014) and high-powered microwave power modules originally developed for tactical aircraft (Tremper and Heyer 2007).

A solar UAV uses solar energy from solar cells on the UAV for longer endurance flights. The solar power can be distributed to the electric motors used to propel the vehicle and for the electronic subsystems used for EW and/or meteorological monitoring. Solar energy is, however, restricted to lower energy density compared to traditional fossil fuels and must therefore be efficiently distributed. An important subsystem in a solar UAV is the power management system consisting typically of the maximum power point tracking, the solar modules, the battery modules and the charging controller. The power required is determined by the weight of the aircraft and its aerodynamic properties, such as the lift coefficient,

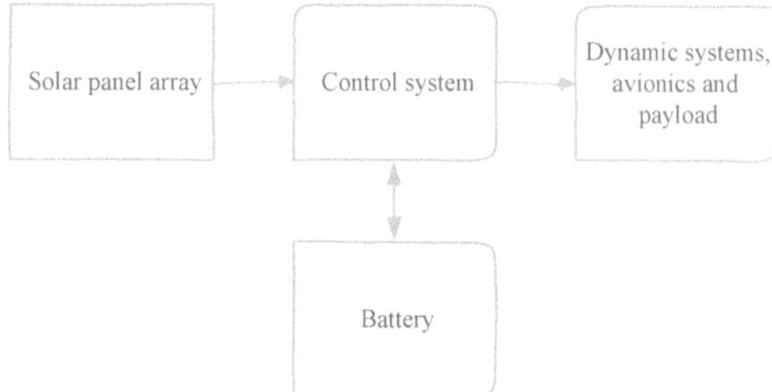

Fig. 9.6 Simplified diagram of a solar-powered energy system for a UAV

drag coefficient, wing area and air density (Lee et al. 2013), as well as the payload power requirements. A block diagram of the working principle for the solar energy system is presented in Feng et al. (2014), adapted and given in Fig. 9.6.

As shown in Fig. 9.6, a solar-powered UAV operates on a similar principle as any other solar-powered device and the control system is used to distribute the available power to the subsystems of the UAV. These subsystems include the dynamic systems and actuators, general avionics and the payload of the UAV. A typical UAV control system is represented in Fig. 9.7, showing the feedback requirements to guide and navigate the UAV.

As shown in Fig. 9.7, the typical UAV control system uses a feedback network to minimize the tracking error between the desired and the estimated states when external disturbances such as wind are accounted for. UAVs are the most difficult systems to design and implement (when compared to ground-based unmanned systems), since aerial vehicles are not in equilibrium with their environment. Ground vehicles are typically in equilibrium with the environment, meaning that if no inputs are given, there is a high probability that the system will not be in an uncontrollable state. Errors are constantly determined and minimized to ensure that the trajectory of the UAV follows the desired direction. External disturbances are not only able to influence the direction of travel of the UAV, but can also cause the aircraft to lose control; sensors and GPS are used for corrective action against external disturbances. The tasks of the control system are complicated multi-input multi-output and time-varying nonlinear systems (Duan and Li 2014). The control system of the UAV is therefore a critical component in UAVs and must be able to process complex algorithms, in real-time, including linear and non-linear dynamic equations to estimate the flight trajectory of the vehicle. For fixed-wing, endurance, military-based UAVs the control system must be able to compute the geometric parameters of the aircraft, basic aerodynamic parameters, longitudinal aerodynamic coefficients, lateral aerodynamic coefficients and fixed-wing UAV control surfaces. For these parameters, flight dynamic models are derived based on the forces and moments applied to the aircraft, atmospheric disturbances and the full non-linear equations of motion (Ahmed et al. 2015). For most UAV flight equations, the earth can be used as the inertial reference frame and the aircraft can be modeled as a rigid body. According to How et al. (2012), the following set of equations typically

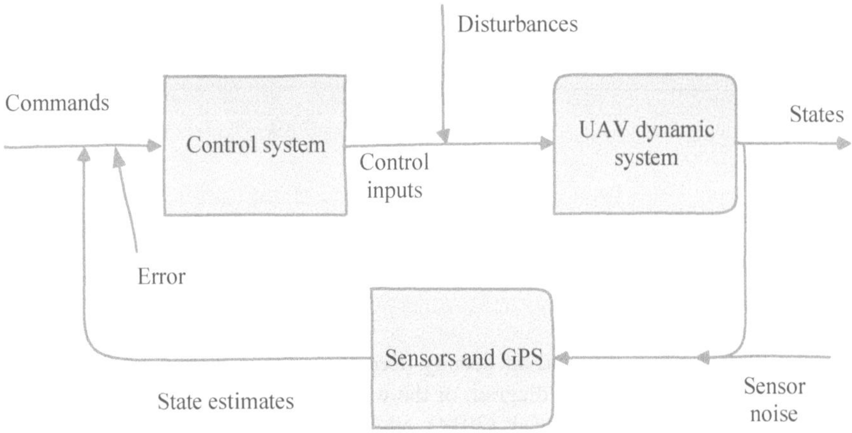

Fig. 9.7 Typical UAV control system

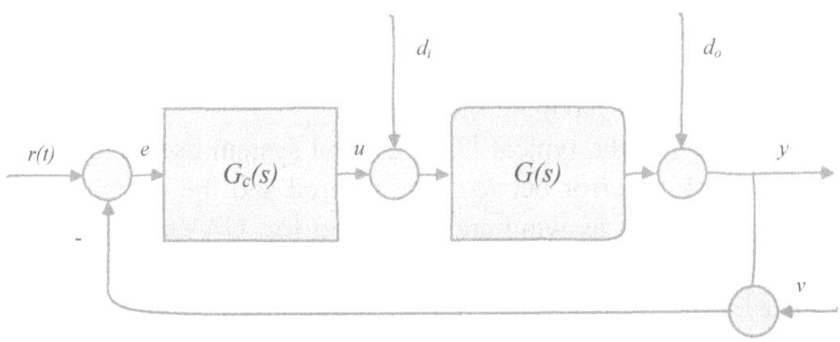

Fig. 9.8 Typical UAV control system showing the signals in the input, output and feedback loop. Adapted from How et al. (2012)

represent the signals processed in the situation presented in Fig. 9.8. The signals presented in Fig. 9.8 are adapted from How et al. (2012).

In Fig. 9.8, adapted from How et al. (2012), the tracking error can be determined by expressing the error e in the form

$$e = r - (y + v) \tag{9.2}$$

which can be expanded upon such that

$$e = S(r - d_o - v) - SGd_i \tag{9.3}$$

where ideally, e should be kept as small as possible. Computational intelligence methods are used to program control systems to execute a set of instructions autonomously. Current and future control systems employ these intelligence methods, or a combination of these methods, typically using the following techniques (Long et al. 2007):

- traditional control,
- fuzzy logic,
- neural networks,
- genetic algorithms,
- rule-based methods, or
- symbolic artificial intelligence.

An alternative consideration in unmanned vehicle autonomous control is considering biological systems such as human behavior. These systems give a holistic view on intelligence and the use of sensing, reasoning, action, learning and collaboration when performing tasks. Applying robotic systems such as UAVs based on intelligent systems creates a path towards cognitive UAVs. These techniques also determine the processing requirements due to the difference in complexity of each method. As current FPGA and DSP technologies improve, the integration of complex, high-speed, real-time and cognitive/neural networks (therefore large availability of memory to store previous events and learn from them) is more realizable compared to earlier systems. According to Long et al. (2007), the most promising approaches to implementing control systems use layered or hierarchical control strategies such as

- subsumption,
- behavior-based models,
- reference models, and
- three-layer models.

In Long et al. (2007), a comprehensive review on the software systems for autonomous vehicles is given. These software systems include

- Java Expert Systems Shell,
- Fuzzy Logic in Integrated Reasoning,
- Subsumption Architecture,
- Autonomous Robotic Architecture (AuRA),
- Applied Research Laboratory at the Pennsylvania State University ARL/PSU Intelligent Controller architecture,
- Jet Propulsion Labs Advanced Sciencecraft Experiment,
- Cognitive Architectures,
  - ACT-R
  - SOAR
- NIST Real-time Control System.

Advanced UAV sensor payloads are able to acquire large amounts of data, including full-motion video and high-definition images. Sharing these data is often bandwidth-limited; although the processor and control system are able to process and store the data relatively easily nowadays, transmission to the ground station is still limited by the bandwidth of the communication link. For this reason, intelligent systems only send mission-critical data to the ground station, and the commands sent to the UAV should also be limited. Efficient data processing on the UAV control system ensures that the communication link is available to send and receive data, prioritized between mission-critical commands and data that can be transmitted when the channel is open. Control compatibility is a challenging issue in military and commercial UAVs. The most prominent and widely debated issues are associated with control compatibility with other UAV ground stations, ability of the UAV to communicate directly with various ground and air support systems in the area (regardless of which architecture they belong to), secure down-link bandwidth limitation as more UAVs are streaming data at the same time and at a higher throughput and timely availability of relevant, post-processed data (Howard 2013). This book does not aim to provide all equations and considerations for implementing autonomous flight, but rather to highlight the complexity and the processing requirements of these equations. Companies specializing in microelectronic integrated systems provide fully integrated systems that achieve high levels of performance in UAVs while reducing size, weight, power, system cost and development time. Such integrated solutions and off-the-shelf components are supplied by various microelectronic integrated system manufacturers. These solutions include

- video transmission with low-power digital signal processors,
- safety and surveillance chipsets to support ADS-B and transponders with IFF and sense and avoid capabilities,
- navigation integrated inertial measurement units for attitude and position sensing, accelerometers, gyroscopes, magnetic sensors, pressure sensors and temperature sensors,
- airframe monitoring systems with precision vibration sensors and ADC converters that support health utilization monitoring systems,
- data transmission and vehicle control ICs for both flight management control and data transmission,
- electrical systems such as motor control, power management, isolation, conditioning, high-quality amplifier components supporting flight control, power management, charging (for solar-powered devices) and propulsion systems,
- control surfaces for flight control sensing and precision measurement through linear variable differential transformers, rotational converters and strain sensor converters,
- payloads that include high-speed and precision converters and DSPs for interfacing to analog sensors and detectors, advanced gyroscopes and accelerometers based on MEMS for payload stabilization, HDMI and DVI switches and buffers,
- SAR signal processing of the successive radio-wave pulses to form images of landscapes and other point of interest,
- collision avoidance and mapping using high-GHz radar solutions for object detection, mapping and ranging, and
- supporting systems for imaging, radar, surveillance, reconnaissance and communications.

Numerous electronic sensors for precise control and feedback are required on UAVs. To control the altitude (pitch, roll and yaw and translation movements) of the UAV, actuators are used to apply forces to the flight mechanical equipment. The precise extent of the position of these actuators is critical in sustaining the correct path of flying. The technology, hardware and software, to realize autonomous algorithms is an important consideration and there are several FPGA and DSP products available that can compute, store and communicate the relevant data of a UAV control system. These digital systems must also be able to accept analog input from all sensors and provide analog output to actuators situated all around the UAV. The vehicle dynamics processor is typically separate from the mission-specific telemetry equipment, although the physical processor could be similar and the workload for each task shared.

Dynamic modeling of autonomous vehicles, like manned vehicles, requires accurate aerodynamics of many types and variations of vehicles. The fundamentals of state estimation and flight control are common between these vehicles and must be simulated, optimized and programmed into the flight controller of the UAV. The rotational state estimation (attitude and attitude rate) and translational state computations (position and velocity) for fixed-wing and rotor-equipped UAVs differ significantly, primarily owing to the additional degrees-of-freedom of the rotor-equipped vehicles. For larger UAVs, the earth's rotation and curvature effects must be accounted for as well, whereas for smaller UAVs, this can be ignored to simplify the initial computations. Depending on the type of UAV, either translational motion (for fixed-wing aircraft) or rotational motion (rotor-equipped aircraft) must be programmed into the flight controller. Translational motion is the movement of the center of mass of an object between two positions. Rotational motion occurs if an object moves about an internal axis continuously and similar quantities and concepts apply, but certain translational quantities must be replaced by their equivalent rotational quantities. If the UAV is subjected to external forces and moments due to gravity, propulsion and/or aerodynamics, the applied forces are combined and expressed in the body frame. For in-depth reviews on the derivation of linear and non-linear translational and rotational motion for UAVs, Table 9.4 lists several references dedicated to explaining the fundamentals of motion-translation in aircrafts. These principles fall outside the scope of this book, which focuses on the computational processing requirements of these techniques.

In order to implement the algorithms required to autonomously fly UAVs, either fixed-wing or rotor-equipped, a platform for developing the systems is required, which reduces the dependence on costly field trials, simulating the flight system entirely in software and ensuring that each flight trial is successful, with only minor adjustments required after each trial. As hardware and software continue to develop and become more cost-effective to implement high-speed computers for model development, this process has become commonplace for UAV simulation and modeling. Chen et al. (2009) state that the fundamental goal of flight dynamics modeling is to represent the flight motion numerically for a given input, as close to the flight motion in the real world as the application requires. As a result of the wide applications of flight models, a large number of implementations and algorithms of

Table 9.4 References related to linear and non-linear translational and rotational motion to model the movement and control of UAVs

Reference	Title
Cooke et al. (1992)	Flight simulation dynamics modeling using quaternions
Ritter and Dillinger (2011)	Nonlinear numerical flight dynamics for the prediction of maneuver loads
Lozano (2013)	Unmanned aerial vehicles: embedded control
Duan and Li (2014)	Bio-inspired computation in unmanned aerial vehicles
Franchi et al. (2012)	Modeling and control of UAV bearing-formations with bilateral high-level steering
Hajiyev et al. (2015)	State estimation and control for low-cost unmanned aerial vehicles
Shima and Rasmussen (2009)	UAV cooperative decision and control: challenges and practical approaches

flight models exist. Essentially, all flight models are based on the mathematical theorems described in Newtonian physics; from Newton's second law, *an aircraft's motion in its six degrees of freedom can be described by a system of non-linear first order differential equations* (Chen et al. 2009). Numerical techniques to solve these differential equations, each defined by its own complexity and accuracy, include

- using Euler angles,
- the Heun method,
- the Bogacki-Shampine method for ordinary differential equations,
- the Runge-Kutta iterative method, and
- the Dormand-Prince method.

The non-linear model of flight-modeling is most commonly implemented in microprocessors where speed is abundant; to simplify the equations for small UAVs where processing power is limited, these non-linear equations are linearized to decrease the required computational load.

Two common orientation representations are typically used to build flight dynamics models, namely the Euler angles and quaternion. Euler angles use yaw, pitch and roll to represent a rigid body's orientation and movement, whereas quaternion is a generalization that extends the complex numbers, *i*, *j* and *k*, where the principal equation in quaternion mathematics still holds as

$$i^2 = j^2 = k^2 = ijk = -1. \tag{9.4}$$

Euler angles are most typically used and implemented in modern computers and microprocessors (Cheng et al. 2009). The DARPA Simulation Networking System (SIMNET) is a networked battlefield simulator that allows multiple user interaction on the battlefield at various levels. SIMNET, developed in 1987, uses quaternions for aircraft orientation representation to avoid singularities and high data rates associated with the more common Euler angle representation of orientation. The use of either Euler angles or quaternions depends strongly on the application, type of aircraft and the expected forces to which the aircraft will be exposed. Quaternions are widely used in two- and three-dimensional rotation in computer graphics and game development. Smooth camera movements can be achieved because of the ease of interpolating between two quaternions and unit normalization of floating point quaternions suffers from fewer rounding defects than conventional matrix representations. The main disadvantages of quaternions are that the computations are more intensive (more complex but generally faster to execute) and also less intuitive compared to Euler angles and quaternions only contain rotation and no translation and scaling.

From the above discussion it becomes evident that UAV applications, vehicles, technologies, applications, sensors, subsystems and topologies vary significantly in literature and in practice. The technologies, in terms of semiconductor processing requirements for each subsystem, differ and it is practically impossible to recommend a single technology to fulfill all the requirements of the UAV. One has to identify and analyze each subsystem, the information and data fed to and from the subsystem, and the requirements of each, to implement a chosen technology efficiently.

For example, in the case of a SAR subsystem, Caris et al. (2014) present a typical SAR scenario and concept. According to Caris et al. (2014), the trajectory knowledge of the UAV is obtained from GPS navigational sensors and an inertial navigation platform from receiving the radar signal using elaborated autofocusing procedures. The SAR system in Caris et al. (2014) uses an implementation of a modern frequency modulated continuous wave generator (FMCW) with a high frequency bandwidth of 1 GHz. To obtain the required resolution, a high-frequency sensor at mm-wave (94 GHz) is implemented. The radar front-end generates chirp signals at a bandwidth of 77 MHz between 87 and 164 MHz, after which the chirp signal is up-converted through frequency multiplication to 7.68 and 7.99 GHz. This signal is then passed to the frequency multiplier to multiply the signal by 12, to 92.16 and 95.88 GHz, respectively. A W-band MMIC is specifically fabricated for this application in InAlAs/InGaAs metamorphic high electron-mobility transistor technology with a 100 nm gate length, which provides transistors with f_T and f_{max} of 220 and 300 GHz respectively at a drain voltage of $V_d = 1$ V. The achieved conversion gain is 7 dB with a noise figure of 4 dB.

Conversely, Jaeschke et al. (2014) implement a 240 GHz radar with 42 GHz bandwidth using a SiGe BiCMOS transceiver with integrated antennas. The 240 GHz MMIC transceiver chip also implements the FMCW wideband modulation technique and achieves a gain of 0.65 dB at 240 GHz in the far-field for the simulated integrated patch antenna.

These two examples already show that for a UAV subsystem, such as radar (or SAR implementations), the choice of technology, in these cases InAlAs/InGaAs and SiGe, depends not only on the implementation frequency but also on the overall gain in the system. It has been established that SiGe offers very high frequency operation, nearing the THz regime, but also has significantly lower gain capabilities as a result of the trade-off between bandwidth and gain. Combining technologies in subsystems in any application aims to provide high-performance electronic systems capable of achieving high-frequency operation with high gain and low noise. As seen in Chap. 8, the more recent optoelectronic systems aim to increase the performance gains further in terms of bandwidth availability, and when integrating microelectronic systems, it is important to consider technologies that are capable of modular integration with optic systems as well.

9.7 Conclusion

This chapter reviews the potential future of EW, placing the focus on the enabling technologies and subsystems that facilitate upgrading of current systems and re-inventing traditional applications with advances in technology. These advances are specifically aimed at technical parameters, process improvements and efficient use of current as well as older (more cost-effective) technologies to benefit EW in various fields. Cognitive EW, the ability of electronic systems to learn from the environment and from historic events, has valuable advantages in warfare as technology advances and the processing performance of miniature processors increases. Cognitive warfare and cognitive EW can be applied in many fields and disciplines, applications and subsystems in electronic equipment and can be implemented in AESA, PGM and SDR. In addition, UAVs have become more popular in recent years and their potential applications are practically limitless, if the enabling technologies continue at the pace they have been developing over the past decades.

In the following chapter, a summary of the advantages and disadvantages of microelectronic technologies such as SiGe, InP, InGaAs, InSb, GaAs and GaN are summarized from a general perspective with reference to the topics discussed in this book. Since each technology presents its own set of strengths and weaknesses and it has been established in this book that most applications and systems require different properties and performance characteristics for efficient operation, a simplified approach to realizing future-generation EW equipment cannot be based solely on a single technology. This book is essentially pro-SiGe, but it is not biased and takes into account the complex nature of electronic equipment, especially in EW. It therefore aims to provide a holistic view on technologies and processes that can contribute to re-imaging the future of EW.

References

Ahmed, A. E., Hafez, A., Ouda, A. N., Ahmed, E. H., & Abd-Elkader, H. M. (2015). Modeling of a small unmanned aerial vehicle. *International Journal of Mechanical, Aerospace, Industrial, Mechatronic and Manufacturing Engineering*, *9*(3), 498–506, 2015.

Altera. (2007). Architecture and Component Selection for SDR Applications. *White Paper*. Retrieved June 9, 2016 from www.altera.com

Analog Devices. (2015). Multifunction: A dilemma or reality? Retrieved June 12, 2016 from http://www.analog.com

BAE. (2014). Digital electronic warfare system (DEWS). Retrieved June 12, 2016 from http://www.baesystems.com

Barton, J. D. (2012). Fundamentals of Small Unmanned Aircraft Flight. *John Hopkins APL Technical Digest, 32*(2).

Baylis, C., Fellows, M., Cohen, L., & Marks, R. J. (2014). Solving the spectrum crisis. *IEEE Microwave Magazine*, 94–107.

Bell, K. L., Baker, C. J., Smith, G. E., Johnson, J. T., & Rangaswamy, M. (2015, December). Cognitive Radar Framework for Target Detection and Tracking. *IEEE Journal of Selected Topics in Signal Processing, 9*(8), 1427–1439.

Caris, M., Stanko, S., Malanowski, M., Samczynski, P., Kulpa, K., Leuther, A., et al. (2014). mm-Wave SAR demonstrator as a test bed for advanced solutions in microwave imaging. *IEEE Aerospace and Electronic Systems Magazine, 29*(7), 8–15.

Chen, X. Q., Ou, Q., Wong, D. R., Li, Y. J., Sinclair, M., & Marburg, A. (2009). Flight dynamics modelling and experimental validation for unmanned aerial vehicles. *INTECH Open Access Publisher. Mobile Robots—State of the Art in Land, Sea, Air, and Collaborative Missions*. May, 2009.

Clarke, B., & Kreitzer, K. (2014). Maximizing the dynamic range of software-defined radio. *Analog Devices Technical Article MS-2735*, 1–4.

Cooke, J. M., Zyda, M. J., Pratt, D. R., & McGhee, R. B. (1992, Fall). NPSNET: Flight simulation dynamic modeling using quaternions. *The Massachusetts Institute of Technology. Naval Postgraduate School: Department of Computer Science, 1*(4).

Das, T. (2013). Practical considerations for low noise amplifier design. *Freescale Semiconductor White Paper*. Rev. 0, 5/2013, RFLNAWP, 2013.

DHL Trend Research. (2014). Unmanned aerial vehicle in logistics: A DHL perspective on implications and use cases for the logistics industry. DHL Customer Solutions and Innovations.

Du Plessis, W. (2014). Software-defined radio (SDR) as a mechanism for exploring cyber-electronic warfare (EW) Collaboration. In *2014 Information Security for South Africa* (pp. 1–6), 2014.

Duan, H., & Li, P. (2014, January 2). *Bio-inspired computation in unmanned aerial vehicles*. Springer Science & Business Media.

Feng, L., Huiming, L., & Xiaoxin, Z. (2014). Research and development of energy monitoring system for solar powered UAV. In *12th International Conference on Signal Processing (ICSP)* (pp. 393–398), 2014.

Franchi, A., Masone, C., Grabe, V., Ryll, M., Bülthoff, H. H., & Giordano, P. R. (2012, September). Modeling and control of UAV bearing-formations with bilateral high-level steering. *The International Journal of Robotics Research*.

Griffiths, H., & Baker, C. J. (2013). Towards the intelligent adaptive radar network. In *2013 IEEE Radar Conference (RadarCon13)* (pp. 1–5), 2013.

Guerci, J. R. (2010). Cognitive radar: A knowledge-aided fully adaptive approach. In *2010 IEEE Radar Conference* (pp. 1365–1370), 2010.

Hajiyev, C., Soken, H. E., & Vural, S. Y. (2015). *State estimation and control for low-cost unmanned aerial vehicles*. Springer. ISBN 3319164171, 2015.

Haykin, S. (2006, January). Cognitive radar: A way of the future. *IEEE Signal Processing Magazine, 23*(1), 30–40.

How, J. P., Frazzoli, E., & Chowdhary, G. (2012, August). *Linear flight control techniques for unmanned aerial vehicles*. Massachusetts Institute of Technology.

Howard, C. (2013). UAV command, control & communications. Retrieved June 12, 2016 from http://www.militaryaerospace.com

Huang, Y., Van Wyk, M. A., & Cilliers, J. E. (2015). On the detectability of multiple input multiple output (MIMO) radar signals using conventional electronic warfare support (ES) receivers. In *2015 IEEE Radar Conference* (pp. 469–474), 2015.

Jaeschke, T., Bredendiek, C., & Pohl, N. (2014). 3D FMCW SAR imaging based on a 240 GHz SiGe transceiver chip with integrated antennas. In *2014 German Microwave Conference (GeMIC)* (pp. 1–4), March 2014.

Keller, J. (2011). SDR and cognitive radio: The new horizons of military communications. *Military & Aerospace Electronics: Executive Briefing*. October 2013.

Keller, J. (2013, April). Programmable radar and adaptive electronic warfare take center stage. *Military & Aerospace Electronics: Executive Briefing*.

Lee, J., Park, H., Jung, G., & Yu, K. (2013). Design of virtual flight system for evaluation of solar powered UAV. In *39th Annual Conference of the IEEE Industrial Electronics Society* (pp. 3463–3467), 2013.

Li, X., Hu, Z., Qiu, R. C., Wu, Z., Browning, J. P., & Wicks, M. C. (2014). Demonstration of cognitive radar for target localization under interference. *IEEE Transactions on Aerospace and Electronic Systems, 50*(4), 2440–2455.

Lichtman, M., Poston, J. D., Amuru, S., Shahriar, C., Clancy, T. C., Buehrer, R. M., et al. (2016, January). A communications jamming taxonomy. *IEEE Security & Privacy, 14*(1), 47–54.

Long, L. N., Hanford, S. D., Janrathitikarn, O., Sinsley, G. L., & Miller, J. A. (2007). A review of intelligent systems software for autonomous vehicles. In *Proceedings of the 2007 IEEE Symposium on Computational Intelligence in Security and Defense Applications* (pp. 69–76), 2007.

Lozano, R. (2013). *Unmanned aerial vehicles: Embedded control*. John Wiley and Sons.

Maurer, D. E., Rogala, E. W., Bankman, I. N., Boone, B. G., Vogel, K. K., & Parris, C. (2001). A low cost gun launched seeker concept design for naval fire support. *John Hopkins APL Technical Digest, 22*(4), 2001.

Potenziani, E. (2006, October). Current and future trends in military electronic warfare systems and the role of thin films and related materials. *Ferroelectronics, 342*(1), 151–161.

Ramasamyraja, R., Pandiguru, M., & Arun, V. (2014). Design of ultra wide band antenna for tactical communication in electronic warfare. In *2014 International Conference on Communications and Signal Processing (ICCSP)* (pp. 1256–1259), April 3–5, 2014.

Ritter, M., & Dillinger, J. (2011). Nonlinear numerical flight dynamics for the prediction of maneuver loads. In *IFASD 2011* (pp. 1–10), 2011.

Scott, A., Hardy, T. J., Martin, R. K., & Thomas, R. W. (2011). What are the roles of electronic and cyber warfare in cognitive radio security? In *2011 IEEE 54th International Midwest Symposium on Circuits and Systems (MWSCAS)* (pp. 1–4), 2011.

Shima, T., & Rasmussen, S. J. (2009). UAV cooperative decision and control: Challenges and practical approaches. *SIAM*.

Silicon Radar. (2016). 122 GHz FMCW-radar-frontend with integrated antennas to cover UAVs sensor requirements. Retrieved June 7, 2016 from http://www.siliconradar.de

Song, P. (2014, May). *Millimeter-Wave integrated circuit design in silicon-germanium technology for next generation radars*. School of Electrical and Computer Engineering, Georgia Institute of Technology.

Sturdivant, R., & Parrish, C. (2015). *Gallium nitride (GaN) enabled C-Band T/R Modules*. Microwave Packaging Technology, Inc.: White Paper.

Thompson, L. (2015). Deft Jam: Electronic Warfare is the face of future combat. Retrieved June 23, 2016 from www.forbes.com

Tremper, D., & Heyer, J. (2007). Unmanned sea surface electronic warfare. In *2007 NRL Review: Electronics and Electromagnetics* (pp. 153–155), 2007.

Tucker, P. (2016). This is the most important technology on the F-35. Retrieved June 7, 2016 from www.defenseone.com

Walton, E., Young, E. L., Gemeny, S., Crowe, D., Bryant, C., Harton, C., & Duale, J. (2008). The software defined antenna; prototype and programming. In *The 30th AMTA Symposium*, 2008.

Wolff, C. (2016). Digital beamforming. Retrieved June 22, 2016 from http://www.radartutorial.eu

Chapter 10
A Review on Si, SiGe, GaN, SiC, InP and GaAs as Enabling Technologies in EW and Space

10.1 Introduction

In May 2016 the United States DoD researchers announced the development of an ADC that processes data at a rate of 60 GHz, fast enough to ensure the potentially uninterrupted operation of spectrum-dependent military systems, including communications, EW signals and radar, also in contested and congested EM environments, to offer situational awareness of the RF spectrum. The amount of data generated reaches nearly a terabyte per second, a data rate so high it requires on-chip data-management circuitry. The circuitry allows signals to be processed locally on-chip, reducing the amount of information passed on to supporting circuitry (DARPA 2016). The researchers plan to develop an ADC that uses 50 % less power than current technology while sampling greater bands of spectrum. The high speeds of these ADCs are in part attributed to performance increases of the operating speed of transistors, the miniaturization of transistors, which enables more transistors to be integrated on a single die, and a parallel-processing architecture where each branch samples the signal at a different time and combines the result at the output. The ADC is manufactured in a 32 nm SOI process from GlobalFoundries® and future improvements are planned on a 14 nm SOI process using fin field-effect transistor (FinFET) technology.

In May 2016, NXP Semiconductors® expanded its portfolio of broadband GaN RF power transistors targeted specifically at EW and battlefield radio applications. This expansion includes final-stage amplifiers with frequency coverage as broad as 1 Hz to 3 GHz (from HF to S-band). GaN transistors on SiC substrates enable such performance and combine high power density, ruggedness and a flat frequency response over a large bandwidth, enabling the devices to cover virtually all frequencies of electronic devices used in EW and other military applications.

W. Lambrechts and S. Sinha, *SiGe-based Re-engineering of Electronic Warfare Subsystems*, Signals and Communication Technology,
DOI 10.1007/978-3-319-47403-8_10

Rapid progress has been achieved in high-voltage transistors such as laterally diffused MOSFETs (LDMOS), high-voltage MESFETs, HBTs and AlGaN/GaN HEMTs, which have an increased breakdown voltage and an increased gate capacitance. Therefore, the power density of these transistors is high, whereas the operating frequency is relatively low because of the increased gate capacitance. Wide-bandgap semiconductors such as GaN and SiC show great promise for advancing high-power microwave electronic devices, where GaN additionally offers relatively high operating frequencies. Primarily, owing to low breakdown voltage, it has not been possible to design and fabricate solid-state transistors in Si and GaAs (Trew 2004) that can yield RF output power necessary to compete with microwave vacuum tubes or GaN. This has severely limited the use of Si and GaAs microwave solid-state transistors and devices in power applications, such as transmitters for wireless communications systems, radars and guidance (Trew 2002). The RF PA is a key component of the transmitter unit in many EW systems, microwave transceivers, radio-frequency identification, radar, satellite transceivers and remote sensing systems. The performance of the RF PA has a direct impact on the performance of the entire system, whereas the performance of the PA is predominantly determined by the power capabilities of the transistors and active devices of the enabling technology.

In military communications there is a shift towards SDR architectures that reduce the number of different radio platforms currently used by the armed forces. New configurable architectures and devices, using SiGe and CMOS processes, drive the objective to software definable systems. Higher integration in SDR applications also enables cognitive functionality, anti-jamming and encryption McHale 2015). SDR, for example, requires high-frequency, low-noise and low-power technologies to realize practical and modular systems that can be used by soldiers on the battlefield. SiGe is one technology that provides these capabilities and is able to process data at the extremely high rates required for real-time SDR communications.

From the above discussion it is evident that no one-size-fits-all approach is possible when choosing a technology for specific applications, whether in commercial, military or space-based applications. Various trade-offs of the potential technologies must be weighed and compared to determine the optimum solution for each application. Even subsystems could benefit from the superior properties provided by each technology and, if possible, more than one technology could be implemented in a system to provide benefits for each subsystem. This chapter highlights the advantages and disadvantages and the most crucial characteristics of technologies such as Si, SiGe, GaN, InP, GaAs and SiC when considering implementation of a system in military application and EW; these are also applicable to commercial and space uses. The following section summarizes these characteristics by individually reviewing each technology/process material.

10.2 Semiconductor Process Highlights

10.2.1 Process Highlights: Si

This section describes the highlights of Si as a semiconductor material as well as its optoelectronic properties, which make the material such a widely used and successful semiconductor material. The properties of Si have been intensively studied and its advantages and disadvantages are well known across the research and development segments. This paragraph briefly highlights these properties to assist in a comparison between Si and other emerging and established/mature semiconductor materials. Table 10.1 lists the parameters of Si used in the comparative study between semiconductor materials used to manufacture integrated circuits, based on its electrical performance. All the parameters presented in this section are obtained from various sources and in the case of slight differences on reported values, an average is taken and used in this book.

It is difficult to highlight the characteristics of Si that make it such a popular semiconductor material without repeating what generations of researchers, scholars, physicists and engineers have discovered and published since the discovery of the *p-n* junction by Russel Ohl in 1940, which led to the development of junction transistors and solar cells. By 1963, Schultz (1963) had already outlined the dominance of Si as a semiconductor used extensively in the fabrication of electronic devices such as diodes, transistors, ICs and solar cells, as well as being a window material for certain portions of the near-IR spectrum, especially the visible band. Schultz (1963) provides a concise summary of the characteristics of Si and the principal features of its behavior as a semiconductor material. Already in this paper, Schultz (1963) refers the readers to existing published material, original papers and review articles to ensure that the published paper only highlights the crucial elements that make Si such a successful material for semiconductor processing and avoid re-iterating the well-known characteristics. From this perspective, this book also does not aim to repeat all the material properties of Si and only highlights important properties that (still) make Si a preferred semiconductor material in the 21st century. Highlights include:

- Si is abundant: SiO_2 is the most common compound in the earth's crust and commonly takes the form of ordinary sand, quartz, rock crystals, amethyst, agate, flint, jasper and opal. Other useful compounds of Si include SiC, used as a semiconductor and as an abrasive since it is nearly as hard as diamond, sodium silicate (Na_2SiO_3), which is known as water-glass and used to produce soap and adhesives, as well as silicon tetrachloride ($SiCl_4$), which is used to create silk screens.

Table 10.1 Material characteristic highlights of Si

Material	Bandgap (eV)	Electron-mobility (cm^2/V s)	Power density (W/mm)	Breakdown voltage (kV/cm)	Thermal conductivity (W/cm K)	Dielectric constant	f_T (GHz)
Si	1.12	1500	0.2	300	1.5	11.8	20

- Extracting and purifying Si from its raw material are both well-established, inexpensive and efficient methods, which involve reducing the silica compounds in furnaces at temperatures exceeding 2000 °C. Carbon reacts with the oxygen in the molten silica to produce CO_2 and Si, which settles at the bottom of these furnaces. The remaining Si is then treated with oxygen to reduce any calcium and aluminum impurities (Plus 2009). The result is metallurgical-grade Si, which is up to 99 % pure. To manufacture electronic-grade Si, which is 99.99999999 % pure, additional processing to refine the metallurgical-grade Si is necessary. This process includes grinding the Si to a fine powder and reacting it with gaseous hydrogen chloride at 300 °C to give a liquid compound of Si called trichlorosilane (HCl_3Si). Impurities such as iron, aluminum, boron and phosphorus are removed with fractional distillation and the purified HCl_3Si is vaporized and reacted with hydrogen gas at 1100 °C to retrieve the elemental Si. During this reaction, Si is deposited at the surface of an electrically heated ultra-pure Si rod to produce the electronic-grade Si (Plus 2009). Electronic-grade Si is therefore a near-perfect material with practically zero impurities and defects, making processing with the material a virtually exact science resulting in high yield and cost-effective processing.
- Mechanical handling of Si wafers is more forgiving than most semiconductor materials, for example InSb. Mechanical handling might be overlooked in many cases as an important characteristic, but in fact, the handling of a semiconductor material through its entire processing procedure typically introduces material cracks and defects into the material. This results in lower yield and higher cost of manufacturing; Si crystallizes in a diamond form with relatively strong bonds holding the crystals together.
- The bandgap of pure Si is moderate at 1.12 eV with an intrinsic carrier concentration of $n_i = 1.5 \times 10^{10}$ cm^{-3} at 300 K. This leads to relatively small leakage currents of the active devices. Doping Si with impurities is relatively easy according to the available research, techniques and parameters defined throughout literature.
- SiO_2 is grown on Si and used as a passivation layer and insulator in CMOS processing and to form the gates of MOSFETs. The advantage of using SiO_2 is that the material is perfectly matched to Si when grown on a pure Si wafer, leading to minimal defects and reducing surface states between layers.

More recently, since approximately the late 1980s, Si photonics have made substantial progress because of its compatibility with mature CMOS manufacturing processes and high-quality SOI wafers, an ideal platform for creating planar waveguide circuits (Jalali and Fathpour 2006). As high-bandwidth data transmission is increasingly becoming limited by the parasitics introduced by the transmitting channel, traditional methods such as copper wires are becoming less effective. Optical transmission is becoming a crucial requirement to reduce the inherent attenuation of high-frequency signals. Traditional copper wires also demonstrate large power dissipation due to the charge and discharge effect of the parasitic capacitance between signal wires. The high level of integration of Si photonics with traditional Si IC manufacturing can offer high stability and low power consumption of interconnects in high-bandwidth applications (Zhou et al. 2015). The critical milestones of Si photonics include successful demonstration and implementation of passive components, modulators and photodetectors. Si-based photonics and OEICs can operate at wavelengths anywhere between 1.2 and 100 μm, depending on the design, material doping and operating temperature (Soref 2006).

By doping Si with Ge, SiGe provides additional benefits over the customary Si process. It is reviewed in the following paragraph.

10.2.2 Process Highlights: SiGe

The competitiveness of Si as a semiconductor material and its inherent limitation to high-power and high-frequency applications have motivated researchers to find alternative alloys, based on Si as an enabling technology to keep the costs low, but with increased performance. One such material is SiGe, which uses similar processing techniques as traditional Si, but adds Ge as a dopant in the base of the transistors to increase frequency performance. This book provides highlights of the performance gains of SiGe in the EW field of applications, which can potentially take advantage of the enhanced performance without sacrifices in other areas, such as cost. Table 10.2 lists the comparative properties and parameters for SiGe.

The following paragraph from Robinson (2014) is cited: "*Si, a material used in the manufacture of most modern microchips, is not competitive with other materials when it comes to the extremely high performance levels needed for certain types of emerging wireless and wired communications, signal processing, radar and other applications. Certain highly specialized and costly materials—such as indium phosphide, gallium arsenide and gallium nitride—presently dominate these highly demanding application areas. But SiGe changes this situation. In SiGe technology, small amounts of Ge are introduced into Si wafers at the atomic scale during the standard manufacturing process, boosting performance substantially.*" Robinson (2014) highlights a new record for transistor speed at an f_{max} of 798 GHz; albeit at a cryogenic temperature (4.3 K), still an impressive accomplishment for the technology and valued insight into its potential capabilities for room temperature operation. The same transistor, achieving 798 GHz at 4.3 K, was operated at 417 GHz at room temperature, a speed already faster than 98 % of all the transistors available currently, irrespective of technology.

Table 10.2 Material characteristic highlights of SiGe

Material	Bandgap (eV)	Electron-mobility (cm^2/V s)	Power density (W/mm)	Breakdown voltage (kV/cm)	Thermal conductivity (W/cm K)	Dielectric constant	f_T (GHz)
SiGe	0.945	7700	1	200	0.083	13.95	285

SiGe, deemed the successor to GaAs since the early 1990s, is characterized by higher integration and reduced cost and is fast becoming the superior technology for high-speed and complex ICs. Key applications where SiGe are dominant are in high-speed optical network transceivers, base stations, wired broadband communications, high-speed cross-point switches, GPS and circuits and subsystems of radar and mm-wave applications. Although unable to outperform technologies such as GaAs in terms of microwave performance parameters such as noise figure, power output or robustness (Schuh et al. 2009), SiGe offers various distinct advantages in its BiCMOS integration capabilities and multiple channel communications infrastructures on single chips. As integrated solutions are shifting the focus from high power output to increased processing capabilities, high integration levels, high volume, low power consumption and low cost, the SiGe HBT has emerged as a prominent candidate for replacing GaAs.

The introduction of a Ge-base in traditional Si technology increases the transit frequency into the high-GHz range and a high base doping reduces the base sheet resistance, which also improves the f_{max} of these transistors. Another important advantage is the excellent passivation by SiO_2, which reduces the $1/f$ corner frequency in SiGe HBTs (König et al. 1995). Logan and Noras (2009) presented a comparison between a SiGe BiCMOS LNA and a Si CMOS LNA with identical topology at 2.1 GHz and concluded that the SiGe LNA presented a better NF with significantly lower current consumption under similar operating conditions. This result shows that in the case of LNAs, operating at relatively low frequencies, SiGe BiCMOS is superior to the traditional Si CMOS implementations. Although it is difficult to ascertain that SiGe BiCMOS technology outperforms CMOS on all FOMs, at all frequencies and under all circumstances, isolating and comparing electronic circuits give a holistic overview of the typical performance gains when implementing SiGe BiCMOS technology.

In RF wireless communications, as presented in Senapati and Maiti (2000), SiGe HBTs are examined and device performance on a radio-frequency IC is compared against Si BJTs, GaAs HBTs and GaAs HEMT. From this comparison, SiGe HBTs presented the lowest power consumption with respectable performance in gain, NF and linearity. GaAs HEMT devices presented the lowest NF and Si BJTs presented the best linearity.

In Long (2005), the impact of SiGe bipolar and BiCMOS technologies on wireless transceivers for portable telephony and data communications is surveyed. Long (2005) also identifies the favorable characteristics of the SiGe technology in terms of turnaround time and overall cost compared to traditional CMOS or III-V technologies. Importantly, as also mentioned in Long 2005, is the fact that SiGe technologies offer space for innovation in circuit and system design (such as direct conversion and homodyning) by offering low-cost and highly integrated devices to overcome the shortfalls of NF and speed when compared to GaAs and even InP technologies. As opposed to increasing the operating frequencies of subsystems towards the limitations of the technology, SiGe allows multiple cascaded circuits (such as multi-core CPUs) to co-exist on a single chip without compromising the yield of the device, through its ability to contain millions and up to billions of transistors on a single die. In addition, as applications start operating in the hundreds of GHz range, the interconnect and parasitic effects become more pronounced and the transistor f_T and f_{max} are not the only limitations in these circuits (Niu 2005). In these cases, the ability to slow down the operating frequency but increase the circuit complexity and functionality becomes more attractive and a technology such as SiGe BiCMOS becomes versatile with its high dynamic range and low power consumption.

For power electronics, GaN is a wide-bandgap material that provides benefits of high operating frequency and high breakdown voltage. It is reviewed in the following paragraph.

10.2.3 Process Highlights: GaN

In terms of wide-bandgap and high-power density semiconductor materials, GaN still remains an important contender in EW and space applications where energy efficiency is a crucial parameter. Although not the fastest material in terms of transistor switching speeds, GaN has a modest and competitive maximum operating frequency, considering its high-power density and efficiency. The properties of GaN related to the comparative discussion in the chapter are presented in Table 10.3.

GaN is a binary III-V direct bandgap semiconductor and its electrical properties make it ideal for optoelectronic devices, high-power and high-frequency applications. Initially, GaN was developed for improvised explosive device jammers used in the military and has evolved to serve many additional applications. GaN was primarily used for L-, S- and C-band applications such as radar, cable television and power management. In 2015, the GaN market share was estimated at US $180 million with approximately 67 % allocated to military applications and a growing demand for more efficient and compact hybrid electric vehicles (Shamsi et al. 2013).

Advances in GaN HEMT technology have made GaN a contender in military and commercial satellite markets and power amplifiers requiring reliability, linearity, power density and energy efficiency (Damian and Gelerman 2012). GaN transistors generally present a high output impedance, making it easier to match and combine devices in cascaded amplifiers. GaN transistors can be modified to produce BJTs, enhanced-mode MOSFETs and more recently MESFETs, HEMTs, FinFET and LDMOS variants to improve performance. In the automotive industry,

Table 10.3 Material characteristic highlights of GaN

Material	Bandgap (eV)	Electron-mobility (cm^2/V s)	Power density (W/mm)	Breakdown voltage (kV/cm)	Thermal conductivity (W/cm K)	Dielectric constant	f_T (GHz)
GaN	3.49	1500	30	3300	1.5	9	150

the performance and cost advantages of SiC have limited the market share of GaN (although integrating GaN on SiC substrates has produced high-frequency and high-powered devices), whereas in higher frequency RF and high-power transmission communications technology such as WiMAX, LTE and legacy military applications, GaN has gained traction against GaAs and LDMOS technologies. This is primarily due to its higher frequency performance and ability to output significant power even in the mm-wave range, wide bandgap, heat capacity and thermal conductivity. These characteristics also open up new opportunities, not only for military applications such as radar and EW, but also in space-based communication applications.

Generally, GaN technologies are placed on SiC substrates to further improve their heat distribution (thermal conductivity) capabilities, while maintaining the high-power and high-frequency advantages of GaN (Runton et al. 2013). GaN provides orders higher power density compared to GaAs and SiGe, which enables it to dissipate heat efficiently while operating in the upper-GHz ranges owing to its high electron mobility. In addition, GaN presents high cut-off frequencies in the 200 GHz range and has a relatively high breakdown voltage (approximately 100 V and higher). A higher electric field breakdown and higher doping density in GaN essentially translate to reduced width of the drift region, reducing its size and effectively increasing its power density (Han et al. 2009).

The biggest disadvantage of GaN and the primary factor inhibiting the technology from being integrated in more commercial products is its cost. The manufacturing process of GaN is complex and the material is unforgiving when compared to Si with its perfectly defined crystal lattice, high yield and reproducibility. The cost of the GaN process will inevitably decrease as volumes increase, but will possibly never be below the cost of bulk Si CMOS or even GaAs.

Another wide-bandgap material with high breakdown voltage is SiC, reviewed in the following paragraph.

10.2.4 Process Highlights: SiC

A wide-bandgap material that competes with (and complements) GaN in the power-electronics field is SiC. Although its power density is not as high as that of GaN, it still presents various advantages in semiconductor processing, such as its ability to conduct heat away from active devices quickly. Table 10.4 lists the properties of SiC, obtained from various sources and used in this comparative study.

Table 10.4 Material characteristic highlights of SiC

Material	Bandgap (eV)	Electron-mobility (cm^2/V s)	Power density (W/mm)	Breakdown voltage (kV/cm)	Thermal conductivity (W/cm K)	Dielectric constant	f_T (GHz)
SiC	3.25	900	10	3500	4.5	10	20

Wide-bandgap technologies such as SiC, along with GaN, are predominantly used in power electronics systems, medium- and high-power (Liang et al. 2014) and high-temperature applications (Singh and Sundaresan 2015). SiC MOSFETs and Schottky barrier diodes have started to enter the market previously dominated by traditional Si because of their high-power and high-voltage capabilities. SiC and GaN present the best trade-offs between high-voltage blocking capability, high-temperature operation, high switching speeds, commercial availability and technology maturity (Millán 2012). SiC presents higher thermal conductivity compared to GaN, but GaN offers better high-frequency and high-voltage performance, although it lacks good quality bulk substrates, which is a disadvantage for vertical devices. The higher thermal conductivity reduces the requirements for large overheads of thermal management systems. The high breakdown electric field of SiC allows for thinner and more highly doped blocking layers. The wide bandgap results in higher temperature operation and radiation-hardened devices, while the high thermal conductivity allows heat to be efficiently extracted away from the device (Singh and Sundaresan 2015). Si insulated-gate bipolar transistor (IGBT) power devices have a limitation on the maximum voltage of up to 6.5 kV (Madhusoodhanan et al. 2015). As a result, for a distribution grid (up to 13.8 kV at 60 Hz) interfacing applications of power electronics converters using Si IGBTs, integrated gate-commuted thyristors, or gate turn-off thyristors for higher order multilevel converters (more than three levels) are mandatory. A 15 kV SiC IGBT (Kadavelugu et al. 2013) has facilitated the use of power converters at distribution voltage level without employing higher order multilevel converter topologies (Madhusoodhanan et al. 2015).

Power semiconductors are required to present low leakage current at high voltage-blocking levels to limit power consumption and prevent thermal runaway leading to device failures. Leakage current is a function of doping concentration in the semiconductor and SiC (and GaN) can operate at much higher blocking voltages with lower doping concentrations as a result of the wide bandgap, when compared to traditional Si, for example. Power devices grown on Si substrates effectively reduce the manufacturing cost but at the expense of crystal defects in the grown layers, which leads to surface effects such as leakage currents.

SiC circuit breakers for DC power systems using 1200 V SiC JFET at DC operating voltages of 400 V with turn-off fault currents of 125 A at response times of 1 μs have been described in Shen et al. (2015). SiC PIN diodes with 6.5 kV IGBT switches in a 4.16 kV, 8 MVA neutral-point clamped voltage source converter with a switching frequency of 465 Hz are presented in Filsecker et al. (2014). The most widely studied configurations of SiC devices are normally-on SiC JFET and cascaded topologies using a normally-on SiC JFET together with a low-voltage Si MOSFET (Alonso et al. 2014). The study carried out by Alonso et al. (2014) confirms that the overall switching losses of the JFET in continuous conduction modes are low, making it suitable for operating in this mode in terms of overall converter efficiency. However, the lower turn-off losses of the cascade configuration point to this configuration as the most appropriate for the discontinuous conduction modes when zero-voltage switching is achieved at the turn-on of the main power.

The cost of SiC material processing has decreased to approximately 10 Euro/cm^2 (13.6 US \$/cm^2) with 100 mm (4 inch) wafers as the standard and 150 mm (6 inch) wafers available (Kaminski and Hilt 2014).

InP is a semiconductor material that provides very high-frequency operation, a relatively narrow bandgap, high-electron mobility and advantages when applied in optoelectronic ICs. It is reviewed in the following paragraph.

10.2.5 Process Highlights: InP

The ability to drive optical signals on-chip using semiconductors that are able to accommodate high-speed analog and digital electronics is becoming more commonplace and a greater requirement in modern electronic communication systems capable of multi-Gbps transfer speeds. Parasitic effects of traditional interconnects have become a limiting factor in high-frequency applications and optical communications aim to overcome these limitations. OEICs can be achieved by materials that are responsive to light waves at practical wavelengths, such as within the 1.3–1.55 μm band. InP, when combined with InGaAs, can produce high-quality and high-speed optoelectronic circuits, which can combine photosensitive detectors, waveguides and high-speed active circuits on a single chip. The properties of InP are summarized in Table 10.5.

A primary motivation for direct bandgap zincblende[1] InP transistor technologies and microelectronic devices is its ability to integrate with optoelectronic and photonic devices such as InGaAs photodiodes in the 1.3–1.55 μm optical band. Semiconductor nanowire heterostructures are increasingly more dominant and crucial building blocks for optoelectronic devices and the optical properties on InP nanowire heterostructures present small lateral size devices and high aspect ratios in one-dimensional quantum system physics (Cirlin et al. 2008). InP-based HBTs typically have a vertical MBE device profile and exhibit low surface recombination and low flicker noise. These devices can produce digital and analog signal processing circuits that exhibit the low phase jitter and high receiver sensitivity characteristics of conventional Si bipolar technologies, but at greater speeds (Cirlin et al. 2008). InP HBTs in addition offer advantages over GaAs technologies such as reduced power consumption due to the lower bandgap of GaInAsP. InP-based HBTs offer high transconductance, good power driving capabilities, the ability to manufacture vertical structures and thus small area requirements with high-speed operation (Topham et al. 1990). In addition, InP presents low parasitic capacitances and ease of device isolation, which make them ideal not only for high-frequency

[1]The indium and phosphide atoms form two interpenetrating face-centered cubic lattices, referred to as the zincblende structure.

Table 10.5 Material characteristic highlights of InP

Material	Bandgap (eV)	Electron-mobility (cm^2/V s)	Power density (W/mm)	Breakdown voltage (kV/cm)	Thermal conductivity (W/cm K)	Dielectric constant	f_T (GHz)
InP	1.35	5400	1	500	0.7	12.5	300

transistors but also for high-frequency optoelectronic devices. An attractive feature of InP HBTs is that they are based on a vertical MBE structure as opposed to a planar FET or HEMT structure and are compatible with the monolithic integration of high-performance PIN diode photodetectors. InP can also be grown epitaxially and lattice-matched to InGaAs to create PIN-structured multi-pixel photodiodes with high-speed capabilities for IR imaging. To manufacture InP HBTs or InP HEMTs effectively, a stable epitaxial material as well as stable top- and bottom-wafer processing is required, making InP more difficult and costly to manufacture compared to more traditional materials.

InP HBTs have been an integral component in applications such as mobile and fiber-optic telecommunications with its high PAE, low leakage current and high breakdown voltage in mm-wave telecommunications owing to its high efficiency and high linearity and in satellite communications where size, weight and power efficiency are crucial. InP HBTs can operate under lower DC currents at high frequencies in comparison with GaAs HBTs, making the technology ideal for high-speed space communications. The turn-on voltage of an InGaAs/InP HBT can be as low as 0.4–0.6 V, whereas AlGaAs/GaAs HBTs have turn-on voltages of typically 1.1–1.3 V.

Mixed signal circuits based on InP HBTs are being challenged by aggressively scaled SiGe bipolar and BiCMOS technologies to meet high bandwidth and high dynamic range requirements (Zopler 2003). InP HBTs are topologically similar to graded-base SiGe devices, which are more commonly available; however, InP devices have distinct characteristics that influence the accurate modelling of these devices (Hitko and Li 2013). InP HBT modelling is approached with a thermal-resistance-first method because of the inherent lower thermal conductivities of III-V semiconductor materials. Self-heating occurs in InP/InGaAs junctions, which also varies the output resistance in DC measurements. As a result of the high levels of base doping, base-width modulation effects are smaller in InP HBTs when compared to SiGe.

From a mathematical and physics perspective, InP model parameters are generally extracted from measurements as opposed to relying on computational estimations and simulations. In III-V materials, electron-velocity overshoot is significant and makes it difficult to project the current gain and transit time based on only the physical dimensions of the device. In addition, high collector voltages, typically applied to InP HBTs, vary the effective base-collector junction capacitance and affect the transistor f_{max}, making frequency parameter estimation for these devices difficult (Hitko and Li 2013).

Reduction of the junction temperature and the intrinsic base-collector capacitance to improve f_{max} of the InP HBTs is commonly achieved by transferring the InP technology to substrates with high thermal conductivity such as SiC, as described in Radisic et al. (2015), implemented in a 200 nm InP technology. In Griffith et al. (2015), a 250 nm InP HBT technology with f_T and f_{max} of 400 GHz and 700 GHz respectively highlights the capabilities of integrating high-speed and high-efficiency power amplifiers in InP technology.

Finally, GaAs, a very high electron mobility semiconductor material with a relatively high operating frequency, is reviewed in the following paragraph.

10.2.6 Process Highlights: GaAs

GaAs also presents a relatively wide bandgap, high mobility and high-frequency operation with the additional advantage of inherently being radiation resistant. More importantly, GaAs possesses ideal characteristics for solar radiation conversion to electrical energy, which when combined with its radiation-hardened characteristic are ideal for space-based and military applications requiring solar-powered supplies. The properties of GaAs are presented briefly in Table 10.6 and the advantages and disadvantages of the material are described in this section.

In an article published by Whitesides (1988), mention of GaAs technologies pushing for higher dominance in the semiconductor manufacturing market is preceded by the observation that Si was already, and still is, the abundant and more inexpensive alternative. GaAs substrates inherently offer high resistance to electron flow, making it a semi-insulating material, whereas Si, for example, is semi-conducting before it is doped with impurities. GaAs material has a wider bandgap and higher drift velocity than traditional Si (Nair 1995) and InP. The higher drift velocity at low electric field leads to higher mobility in GaAs devices.

The usefulness of GaAs becomes apparent in applications requiring radiation-hardened and high-temperature operation. In addition, GaAs can relatively easily be doped to emit light, making it useful for lasers and LEDs. The crystal structure of GaAs is essentially its most important contribution to its high

Table 10.6 Material characteristic highlights of GaAs

Material	Bandgap (eV)	Electron-mobility (cm^2/V s)	Power density (W/mm)	Breakdown voltage (kV/cm)	Thermal conductivity (W/cm K)	Dielectric constant	f_T (GHz)
GaAs	1.42	8500	0.5	400	0.5	12.8	150

cost, low reliability and difficulty of processing. Growing large and crystal-defect free GaAs materials is an exact and labor-intensive process compared to Si, which is relatively easy to grow. Moreover, arsenic is a toxic and volatile material. Ideally, by harnessing the advantages of the speed of GaAs with the efficiency and yield of Si, manufacturing GaAs and Si devices on a single substrate can be achieved (Yamaguchi et al. 2000); however, matching between these materials is not perfect and depositing GaAs on Si without spreading the crystal defects is another costly and challenging process (Whitesides 1988).

Abate (2015) presents a similar trend in terms of the cost and availability of GaAs. Quoting from the 2015 publication: "*Si is typically used in solar cells and computer chips. Gallium arsenide is an alternative material with many advantages*". It is still widely accepted that GaAs is a much faster (in terms of operating frequency) material compared to Si, but it is only used in specialized applications where the higher cost can be justified. In Lin et al. (2013), GaN is proposed as an alternative to earlier-generation GaAs-based transceiver modules in phased array antennas and AESA radar (Schuh et al. 2009) for EW applications. This is primarily due to the fact that GaAs presents much lower power output capabilities (5–15 W) compared to the much wider bandgap material GaN, with output powers at S-band in the 85 W region.

Kuzuhara and Tanaka (2003) published transistor performance of GaAs-based high-frequency and high-speed devices and explicitly highlighted the fact that GaAs has inferior electron mobility compared to InGaAs and lower peak drift velocity compared to InP. It has a wider bandgap compared to Si, but smaller than SiC and GaN. GaAs remains an excellent candidate for high-frequency electronic devices and with the increasing popularity of the mm-wave domain, GaAs still proves its worth as an enabling technology. Firstenberg and Tomasetta (1984) highlight several military applications where radiation hardness, high throughput and environment severity are stringent process requirements, where GaAs shows innate advantages. The list of applications in the 1984 article include:

- EW (early warning receivers),
- secure and satellite communications,
- air defense,
- space and ocean surveillance,
- intelligence, and
- image processing.

Firstenberg and Tomasetta (1984) also concluded that in advanced defense systems where functional performance is limited by the speed of the electronic components, GaAs can provide the benefits of introducing higher-speed electronics in systems where the complexity can be reduced to improve on yield and cost.

The following section reviews the semiconductor materials based on performance parameters such as electron bandgap, electron (carrier) mobility, power density, breakdown voltage, thermal conductivity and cut-off frequency.

10.3 Material Performance: Si, SiGe, GaN, SiC, InP and GaAs

Reviewing the advantages and disadvantages of semiconductor materials used in EW, space and commercial applications in isolation can also be presented as a comparative study in terms of material parameters. This section summarizes these properties and parameters of Si, SiGe, GaN, SiC, InP and GaAs with regard to their electrical and physical parameters: their electron bandgap, electron mobility, power density, breakdown voltage, thermal conductivity and transistor cut-off frequency. Each paragraph presents a chart comparing the parameters of each material and a discussion on the performance variations attributed to each parameter.

From these data, an overall performance comparison equation is presented with weighting factors attached to each parameter, which can be altered depending on the requirements of the application. By increasing or decreasing each weighting factor, the optimal material for a proposed application can be determined and two such examples are given.

The first parameter presented is the electron bandgap of each material, specified in eV.

10.3.1 Performance Comparison: Electron Bandgap (EV)

The electron bandgap of a solid insulator or semiconductor material is the energy difference between the top of the valence band and the bottom of the conduction band, which must be overcome to allow free conduction of electrons. Essentially, this difference determines the amount of energy (electric field) that should be applied across the terminals of the material in order to change its properties from non-conducting to conducting. Insulators have larger electron bandgaps, meaning that it takes a larger electric field to induce the flow of current in the material. Conductive materials have very low electron bandgap where the ambient energy such as room-temperature radiation or incident light are enough to facilitate current flow through the material. Semiconductor materials have a small but non-zero bandgap, which requires some external energy applied to the material in order to facilitate the flow of electric current. If no energy is applied, ambient energy is not enough to overcome the energy difference between the valence and conduction band and the flow of current is not possible.

The conductivity of intrinsic semiconductors is strongly dependent on the bandgap of the material. The introduction of alloys such as Ge into traditional materials such as Si intentionally changes the bandgap of the material to control the electrical properties of the device. Temperature also plays a significant role in the determination of the bandgap properties of a semiconductor material, where a decrease in temperature decreases the bandgap energy of the material. An increase in temperature can also add enough energy to the material to overcome the inherent energy differences and facilitate current flow. In such applications (high temperature), a larger bandgap is preferred to ensure that the semiconductor material properties are controlled only by the energy supplied in the form of an electric field as opposed to variations in temperature or radiation. Military applications often opt for larger bandgap materials for this reason and can be operated at much higher energy levels under typical conditions. Si, for example, has a bandgap of approximately 1.12 eV, a relatively low value, whereas if it is doped with Ge, the bandgap is decreased but the electron mobility is increased; electron mobility also plays a role in the characteristics and current-flowing capabilities of a material. In solar cells, the Shockley-Queisser limit refers to the maximum theoretical efficiency of a solar cell using a single *p-n* junction to collect power from the cell (Shockley and Queisser 1961). The limit places an inherent maximum solar conversion efficiency based on the bandgap of the *p-n* junction, where a lower bandgap relates to a lower solar conversion efficiency.

The bandgaps of the materials discussed in this chapter are shown in Fig. 10.1.

Figure 10.1 shows the differences in bandgaps of the materials SiGe, GaN, SiC, InP, GaAs and Si, in no particular order. From Fig. 10.1 it can be seen that Si presents the smallest bandgap apart from its alloy, SiGe, and is the most commonly used material not only in IC fabrication, but also in solar panels. InP and GaAs have comparable bandgaps around 1.35 and 1.42 eV, respectively. GaN and SiC have the widest electron bandgaps at 3.49 and 3.25 eV, respectively. These wide bandgaps of GaN and SiC make them ideal for use in high-temperature and radiation-prone environments such as space and in the military, where they can be intentionally radiated with levels of energy in EW. SiGe has the lowest bandgap, where a direct result of the Ge grading is an increase in operating frequency; since a smaller bandgap increases the electron injection speed (resulting in higher gain), the field from the graded base bandgap decreases the base transit time, produces a higher Early voltage and eliminates the base profile from performance figures. SiGe

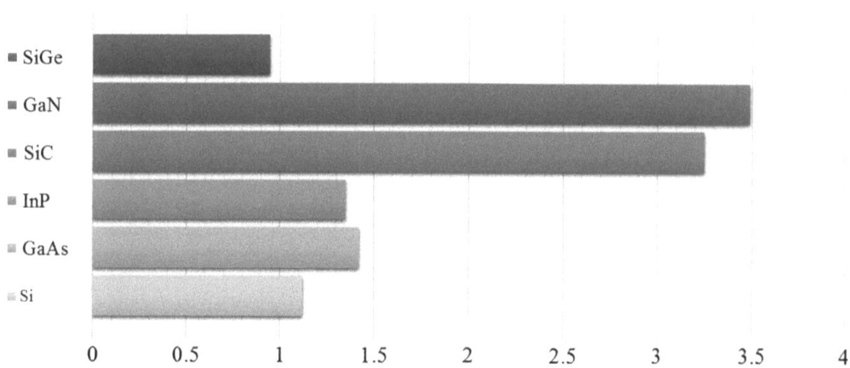

Fig. 10.1 Performance comparison between SiGe, GaN, SiC, InP, GaAs and Si semiconductor processes in terms of electron bandgap and specified in eV

is the first applied bandgap-engineered device in Si technology and offers better performance without the trade-off of additional cost for process alterations.

Electron mobility, specified in cm^2/V-s, is reviewed in the following paragraph.

10.3.2 Performance Comparison: Electron Mobility (cm^2/V-s)

An electric field applied across a semiconductor material causes the carriers to drift or move at an average velocity between the positive and negative electrodes. The induced current is the ratio of total charge and the time interval needed to move between the electrodes. This effect results in the current density performance characteristics of the material.

Closely related is the carrier mobility. The carrier mobility is directly linked to the conductivity and resistivity of the semiconductor material. For low doping concentrations the mobility in a semiconductor is moderately constant and largely restricted by phonon scattering. As the doping is increased, the mobility in the semiconductor decreases due to ionized impurity scattering. The type of dopant determines the mobility in the semiconductor material.

The conductivity of a material is defined as the current density divided by the applied electric field and since current density equals the product of the charge of the mobile carriers, its density and velocity are expressed as a function of the electric field using mobility (Van Zeghbroeck 2011). Resistivity is the inverse of conductivity; the resistivity of a material decreases with an increase in doping density. The sheet resistance of a semiconductor material is an alternative, somewhat easier, method to determine the resistance to the flow of current and is expressed in Ω/square.

The electron mobility of the materials discussed in this chapter is given in Fig. 10.2.

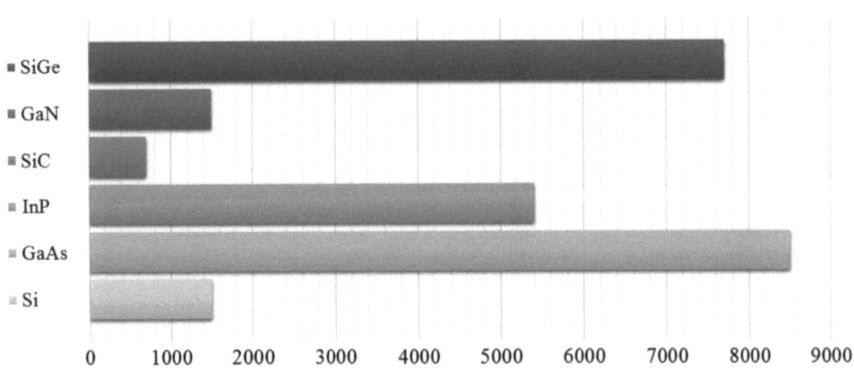

Fig. 10.2 Performance comparison between SiGe, GaN, SiC, InP, GaAs and Si semiconductor processes in terms of electron mobility and specified in cm^2/V-s

From Fig. 10.2, it can be seen that SiGe outperforms Si in terms of electron mobility because of the Ge-graded base doping profile introduced in the Si substrate. GaAs has the highest electron mobility at approximately 8500 cm^2/V-s and can be incorporated with AlGaAs to from a HEMT, which has superior performance in terms of its power capabilities through bandgap engineering. Other notable HEMT technologies include AlGaN/GaN compounds, which exhibit not only high power performance, but also high-frequency operation. GaN and SiC present the lowest electron mobility in this comparison, sacrifice their high-power capabilities for operating speed and are typically used in power-switching applications where a rapid transition between states is not required, although GaN also exhibits relatively high-frequency performance as well.

The following paragraph reviews the power density of the semiconductor materials, specified in W/mm^2.

10.3.3 Performance Comparison: Power Density (W/mm^2)

High-power density of semiconductor materials is one of the key drivers and metrics for the advancement of power conversion technologies for applications in the military and space, which require high power density devices to take advantage of the reduced size of ICs while retaining high-power conversion efficiency. Generally, materials with wider bandgaps offer the highest power density, but bandgap engineering of traditional materials such as Si has improved upon the power density of the alloys, similar to SiGe, for example.

Advances in microelectronic engineering aim to improve the power density of materials while maintaining high efficiency, high reliability and low cost. A critical challenge that remains is to provide long-term reliability of these components. Most critically, power density allows the generation or distribution of power in small spaces, inherently important in applications where space is limited. A power density comparison of the materials discussed in this chapter is presented in Fig. 10.3.

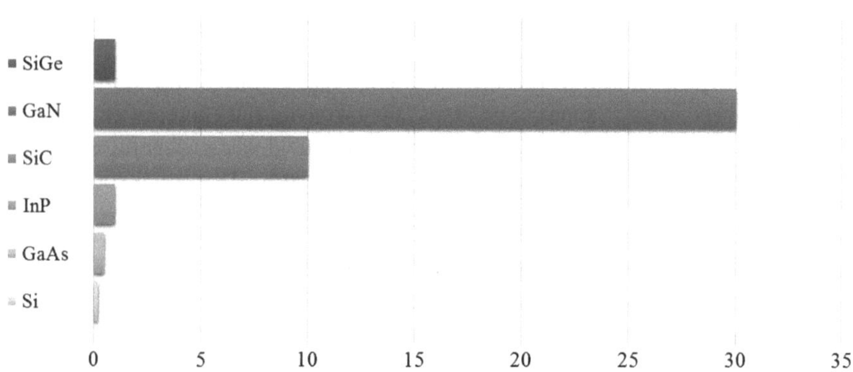

Fig. 10.3 Performance comparison between SiGe, GaN, SiC, InP, GaAs and Si semiconductor processes in terms of power density and specified in W/mm^2

Figure 10.3 shows the differences in power densities between the materials discussed in this chapter. GaN, followed by SiC, presents the highest power density, as seen in Fig. 10.3, and is therefore an ideal semiconductor material used in high-power mobile devices, optical and radio networks as well as in defense and aerospace applications such as radar, covert communications and EW. SiGe presents improved power density, considering the low power densities of InP, GaAs and Si. In Harir and Socher (2013), for example, a SiGe HBT with power density of 1 W/mm^2 is achieved, which shows a significant improvement for high-frequency and relatively low breakdown voltage semiconductors.

In PAs, a higher power density results in less combining loss, an increased PAE and cost-effectiveness of PA solutions. High breakdown voltages and high current enable high voltage and high power switches. SiC, for example, enables high voltage blocking, low switching losses, a low on-resistance and adequate frequency operation. SiC furthermore has high thermal conductivity, which is ideal for power devices and thermal transfer and complements the increased power density.

Related to power density is the breakdown voltage of active semiconductors specified in kV/cm, which is reviewed in the following paragraph.

10.3.4 Performance Comparison: Breakdown Voltage (kV/cm)

Considering the collector to base junction of a BJT, similar to a *p-n* junction diode, an avalanche process limits the collector-base voltage the transistor can sustain (Bairanzade 2003). If the applied voltage is large enough, the *n*-side depletion region will reach the n^+ contact diffusion and, if the voltage increases further, the contact will supply electrons to the *p-n* junction. At this point, the collector-base junction is essentially shorted and the current is limited by the external resistances (Bairanzade 2003). Essentially, three mechanisms control the breakdown voltage of a transistor, these being thermal instability, quantum mechanical tunneling of carriers across the bandgap and avalanche multiplication. Thermal instability occurs with an increase in leakage current due to an increase in temperature. These mechanisms are not directly destructive; however, heating caused by the large breakdown current and high breakdown voltage causes the devices to be permanently damaged unless appropriate steps are taken to sink heat away from the device. The tunneling effect occurs in a high electric field when a significant current flows by means of the band-to-band tunneling process. At even higher electric fields, avalanche multiplication occurs, also called impact ionization, and the upper limit of the breakdown voltage of the transistor is imposed by avalanche multiplication. Large current in reverse-biased conditions flow when multiplication of carriers in the space-charge region break covalent bonds due to the high electric field. Power transistors have high breakdown voltage specifications and are able to sustain much higher electric fields compared to traditional transistors.

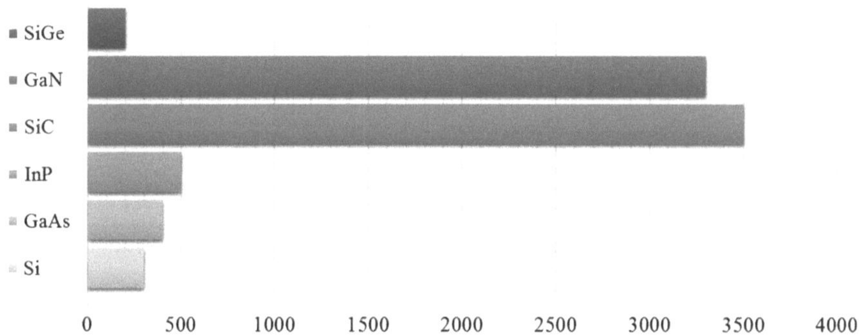

Fig. 10.4 Performance comparison between SiGe, GaN, SiC, InP, GaAs and Si semiconductor processes in terms of breakdown voltage and specified in kV/cm

Breakdown voltage is a key parameter of power switching devices, which must be high and stable with low leakage current, but for logic devices it is equally important, since technology typically decreases in size but the applied voltages remain constant (to avoid redesigning of power supplies and supporting circuitry). Figure 10.4 shows the breakdown voltages of the materials discussed in this chapter.

From Fig. 10.4 it can be seen that GaN and SiC present the largest breakdown voltages by a fair margin, followed by InP, GaAs, Si and SiGe. For this reason, GaN and SiC are the preferred technologies for power-switching devices, similar to the argument regarding power density. Direct bandgap III-V GaN-transistors are ideal for higher frequency ICs up to 1–2 kV and a maximum current rating of a few 10's A. Indirect bandgap SiC transistors are better suited for discrete devices or modules blocking 1 kV and above and virtually no limit in the current (Kaminski and Hilt 2014). Si IGBTs also have the capability to operate at similar specifications as SiC devices. InP, GaAs, Si and SiGe transistors are not commonly used in power-switching applications because of their low breakdown voltages.

Thermal conductivity, the ability to sink heat away from the device, specified in W/cm.K, is reviewed in the following paragraph.

10.3.5 Performance Comparison: Thermal Conductivity (W/cm K)

Thermal conductivity, the inverse of thermal resistance, is a function of temperature and is determined by various physical processes and material characteristics. For relatively pure semiconductors and temperatures well below the melting point of the material(s), generated heat is conducted almost exclusively by lattice vibrations/phonons (Reisch 2012). Only at very high temperatures do mobile

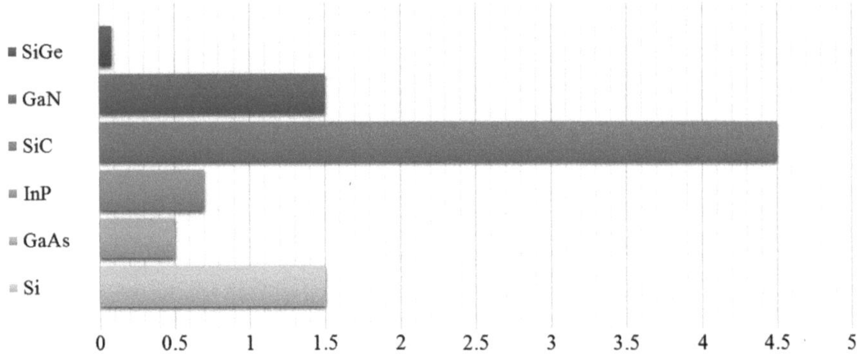

Fig. 10.5 Performance comparison between SiGe, GaN, SiC, InP, GaAs and Si semiconductor processes in terms of thermal conductivity and specified in W/cm K

carriers and photons make a substantial contribution to the heat transport in the crystal lattice of the semiconductor material. Contrary to the expected result that higher doped materials have a better ability to transport heat because of the additional electrons, in actual fact thermal conductivity is lowered at high doping levels, as can be seen in Fig. 10.5, with SiGe compared to Si. Thermal conductivity is also a temperature-dependent attribute of the material and generally decreases as the temperature rises. Figure 10.5 shows the thermal conductivity of the materials discussed in this chapter.

Figure 10.5 shows that SiC has the largest thermal conductivity when compared to SiGe, GaN, InP, GaAs and Si. SiC therefore offers the best ability to conduct heat away from the active devices, an ideal characteristic when considering its wide bandgap and its use in high-power and high-temperature and/or high-radiation switching power electronics, offering a reduction in physical size of these components if integrated into supporting circuitry. Materials such as GaAs and SiGe are rarely used in high-power and high-temperature applications because of their low thermal conductivity, leading to low reliability in such systems, especially crucial in military-based environments such as the seeker-head of a missile. At low operating voltages and high-speed applications, thermal conductivity is often achieved by external methods such as heatsinks attached to ICs (evident in for example PC CPUs). The cut-off frequency of a transistor is a FOM that determines its maximum operating frequency and is reviewed in the following paragraph.

10.3.6 Performance Comparison: Cut-off Frequency F_T (GHz)

A measure of the transistor current-gain bandwidth product is the cut-off frequency, a frequency at which the common-emitter, short-circuit current gain is unity (Kirk

1962). Transistor cut-off frequency is dependent on the current density flowing through the transistor and falls off with an increase in current density (Kirk 1962). The cut-off frequency of a transistor amplifier is greatly affected by its feedback network (Thomas 1952). The most common description of the cut-off frequency is the frequency where the output signal has experienced 3 dB attenuation due to the parasitic losses in the transistor resulting from high-frequency operation. The cut-off frequency is therefore an ideal parameter to describe the frequency response, thus the maximum operating speed, of the transistor and has been a FOM of transistor performance since its invention. f_T is strongly dependent on the material of the semiconductor, the inherent parasitic capacitances between the gate-source in CMOS (or base-emitter in BJTs) and gate-drain in CMOS (or base-collector in BJTs) junctions and the intrinsic gain of the transistor. Traditionally, III-V HEMT and III-V HBTs have outperformed traditional Si CMOS transistors by a fair margin, considering purely their f_T, but SiGe has steadily increased its operating frequency capabilities to compete directly with these materials, at the benefit of lower integration cost into standard Si processes and lower power consumption. Although transistors are rarely used in circuits as stand-alone components, but are connected in amplifier configurations, the cut-off frequency does not determine the overall performance of the IC; however, the cut-off frequency is a good indication of the maximum attainable speed performance from the IC. Typically, trade-offs among the speed of the transistor, its power capabilities and gain are compared when choosing a technology for a specific application. Figure 10.6 presents the cut-off frequencies of the materials discussed in this chapter.

As seen in Fig. 10.6, SiGe and InP present the highest cut-off frequencies currently available, operating at nearly 300 GHz. SiGe is however by far the most cost-effective solution for high-frequency applications, whereas InP offers convenience in its optoelectronic capabilities in the 1.3–1.55 μm spectrum. GaN and GaAs also yield relatively high-frequency performance when compared to SiGe and InP, although they are more expensive to manufacture. GaN additionally presents

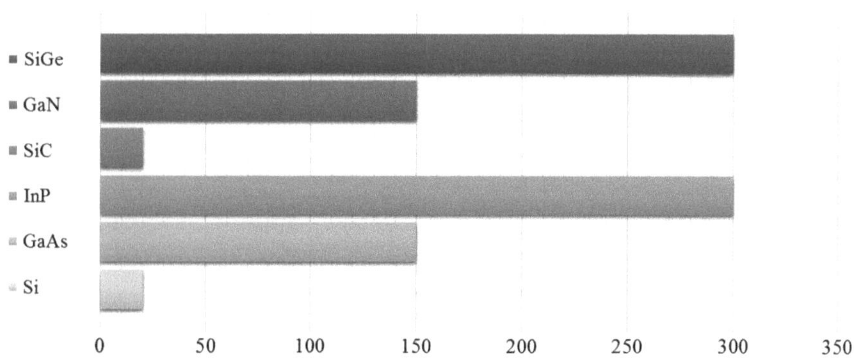

Fig. 10.6 Performance comparison between SiGe, GaN, SiC, InP, GaAs and Si semiconductor processes in terms of cut-off frequency and specified in GHz

high-power capabilities due to its wide bandgap, but power consumption and heat generation of these components increase the complexity and cost of integration. EW uses broadband and high-power emissions to disrupt and jam RF signals, traditionally performed by GaN. Similarly, AESA radar requires optimum power and size trade-offs, another discipline dominated by GaN. SiGe, operating at high-GHz frequencies, is advantageous because of its low power consumption, which is required for most mobile communications and military devices situated in power-limited environments. SiC presents the lowest cut-off frequency, somewhat lower than traditional Si, and is therefore ideally used in low-frequency, high-power switching circuits operating at high-temperature and/or high-radiation environments.

Based on the individual parameters reviewed in this section, the following section determines the desirability of each semiconductor technology based on example applications.

10.4 Semiconductor Material Desirability Based on Application Requirements

10.4.1 Performance Comparison: Overall Desirability

Two examples of transistor desirability depending on its application are presented in this section. Each parameter as discussed in the previous section is given a weight factor based on the application: high-power, high-temperature low-frequency switching applications (Table 10.7) and high-frequency, low-power and cost-effective applications (Table 10.8). For each of these proposed applications, the resulting semiconductor desirability is given as a normalized value with a maximum of 1 (100 %).

The weighting factors in Table 10.7, assigned to the parameters based on its importance to realize a high-power, high-temperature and low-frequency switching application, are used to calculate the desirability of each technology and are presented in Fig. 10.7.

As seen in Fig. 10.7, not surprisingly, GaN and SiC are the most desirable semiconductor materials based on the weighting factors in Table 10.7 for

Table 10.7 Weighting factors for high-power, high-temperature low-frequency switching transistors

Parameter	Unit	Weight
Electron bandgap	eV	5
Electron mobility	$cm^2/V\text{-}s$	1
Power density	W/mm^2	3
Breakdown voltage	kV/cm	3
Thermal conductivity	W/cm.K	2
Cut-off frequency	GHz	1

Table 10.8 Weighting factors for high-power, high-temperature, low-frequency switching transistors

Parameter	Unit	Weight
Electron bandgap	eV	1
Electron mobility	cm^2/V-s	3
Power density	W/mm^2	1
Breakdown voltage	kV/cm	1
Thermal conductivity	W/cm K	1
Cut-off frequency	GHz	5

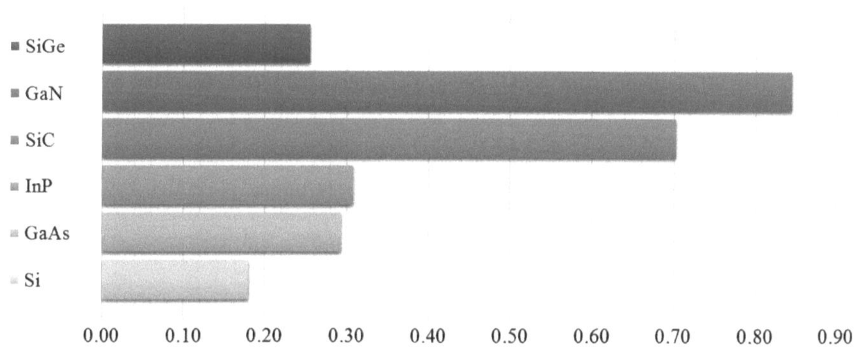

Fig. 10.7 Overall desirability of SiGe, GaN, SiC, InP, GaAs and Si when considering a high-power, high-temperature and low-frequency switching application

high-power applications. Based on these weighting factors, it is noticeable that there is no significant difference among SiGe, InP and GaAs, which shows good promise for integrating SiGe as an alternative technology with its added advantages of high frequency and low power consumption. Considering a high-frequency application, Table 10.8 presents weighting factors to achieve high-frequency operation, while maintaining low power consumption.

The resulting desirability of the semiconductor materials based on the weighting factors presented in Table 10.8 is given in Fig. 10.8.

According to Fig. 10.8, SiGe is the most desirable material if high-frequency and low-power applications are required with emphasis on cost-effectiveness. InP, GaN and GaAs also present significant performance capabilities in terms of operating speed, with the trade-offs being its higher cost and its relatively marginal superiority in electron mobility when compared to SiGe. SiC and traditional Si present the lowest desirability when high-speed operation is critical, such as in high-speed communications, radar signal processing, EW, data transfer, imaging and digital processing.

Cost is a major deciding factor in choosing an enabling technology for specific applications and semiconductor manufacturing facilities are typically extremely expensive to realize, maintain and operate. The following section summarizes some of the costs involved in semiconductor processing and the reasons why mature

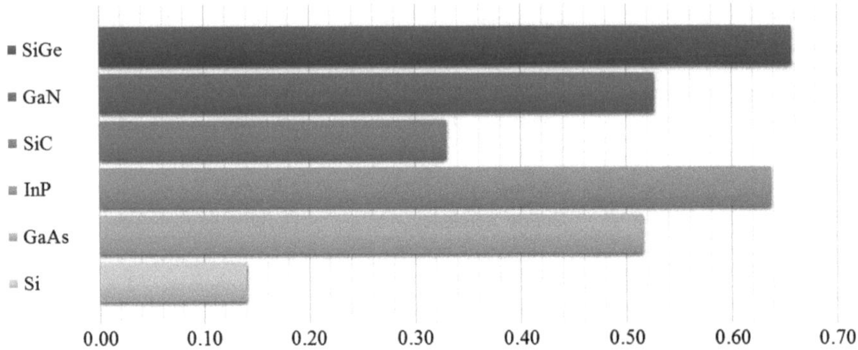

Fig. 10.8 Overall desirability of SiGe, GaN, SiC, InP, GaAs and Si when considering a high-frequency, low-power and cost-effective application

technologies are less costly, although system performance is reduced when using such technologies.

10.5 Cost of Semiconductor Processing

A crucial limitation in adopting any enabling technology to produce large volumes of devices for specific applications is the cost, directly related to the manufacturing cost, yield, efficiency and stability. These factors are typically collectively grouped into the category of technology maturity and determine the viability of using a sub-performing technology in favor of lower cost during mass production. Foundries capable of producing large volumes of semiconductor applications and ICs most commonly opt for mature technologies such as Si, or more prominent ones in recent years such as SiGe, primarily because of its low integration cost with existing Si processes. The total cost of a semiconductor process consists of various factors, including the cost of the facility, consumables, equipment, labor and other costs (such as maintenance and running costs). The percentage of the total cost of each of these factors is summarized in Fig. 10.9 (adapted from Integrated Circuit Engineering Corporation 1996).

According to Fig. 10.9, equipment and labor in a semiconductor fabrication facility are the largest expenses at 31 and 26 %, respectively. Semiconductor manufacturing equipment is extremely expensive to purchase, maintain and repair. Moreover, with the rapid enhancements in the semiconductor industry, a large percentage of the equipment should be replaced or upgraded at relatively short intervals. Photolithography equipment, for example, should be kept up to date with the latest advances in node size and resolution for a foundry to remain competitive and produce state-of-the art devices. Similar to photolithography, ion-implantation requires large and expensive equipment which must be maintained and efficiently used, with the cost of manufacturing masks for photolithography and the cost of

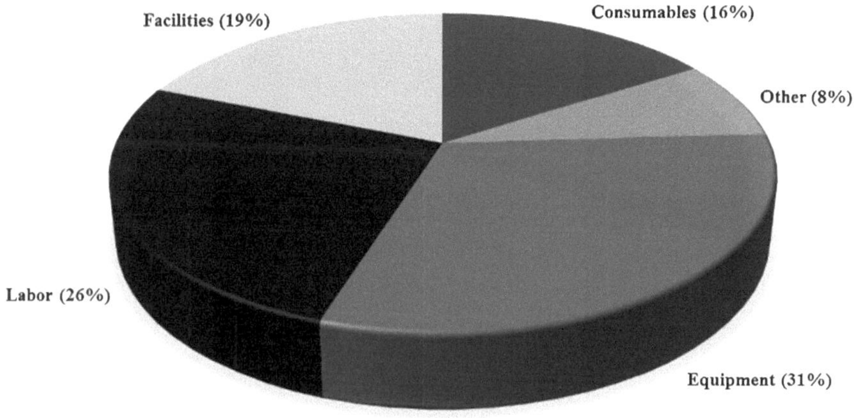

Fig. 10.9 Summary of estimated semiconductor technology development costs (adapted from Integrated Circuit Engineering Corporation 1996)

targets (used in ion-implantation) being large contributors to the high cost of running these facilities.

Skilled workers are required to operate the expensive and complex machinery and in a typical scenario individual workers are only responsible for a fraction of the entire process, therefore requiring a large number of workers to realize a complex process. This pushes labor costs up (approximately 26 % of the total cost) to run and maintain such a facility.

Facilities, accounting for approximately 19 % of the overall cost, are typically large open spaces (such as warehouse-type buildings) with specialized lighting, cleanliness control, vibration reduction and temperature and humidity control. Because of the specialized nature of semiconductor processing, extreme control over the immediate environment and space requirements, a large capital investment is required to erect such a facility initially and to maintain it.

In addition, many consumables are required to manufacture semiconductor devices, including expensive chemicals that are used in large quantities, various gas sources such as silane (SiH_4) and nitrogen, large amounts of liquid nitrogen and costly materials such as lint-free cloth, gloves and cleaning materials. Because of the cost of chemicals used for etching and cleaning and the large quantities used during a process-run, consumables also make up a large portion of the overall cost, approximately 16 %, depending on the process. Among the other expenses (8 % of the total cost) are the maintenance and running costs of such a facility. A further breakdown of the costs involved to do a single process-run, assuming the facility is able to manufacture state-of-the-art devices, is given in Fig. 10.10 (adapted from Integrated Circuit Engineering Corporation 1996).

As shown in Fig. 10.10, a typical process requires various steps of photolithography, diffusion, etching, depositing of thin films and ion implantation. The

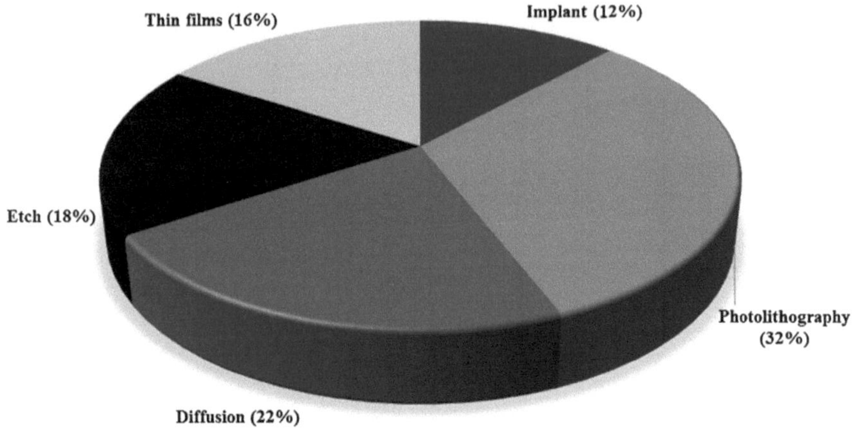

Fig. 10.10 Summary of estimated semiconductor technology development costs (adapted from Integrated Circuit Engineering Corporation 1996)

approximate percentages of the total costs of each of these steps are indicated in Fig. 10.10.

Figures 10.9 and 10.10 present a brief look into the typical costs involved in semiconductor processing. These cost are commonly deciding factors when choosing a technology to realize subsystems for specific application. In military and space-based applications, higher processing cost can often be justified by the lower volume of device manufacturing and using state-of-the-art technologies frequently results in performance breakthroughs. In the commercial sector, high-volume production often leads to a decrease in overall cost, as is the case with personal computer processing units, whereas smaller volume production often opts for older, more mature technologies to limit the cost-to-market.

10.6 Conclusion

This chapter reviews the performance characteristics of various technologies used in commercial, military and space-based systems. Military and space technologies often have stringent requirements based on power consumption, size, weight, frequency performance and output power and it is critical to consider all performance factors when choosing a technology for a specific application. Performance characteristics reviewed in this chapter include electron bandgap, electron mobility, power density, breakdown voltage, thermal conductivity and cut-off frequency. Each parameter is reviewed in terms of its value in the semiconductor processes as well as the performance advantages and disadvantages of each. These performance comparisons are used to determine the viability and desirability of each technology based on specific applications such as high-power or high-frequency operation.

Radiation-hardened technologies are also becoming more crucial in military and space-based applications. Although these have been used for many decades, research focus on inherent radiation-hardened properties is receiving increased attention as space application are becoming more commonplace and less costly to realize.

References

Abate, T. (2015). New Stanford manufacturing process could yield better solar cells, faster chips. Retrieved June 28, 2016 from news.stanford.edu

Alonso, A. R., Diaz, M. F., Lamar, D. G., De Azpeitia, M. A. P., Hernando, M. M., & Sebastian, J. (2014). Switching performance comparison of the SiC JFET and SiC JFET/Si MOSFET cascode configuration. *IEEE Transactions on Power Electronics, 29*(5), 2428–2440.

Bairanzade, M. (2003). Understanding power transistors breakdown parameters. *Application Note: AN1628/D*. 1–16, August 2003.

Cirlin, G. E., Dubrovskii, V. G., Harmand, J. C., Patriarche, G., Glas, F., Tchernycheva, M., et al. (2008). Semiconductor nanowires in InP and related material systems: MBE growth and properties. In *20th International Conference on Indium Phosphide and Related Materials* (pp. 1–5).

Damian, C., & Gelerman, D. (2012). A new generation of gallium nitride (GaN) based solid state power amplifiers for satellite communication. *Advantech Wireless Inc.* 1–4.

DARPA. (2016). New chips ease operations in electromagnetic environs. Retrieved July 11, 2016 from www.darpa.mil

Filsecker, F., Alvarez, R., & Bernet, S. (2014). Evaluation of 6.5-kV SiC p-i-n diodes in a medium-voltage, high-power 3L-NPC converter. *IEEE Transactions on Power Electronics, 29* (10), 5148–5156.

Firstenberg, A., & Tomasetta, L. (1984). Defense applications of GaAs Digital integrated circuits. In *1984 IEEE Military Communications Conference, 1*, pp. 49–53.

Griffith, Z., Urteaga, M., Rowell, P., & Pierson, R. (2015). 340–440 mW broadband, high-efficiency E-Band PA's in InP HBT. In *2015 IEEE Compound Semiconductor Integrated Circuit Symposium (CSICS)* (pp. 1–4).

Han, L., Xin, Z., & Zhen-kun, Y. (2009). The application analysis of GaN power devices in radar transmitter. In *2009 IET International Radar Conference* (pp. 1–5).

Harir, E., & Socher, E. (2013). 0.5 W X-band sige PA with integrated double-tuned transformers. In *2013 IEEE MTT-S International Microwave Symposium Digest (IMS)* (pp. 1–3).

Hitko, D. A., & Li, C. (2013). InP HBT technology and modeling. In *2013 IEEE Bipolar/BiCMOS Circuits and Technology Meeting* (pp. 123–126).

Integrated Circuit Engineering Corporation. (1996). Chapter 2: Cost per wafer. In *NMAH Accession 1996.0089, 1996.3017 & NMAH Archives Center #600*. Retrieved July 11, 2016 from http://smithsonianchips.si.edu/

Jalali, B., & Fathpour, S. (2006). Si photonics. *Journal of Lightwave Technology, 24*(12), 4600–4615, Dec 2006.

Kadavelugu, A., Bhattacharya, S., Ryu, S., Van Brunt, E., Agarwal, A., & Leslie, S. (2013). Characterization of 15 kV SiC n-IGBT and its application considerations for high power converters. In *2013 IEEE energy conversion congress and exposition* (pp. 2528–2535), 15–19 Sept 2013.

Kaminski, N., & Hilt, O. (2014). SiC and GaN devices—wide bandgap is not all the same. *IET Circuits, Devices and Systems, 8*(3), 227–236.

Kirk, C. T. (1962). A theory of transistor cutoff frequency (f_T) falloff at high current densities. *IRE Transactions on Electron Devices, 9*(2), 164–174.

König, U., Gruhle, A., & Schüppen, A. (1995). SiGe devices and circuits: Where are advantages over III/V? In *Technical Gallium Arsenide Integrated Circuit (GaAs IC) Symposium* (pp. 14–17).

Kuzuhara, M., & Tanaka, S. (2003). GaAs-based high-frequency and high-speed devices. *Japan Society of Applied Physics (JSAP)., 7*, 4–11.

Liang, Z., Ning, P., & Wang, F. (2014). Development of advanced All-SiC power modules. *IEEE Transactions on Power Electronics, 29*(5), 2289–2295.

Lin, C., Liu, H., & Zheng, Y. (2013). Application of GaN high power chips in T/R modules. In *2013 Proceedings of the International Symposium on Antennas & Propagation, 01*, 515–518.

Logan, N., & Noras, J. M. (2009). Advantages of bipolar SiGe over Si CMOS for a 2.1 GHz LNA. In *9th International Conference on Telecommunication in Modern Satellite, Cable, and Broadcasting Services* (pp. 29–31), 7–9 Oct 2009.

Long, J. R. (2005). SiGe radio frequency ICs for low-power portable communications. *Proceedings of the IEEE, 93*(9), 1598–1623.

Madhusoodhanan, S., Tripathi, A., Patel, D., Mainali, K., Kadavelugu, A., Hazra, S., et al. (2015). Solid-state transformer and MV grid tie applications enabled by 15 kV SiC IGBTs and 10 kV SiC MOSFETs based multilevel converters. *IEEE Transactions on Industry Applications, 51* (4), 3343–3360.

McHale, J. (2015). RF and microwave innovation drives military radar and electronic warfare applications. Retrieved July 11, 2016 from http://mil-embedded.com/

Millán, J. (2012). A review of WBG power semiconductor devices. *International Semiconductor Conference (CAS), 1*, 57–66.

Nair, V. (1995). GaAs IC technology for wireless communications applications. *WESCON/'95*. 475–482, 1995.

Niu, G. (2005). Noise in SiGe HBT RF technology: Physics, modeling and circuit implications. In *Proceedings of the IEEE: Invited Paper* (Vol. 93, no. 9, pp 1583–1597).

Plus, P. C. (2009). How sand is transformed into silicon chips. Retrieved July 3, 2016 from http://www.techradar.com

Radisic, V., Scott, D. W., Monier, C., Wang, S., Cavus, A., Gutierrez-Aitken, A., et al. (2015). InP HBT transferred substrate amplifiers operating to 600 GHz. In *2015 IEEE MTT-S International Microwave Symposium* (pp. 1–3).

Reisch, M. (2012). High-frequency bipolar transistors. In *Springer Science & Business Media—Volume 11 of Springer Series in Advanced Microelectronics*. ISBN: 364255900X, 658 p.

Robinson, R. (2014). Si-germanium chip sets new speed record. Retrieved July 5, 2016 from www.phys.org

Runton, D. W., Trabert, B., Shealy, J. B., & Vetury, R. (2013). History of GaN: High-Power RF Gallium Nitride (GaN) from infancy to manufacturable process and beyond. *IEEE Microwave Magazine, 14*(3), 82–93.

Schokley, W., & Queisser, H. J. (1961). Detailed balance limit of efficiency of p-n junction solar. *Journal of Applied Physics, 32*, 510–519.

Schuh, P., Rieger, R., Fleckenstein, A., Opperman, M., Adelseck, B., Mussig, H., et al. (2009). T/R-module technologies today and possible evolutions. In *2009 International Radar Conference* (pp. 1–5).

Schultz, M. L. (1963). Si: Semiconductor Properties. *Infrared Physics, 4*(2), 93–112.

Senapati, B., & Maiti, C. K. (2000). Advantages of SiGe-HBTs for RF wireless communication. In *2000 IEEE International Conference on Personal Wireless Communications* (pp. 9–13).

Shamsi, P., McDonough, M., & Fahimi, B. (2013). Performance evaluation of wide bandgap semiconductor technologies in automotive applications. In *2013 IEEE workshop on wide bandgap power devices and applications (WiPDA)* (pp. 115–118).

Shen, Z. J., Sabui, G., Miao, Z., & Shuai, Z. (2015). Wide-bandgap solid-state circuit breakers for dc power systems: Device and circuit considerations. *IEEE Transactions on Electron Devices, 62*(2), 294–300.

Singh, R., & Sundaresan, S. (2015). High-temperature operation with Si carbide devices. In *IEEE Power Electronics Magazine* (pp. 27–35).

Soref, R. (2006). The past, present, and future of Si photonics. *IEEE Journal of Selected Topics in Quantum Electronics, 12*(6), 1678–1687.

Thomas, D. E. (1952). Transistor amplifier-cutoff frequency. *Proceedings of the IRE., 40*(11), 1481–1483.

Topham, P. J., Griffith, I., & Shaw, N. (1990). InP based HBT integrated circuits. In *IEE Colloquium on InP Based Materials, Devices and Integrated Circuits*. 10/1–10/4, 1990.

Trew, R. J. (2002). SiC and GaN transistors—Is there one winner for microwave power applications? *Proceedings of the IEEE, 90*(6), 1032–1047.

Trew, R. J. (2004). Wide bandgap transistor amplifiers for improved performance microwave power and radar applications. In *15th International Conference on Microwaves, Radar and Wireless Communications: Invited Paper*, 1, 18–23.

Van Zeghbroeck, B. (2011). Chapter 2: Semiconductor fundamentals. Retrieved July 2, 2016 from http://eece.colorado.edu

Whitesides, G. (1988). Gallium arsenide: Key to faster, better computing. Retrieved June 28, 2016 from www.the-scientist.com

Yamaguchi, M., Ohmachi, Y., Oh'hara, T., Kadota, Y., Imaizumi, M., & Matsuda, S. (2000). GaAs-on-Si solar cells for space use. In *Conference Record of the Twenty-Eight IEEE Photovoltaic Specialists Conference* (pp. 1012–1015).

Zhou, Z., Tu, Z., Li, T., & Wang, X. (2015). Si photonics for advanced optical interconnects. *Journal of Lightwave Technology, 33*(4), 928–933.

Zopler, J. C. (2003). Challenges and opportunities for InP HBT mixed signal circuit technology. In *2003 International Conference on Indium Phosphide and Related Materials* (pp. 8–11).

MIX
Papier aus verantwortungsvollen Quellen
Paper from responsible sources
FSC® C105338

If you have any concerns about our products,
you can contact us on
ProductSafety@springernature.com

In case Publisher is established outside the EU,
the EU authorized representative is:
Springer Nature Customer Service Center GmbH
Europaplatz 3, 69115 Heidelberg, Germany

Printed by Libri Plureos GmbH
in Hamburg, Germany